A Course in Analytic Number Theory

To the memory of my father Finn and mother Liv

Contents

Preface

This book was written for graduate students looking for an introduction to some basic methods of analytic number theory. It is suitable as a textbook for an introductory one-semester course at the beginning graduate level, but contains more material than can be comfortably covered in such a course. However, by suitably selecting chapters, it is possible to teach courses going in various directions.

Readers should be familiar with ε-δ calculus, have completed an undergraduate course in complex analysis, and possess the proficiency in abstract and linear algebra to be expected of a beginning graduate student. No familiarity with graduate-level analysis is assumed. The first four chapters presuppose no complex analysis beyond simple properties of the exponential function.

Each chapter is followed by notes with references and historical remarks. At the end of many of the sections there are references to more detailed treatments of the topic under consideration.

I wish to thank anonymous referees for suggestions that have improved the book. Naturally I alone remain responsible for all errors and imperfections that remain.

This seems an appropriate place to express my gratitude to Hugh Montgomery, Imre Ruzsa, and the late Sigmund Selberg, from whose teaching of analytic number theory I have benefited.

Marius Overholt

Acknowledgments

Except for some exercises, I am indebted to the literature of analytic number theory for all the material in this textbook.

As references I have chiefly relied on the following works: For arithmetic functions *Introduction to Analytic and Probabilistic Number Theory* by Gérald Tenenbaum, and for prime number theory *Multiplicative Number Theory I. Classical Theory* by Hugh L. Montgomery and Robert C. Vaughan have been the main references. But for an easy proof of the Prime Number Theorem with an error term I have followed *The Distribution of Prime Numbers* [**Ing90**] by A. E. Ingham. For my very modest account of the analytic properties of the Riemann zeta function I am indebted to *The Theory of the Riemann Zeta-function* [**Tit86**] by E. C. Titchmarsh. The chapter on the Circle Method owes the most to the second edition of *Analytic Methods for Diophantine Equations and Diophantine Inequalities* [**Dav05**] by H. Davenport, edited by T. D. Browning and with a Foreword by D. E. Freeman, D. R. Heath-Brown and R. C. Vaughan, though in a few particulars I have followed the treatment in the second edition of *The Hardy-Littlewood Method* [**Vau97**] by R. C. Vaughan. For the chapter on the Dedekind zeta function my main sources have been *Lectures on Algebraic and Analytic Number Theory* [**Gál61**] by I. S. Gál, the contribution by H. A. Heilbronn in the collection *Algebraic Number Theory* [**Hei67**] edited by J. W. S. Cassels and A. Frölich, *Algebraic Number Theory* [**Lan70**] by Serge Lang, *Elementary and Analytic Theory of Algebraic Numbers* [**Nar00b**] by W. Narkiewicz, *Algebraic Number Theory* by Jürgen Neukirch, and especially the expository articles *The Analytic Theory of Algebraic Numbers* [**Sta75**] and *Galois Theory, Algebraic Number Theory and Zeta Functions* [**Sta95**] by H. M. Stark.

Beyond these works that I have mainly relied on for this book, there are many other excellent treatments of analytic number theory. It may be appropriate at this point to mention a few that are particularly important for one reason or another. *Introduction to Analytic Number Theory* by Tom M. Apostol [**Apo76**] has for decades been the most widely used introductory text. It carefully develops material needed from elementary number theory rather than assuming it as a prerequisite, and has many exercises. *Multiplicative Number Theory* by Harold Davenport [**Dav00**] is a classic account of the distribution of primes in arithmetic progressions. It has been in print for more than forty years, and is still one of the more frequently assigned textbooks for courses in analytic number theory. *Analytic Number Theory* by Henryk Iwaniec [**IK04**] and Emmanuel Kowalski is a broad, deep and modern treatment. Of outstanding importance to the development of analytic number theory in its early stages was *Handbuch Der Lehre von der Verteilung der Primzahlen* [**Lan74**] by Edmund Landau.

For information on the historical background I have in addition relied on these sources: *Gauss and Jacobi Sums* [**BEW98**] by B. C. Berndt, R. J. Evans and K. S. Williams, *Pioneers of Representation Theory: Frobenius, Burnside, Schur, and Brauer* [**Cur99**] by C. W. Curtis, the survey paper [**DE80**] by H. G. Diamond on elementary methods in prime number theory, *History of the Theory of Numbers* [**Dic34**] by L. E. Dickson, *Riemann's Zeta Function* [**Edw01**] by H. M. Edwards, the sixth edition of *An Introduction to the Theory of Numbers* [**HW08**] by G. H. Hardy and E. M. Wright and revised by D. R. Heath-Brown and J. H. Silverman, the introductory notes by H. A. Heilbronn in volume I of the Collected Papers of G. H. Hardy [**Har66**], the 1990 edition of *The Distribution of Prime Numbers* [**Ing90**] by A. E. Ingham and with a Foreword by R. C. Vaughan, *Multiplicative Number Theory I. Classical Theory* [**MV07**] by H. L. Montgomery and R. C. Vaughan, and *The Development of Prime Number Theory* [**Nar00a**] by W. Narkiewicz, and the survey paper [**VW02**] on Waring's Problem by R. C. Vaughan and T. D. Wooley.

All figures have been made with Inkscape™ and Mathematica™.

Marius Overholt

How to use this text

Chapters 1, 2, 3, 5 and 6 are suitable for a course emphasizing arithmetic functions and the two classical highlights of analytic number theory: Dirichlet's theorem on primes in arithmetic progressions, and the Prime Number Theorem.

Chapters 1, 2, 3 and 4 are suitable for a syllabus with an emphasis on elementary methods, for students with little knowledge of analysis. Unfortunately the Prime Number Theorem is not covered.

Chapters 1, 3, 4, 5 and 6 are suitable for a syllabus with a Diophantine emphasis, but also including a proof of the Prime Number Theorem.

Chapters 1, 3, 5, 6, 8 and 10 are suitable for a syllabus with an emphasis on analytic methods, concentrating on the Riemann zeta function and the distribution of primes.

Chapters 1, 3, 5, 8 and 9 are suitable for a syllabus with an emphasis on the analytic theory of number fields. Students following such a syllabus should either have some knowledge of algebraic number theory, or else have a good knowledge of abstract algebra and do some reading. Here the Prime Number Theorem is established by means of the Ikehara theorem.

Chapters 1, 3, 5 and 7 are suitable for a syllabus aiming at the Prime Number Theorem for arithmetic progressions. The ordinary Prime Number Theorem is established as a corollary at the end of the course.

To help with the planning of syllabi a diagram of dependencies between chapters has been provided; see Figure 1 on page xvi. To supplement this, more specific comments may prove useful: Unless the material on Artin L-functions in Chapter 9 is covered, Section 3.8 may be substituted for Sections 3.4 and 3.5. This saves a little time, and the construction of Dirichlet

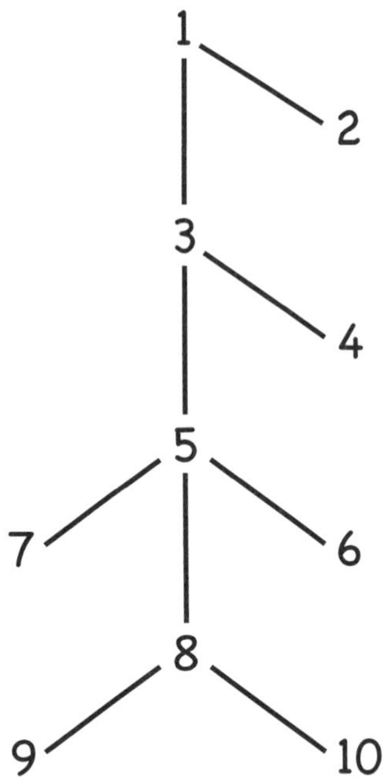

Figure 1. Diagram of chapter dependencies

characters by means of primitive roots seems less conceptually demanding than even an epsilon of representation theory. The theorem of Siegel from Section 7.3 is also needed in Section 9.3. The Borel-Carathéodory lemma of Section 7.4 is also used to prove the Hadamard factorization theorem of Section 8.6. The Hadamard factorization theorem is needed in Section 9.4 as well as in Chapter 10. As usual, starred material may be omitted without loss of continuity.

There are exercises at the end of the chapters, for which solutions are not provided. In addition to the end-of-chapter exercises, there is a chapter with a selection of other exercises with solutions. As a caution, some exercises have been marked with a dagger, because they require more work than the remainder. But they are for the most part not more difficult in the sense of it being harder to see what to do.

A *Summary of Elementary and Algebraic Number Theory* with a condensed exposition of those concepts on which the book draws is available on the web. The Summary presupposes familiarity with groups, rings and fields. All results from elementary and algebraic number theory that are actually needed in this book are proved in the Summary.

Introduction

Analytic number theory is mainly devoted to finding approximate counts of number theoretical objects in situations where exact counts are out of reach. Primes, divisors, solutions of Diophantine equations, lattice points within contours, partitions of integers and ideal classes of algebraic number fields are some of the objects that have been counted. The prototypical approximate count in number theory is the Prime Number Theorem (PNT), stating that

$$\lim_{x \to +\infty} \frac{\pi(x)}{\int_2^x \frac{du}{\log(u)}} = 1$$

where $\pi(x)$ is the number of primes $p \leq x$. This was proved independently in 1896 by Jacques Hadamard and Charles de la Vallée Poussin, building on ideas of Bernhard Riemann, and applying complex analysis to the Riemann zeta function

$$\zeta(s) = \sum_{n=1}^{\infty} n^{-s}$$

to establish the result. An asymptotic count like the PNT usually attracts attention with a view to improve it. As the distribution of prime numbers is one of the central topics in number theory, much effort has been expended to obtain improvements to the Prime Number Theorem. We shall prove one of the weaker ones, to the effect that there exist positive constants c, C, x_0 such that

$$\left| \pi(x) - \int_2^x \frac{du}{\log(u)} \right| \leq Cxe^{-c \log^{1/10}(x)} \quad \text{for} \quad x \geq x_0.$$

This is more precise, though also more complicated to state, than the asymptotic form of the PNT.

n	1	2	3	4	5	6	7	8	9	10	11	12	13	14	15
$d(n)$	1	2	2	3	2	4	2	4	3	4	2	6	2	4	4

Table 1. Values of the divisor function

Counting the number of divisors of positive integers leads to a difficult problem known as the Dirichlet Divisor Problem that is still unsolved today. Denoting the number of divisors of n by $d(n)$, Table 1 shows that these counts fluctuate a good deal. But much more regular behavior is revealed by averaging $d(n)$, and in fact

$$\frac{1}{x}\sum_{n\leq x} d(n) \approx \log(x) + 2\gamma - 1, \qquad \gamma = 0.5772\ldots,$$

with an absolute error that tends to zero as $x \to +\infty$. To determine how fast the error tends to zero is the divisor problem of Dirichlet.

Divisors and primes are the stuff of multiplicative number theory. But there are also interesting counting problems connected with additive questions. The eighteenth-century English algebraist Edward Waring stated that every positive integer may be expressed as a sum of a limited number of k-th powers of nonnegative integers, the number required depending on k only. We shall count the number of such representations for large integers when the number of powers allowed is sufficiently large, finding an asymptotic formula by means of the Circle Method and establishing Waring's claim. This was first achieved by David Hilbert by a method different from the one used here. The proof is the most elaborate in the book, though the prerequisites are surprisingly modest. The Circle Method is related to Fourier theory, but involves only Fourier series with finitely many terms so convergence issues do not arise.

The achievements of analytic number theory are not entirely limited to approximate counts. Some of the quantities estimated are not counting numbers, and for a few problems exact rather than approximate results have been attained. We shall cover one such case from algebraic number theory, that of the analytic class number formula

$$h_K = \frac{w_K|d_K|^{1/2}}{2^{r_1(K)+r_2(K)}\pi^{r_2(K)}R_K}\lim_{s\to 1}\frac{\zeta_K(s)}{\zeta(s)}$$

that expresses the number h_K of ideal classes of the ring of algebraic integers of a number field K in terms of other arithmetic data. This formula is due to Dirichlet and Richard Dedekind.

Chapter 1

Arithmetic Functions

1.1. The method of Chebyshev

Around 1792 J. K. F. Gauss counted primes in successive blocks of a thousand integers, and noticed that the sequence of primes seems to thin out according to a definite law. One formulation of his law is that the intervals $(x-h, x]$ contain about $h/\log(x)$ primes when x is large and h is not too small. Another way to express the same empirical observation is that the chance of a randomly chosen large integer n being prime is approximately $1/\log(n)$. Gauss went on to integrate this density to obtain the approximation

$$\pi(x) \stackrel{\text{def}}{=} \sum_{p\le x} 1 \approx \mathrm{li}(x) \stackrel{\text{def}}{=} \int_2^x \frac{du}{\log(u)}$$

for the counting function $\pi(x)$ of the primes. Here $\mathrm{li}(x)$ is the *integral logarithm.* By his counts of primes, Gauss guessed that $\lim_{x\to+\infty} \pi(x)/\mathrm{li}(x) = 1$. This is simply the statement that the relative error $|\pi(x) - \mathrm{li}(x)|/\pi(x)$ in the approximation goes to zero as $x \to +\infty$. Using the notation

$$f(x) \sim g(x) \quad \stackrel{\text{def}}{\Longleftrightarrow} \quad \lim_{x\to+\infty} \frac{f(x)}{g(x)} = 1$$

of asymptotic equality familiar from analysis, the conjecture can be expressed as $\pi(x) \sim \mathrm{li}(x)$. This is the Prime Number Theorem (abbreviated PNT) first proved by J. S. Hadamard and C. G. J. N. de la Vallée Poussin in 1896. Since $\mathrm{li}(x) \sim x/\log(x)$ by l'Hôpital's rule, the PNT also has the formulation $\pi(x) \sim x/\log(x)$, showing that about $1/\log(x)$ of the positive integers up to x are prime.

Because the density of the primes near n is approximately $1/\log(n)$, it may be more natural to count each prime p with weight $\log(p)$ rather than with weight 1. This gives the weighted counting function

$$\vartheta(x) \stackrel{\text{def}}{=} \sum_{p\leq x} \log(p)$$

introduced by P. L. Chebyshev. To obtain nice formulas that are easier to analyze, it is advantageous also to count prime powers p^k with weight $\log(p)$. The *von Mangoldt function* Λ given by $\Lambda(p^k) = \log(p)$ when the argument is a prime power, and zero otherwise, serves this purpose. The weighted counting function

$$\psi(x) \stackrel{\text{def}}{=} \sum_{p^k\leq x} \log(p) = \sum_{n\leq x} \Lambda(n)$$

was also introduced by Chebyshev. Since prime powers with exponent higher than one are quite sparse, $\psi(x)$ mainly counts primes with weight $\log(p)$. The functions $\psi(x)$ and $\vartheta(x)$ are thus approximately equal, and both are closely related to the counting function $\pi(x)$ of the primes. In particular it will turn out that the Prime Number Theorem may equally well be expressed as one of the asymptotic relations $\psi(x) \sim x$ or $\vartheta(x) \sim x$. These formulations are often more convenient.

The integers $n = pm$ in the interval $0 < n \leq N$ that are divisible by a prescribed prime p are given by the integer solutions m of the inequality $0 < m \leq N/p$. The largest integer less than or equal to a real number x is denoted by $[x]$. It is called the *integer part* of x or the *Gauss bracket.* Clearly $[N/p]$ is the number of integers n as above. The same reasoning shows that, of these integers, exactly $[N/p^k]$ are divisible by p^k. This observation allows us to write down the prime factorization

$$N! = \prod_{p^k} p^{[N/p^k]}$$

of the factorial, due to A.-M. Legendre. The product is taken over all prime powers, but has only finitely many factors different from 1 because $[N/p^k] = 0$ when $p^k > N$. The importance of the identity lies in the fact that the left-hand side does not contain the primes explicitly, and is susceptible of being estimated analytically. Taking the logarithm on both sides of the Legendre identity yields

$$\sum_{n\leq N} \Lambda(n) \left[\frac{N}{n}\right] = \sum_{p^k} \log(p) \left[\frac{N}{p^k}\right] = \log(N!).$$

Now

$$\sum_{p^k} \log(p) \left[\frac{N}{p^k}\right] = \sum_{p^k} \log(p) \sum_{mp^k\leq N} 1 = \sum_{m\leq N} \sum_{p^k\leq N/m} \log(p)$$

by interchanging the order of summation in the double sum. Then

$$\sum_{m\le N}\sum_{n\le N/m}\Lambda(n)=\log(N!)=\sum_{n\le N}\log(n).$$

Any mapping $f:\mathbb{N}\to\mathbb{C}$ from the positive integers into the complex numbers is called an *arithmetic function.* The functions Λ and log are examples of arithmetic functions. Every arithmetic function f has a *summatory function*

$$F(x)=\sum_{n\le x}f(n).$$

Thus ψ is the summatory function of Λ. The summatory function

$$T(x)\stackrel{\text{def}}{=}\sum_{n\le x}\log(n)$$

of log is also important. The identity

$$\sum_{m\le x}\psi\Big(\frac{x}{m}\Big)=\sum_{n\le x}\Lambda(n)\Big[\frac{x}{n}\Big]=T(x)$$

holds for nonnegative x since $\log(N!)=T(x)$ where $N=[x]$. This identity is the starting point for the method of Chebyshev.

Proposition 1.1. *The inequalities*

$$\log(2)x-\log(4x)\le\psi(x)\le 2\log(2)x+\frac{\log^2(x)}{\log(2)}$$

hold for $x\ge 1$.

Proof. The terms in the last sum in the computation

$$\begin{aligned}T(x)-2T\Big(\frac{x}{2}\Big)&=\sum_{n\le x}\log(n)-2\sum_{m\le x/2}\log(m)\\&=\sum_{n\le x}\log(n)-2\sum_{2m\le x}\log(2m)+2\sum_{2m\le x}\log(2)\\&=\sum_{n\le x}(-1)^{n-1}\log(n)+2\Big[\frac{x}{2}\Big]\log(2)\end{aligned}$$

alternate in sign and increase in magnitude. So

$$\Big|T(x)-2T\Big(\frac{x}{2}\Big)-2\Big[\frac{x}{2}\Big]\log(2)\Big|\le\log([x])$$

for $x\ge 1$. Thus

$$\log(2)x-\log(4x)\le T(x)-2T\Big(\frac{x}{2}\Big)\le\log(2)x+\log(x).$$

Substituting the expression

$$T(x)=\sum_{n\le x}\psi\Big(\frac{x}{n}\Big)$$

into $T(x) - 2T(x/2)$ yields

$$\psi(x) - \psi\Big(\frac{x}{2}\Big) + \psi\Big(\frac{x}{3}\Big) - \cdots = T(x) - 2T\Big(\frac{x}{2}\Big).$$

Then

$$\psi(x) \geq \log(2)x - \log(4x)$$

since ψ is an increasing and nonnegative function, and

$$\psi(x) - \psi\Big(\frac{x}{2}\Big) \leq \log(2)x + \log(x)$$

for the same reason. Adding up the inequalities

$$\psi\Big(\frac{x}{2^j}\Big) - \psi\Big(\frac{x}{2^{j+1}}\Big) \leq \log(2)2^{-j}x + \log(x)$$

for $j = 0, 1, 2, \ldots, [\log(x)/\log(2)] - 1$ yields

$$\psi(x) \leq 2\log(2)x + \left[\frac{\log(x)}{\log(2)}\right]\log(x) \leq 2\log(2)x + \frac{\log^2(x)}{\log(2)}$$

since $\psi(x/2^{j+1}) = 0$ when $x/2^{j+1} < 2$. $\square$

The inequalities in Proposition 1.1 and the limits $\log^2(x)/(x\log(2)) \to 0$ and $\log(4x)/x \to 0$ as $x \to +\infty$ imply that for every $\varepsilon > 0$ there exists some $x_0(\varepsilon)$ so that $\log(2) - \varepsilon < \psi(x)/x < 2\log(2) + \varepsilon$ for $x \geq x_0(\varepsilon)$. Such ε-$x_0(\varepsilon)$ inequalities are often expressed in a somewhat different but equivalent way, using concepts from analysis. The *limit superior* $\limsup_{x\to+\infty} f(x)$ of a bounded real function $f(x)$ on an interval $[a, \infty)$ is the unique real number σ such that $f(x) < \sigma+\varepsilon$ holds for all x sufficiently large, while $f(x) < \sigma-\varepsilon$ fails for some x arbitrarily large, no matter how small $\varepsilon > 0$ is taken. Similarly the *limit inferior* $\liminf_{x\to+\infty} f(x)$ is the unique real number ι such that $f(x) > \iota - \varepsilon$ holds for all x sufficiently large, while $f(x) > \iota + \varepsilon$ fails for some x arbitrarily large, no matter how small $\varepsilon > 0$ is taken. That σ and ι must necessarily exist is a consequence of the completeness property of the real number system. Define

$$\mathbf{a} \stackrel{\text{def}}{=} \liminf_{x\to+\infty} \frac{\psi(x)}{x} \quad \text{and} \quad \mathbf{A} \stackrel{\text{def}}{=} \limsup_{x\to+\infty} \frac{\psi(x)}{x}.$$

Then the ε-$x_0(\varepsilon)$ bounds for $\psi(x)/x$ can be reformulated as the statement that $\log(2) \leq \mathbf{a} \leq \mathbf{A} \leq 2\log(2)$.

The definitions of ψ and ϑ yield

$$\psi(x) = \sum_{p^k \leq x} \log(p) = \sum_{k=1}^{\infty} \sum_{p \leq x^{1/k}} \log(p) = \vartheta(x) + \vartheta(x^{1/2}) + \vartheta(x^{1/3}) + \cdots,$$

and so

$$\psi(x) - 2\psi(x^{1/2}) = \vartheta(x) - \vartheta(x^{1/2}) + \vartheta(x^{1/3}) - \cdots.$$

The inequality $\psi(x)-2\psi(x^{1/2}) \leq \vartheta(x) \leq \psi(x)$ follows since ϑ is an increasing and nonnegative function, and then

$$\log(2) \leq \liminf_{x\to+\infty} \frac{\vartheta(x)}{x} \leq \limsup_{x\to+\infty} \frac{\vartheta(x)}{x} \leq 2\log(2)$$

by our estimates on $\psi(x)$.

Clearly $\vartheta(x) \leq \pi(x)\log(x)$, and so $\pi(x)/(x/\log(x)) \geq \vartheta(x)/x$. Finding an upper bound for $\pi(x)$ is only slightly more challenging. The inequality

$$\vartheta(x) \geq \sum_{y<p\leq x} \log(p) \geq (\pi(x)-\pi(y))\log(y)$$

yields

$$\frac{\pi(x)}{x/\log(y)} \leq \frac{\vartheta(x)}{x} + \frac{\pi(y)}{x/\log(y)} \leq \frac{\vartheta(x)}{x} + \frac{y}{x/\log(y)}.$$

To obtain the desired upper bound, we must let y increase fast enough with x so that the left-hand side of the inequality is close to $\pi(x)/(x/\log(x))$ for large x, while the second term on the right-hand side should become negligible in comparison. The choice $y = x/\log^2(x)$ works well, giving

$$\frac{\pi(x)}{x/(\log(x)-2\log\log(x))} \leq \frac{\vartheta(x)}{x} + \frac{x/\log^2(x)}{x/(\log(x)-2\log\log(x))}.$$

The second term on the right-hand side tends to zero, and

$$\frac{\frac{\pi(x)}{x/(\log(x)-2\log\log(x))}}{\frac{\pi(x)}{x/\log(x)}} = \frac{\log(x)-2\log\log(x)}{\log(x)} \to 1$$

as $x \to +\infty$. Thus

$$\liminf_{x\to+\infty} \frac{\vartheta(x)}{x} \leq \liminf_{x\to+\infty} \frac{\pi(x)}{x/\log(x)} \leq \limsup_{x\to+\infty} \frac{\pi(x)}{x/\log(x)} \leq \limsup_{x\to+\infty} \frac{\vartheta(x)}{x}$$

and so

$$\log(2) \leq \liminf_{x\to+\infty} \frac{\pi(x)}{x/\log(x)} \leq \limsup_{x\to+\infty} \frac{\pi(x)}{x/\log(x)} \leq 2\log(2).$$

Since $\lim_{x\to+\infty} \mathrm{li}(x)/(x\log(x)) = 1$, the last inequality shows that $\pi(x)$ and $\mathrm{li}(x)$ have the same order of growth.

Choosing $y = x^a$ with $0 < a < 1$ and a otherwise, arbitrarily yields

$$\frac{\vartheta(x)}{x} \leq \frac{\pi(x)}{x/\log(x)} \leq \frac{1}{a}\frac{\vartheta(x)}{x} + \frac{x^a}{x/\log(x)};$$

thus we see that $\pi(x) \sim x/\log(x)$, and $\pi(x) \sim \mathrm{li}(x)$, and $\vartheta(x) \sim x$, and $\psi(x) \sim x$, and $\mathbf{a} = \mathbf{A} = 1$, are equivalent formulations of the Prime Number Theorem.

1.2. Bertrand's Postulate

The method of Chebyshev does not prove the Prime Number Theorem, but it does show that $\psi(x)$ has the expected order of growth. By a more elaborate version of his method Chebyshev obtained bounds such as $.921\cdot x \leq \psi(x) \leq 1.106\cdot x$ for all x sufficiently large. He used these bounds to give the first proof of Bertrand's Postulate. Today Bertrand's Postulate is taken as the statement that for all $x \geq 2$ there is at least one prime in the interval $(x/2, x]$. The weaker bounds in Proposition 1.1 do not suffice to prove this with Chebyshev's approach. But we shall obtain the result anyway using an idea of S. A. Ramanujan.

Proposition 1.2 (Bertrand's Postulate). *For every $x \geq 2$ there exists at least one prime p with $x/2 < p \leq x$.*

Proof. For $2 \leq x \leq 797$ the interval $(x/2, x]$ contains a prime by a trick of E. G. H. Landau: The chain $2, 3, 5, 7, 11, 17, 31, 59, 107, 211, 401, 797$ consists of primes and each is smaller than twice its predecessor. To detect primes in the intervals $(x/2, x]$ for $x \geq 797$ we show that the difference $\vartheta(x) - \vartheta(x/2)$ is positive.

Fetch the inequality

$$\psi(x) - \psi\Big(\frac{x}{2}\Big) + \psi\Big(\frac{x}{3}\Big) \geq T(x) - 2T\Big(\frac{x}{2}\Big)$$

from the proof of Proposition 1.1. Retention of the term $\psi(x/3)$ is an idea due to Ramanujan. Now

$$\begin{aligned}\vartheta(x) - \vartheta\Big(\frac{x}{2}\Big) &\geq \psi(x) - 2\psi(x^{1/2}) - \psi\Big(\frac{x}{2}\Big)\\ &\geq T(x) - 2T\Big(\frac{x}{2}\Big) - \psi\Big(\frac{x}{3}\Big) - 2\psi(x^{1/2})\\ &\geq \log(2)x - \log(4x) - 2\log(2)\frac{x}{3}\\ &\quad - \frac{\log^2(x/3)}{\log(2)} - 4\log(2)x^{1/2} - 2\frac{\log^2(x^{1/2})}{\log(2)} = f(x)\end{aligned}$$

with

$$f(x) = \frac{\log(2)}{3}x - \log(4x) - \frac{3\log^2(x)}{2\log(2)} - 4\log(2)x^{1/2},$$

by Proposition 1.1 and its proof. The derivative

$$f'(x) = \frac{\log(2)}{3} - \frac{1}{x} - \frac{3\log(x)}{x\log(2)} - \frac{2\log(2)}{x^{1/2}}$$

is an increasing function on $x \geq 797$ since $\log(x)/x$ is decreasing on $x \geq e$. Then $f'(x) > 0$ on $x \geq 797$ because $f'(797) = 0.14$. Thus $f(x)$ is increasing on this interval and hence positive there because $f(797) = 1.2$. □

1.3. Simple estimation techniques

In analytic number theory arithmetical questions are characteristically answered by finding estimates that imply the desired conclusions. The proof of Bertrand's Postulate is typical; we wanted to know that every interval $(x/2, x]$ for $x \geq 2$ contains at least one prime, but to obtain this result we detoured through a lower bound for $\vartheta(x) - \vartheta(x/2)$. Carrying out an estimate often requires long calculations with inequalities, for which the usual notation proves cumbersome. The big-O notation of P. G. H. Bachmann is well adapted for efficient calculations with long chains of inequalities. The statement $f(x) = O(g(x))$ means that there exist some unspecified constants $C > 0$ and x_0 such that $|f(x)| \leq Cg(x)$ on the interval $x \geq x_0$. The bound $\psi(x) \leq 2\log(2)x + \log^2(x)/\log(2)$ for $x \geq 1$ would be expressed as $\psi(x) = O(x)$ in the big-O notation. The latter statement is less precise, but the kind of information suppressed in the big-O notation is often unimportant anyway. The following result is the key to labor-saving calculations using the big-O notation.

Proposition 1.3. *If $f_1(x) = O(g(x))$ and $f_2(x) = O(g(x))$ then $f_1(x) + f_2(x) = O(g(x))$. If also $f_3(x) = O(h(x))$ then $f_1(x)f_3(x) = O(g(x)h(x))$.*

Proof. There are $C_1, C_2, C_3 > 0$ and x_1, x_2, x_3 such that $|f_1(x)| \leq C_1 g(x)$ for $x \geq x_1$, $|f_2(x)| \leq C_2 g(x)$ for $x \geq x_2$ and $|f_3(x)| \leq C_3 h(x)$ for $x \geq x_3$. Then

$$|f_1(x) + f_2(x)| \leq |f_1(x)| + |f_2(x)| \leq C_1 g(x) + C_2 g(x) = (C_1 + C_2)g(x)$$

for $x \geq \max(x_1, x_2)$ and

$$|f_1(x)f_3(x)| = |f_1(x)||f_2(x)| \leq C_1 g(x) C_3 h(x) = (C_1 C_3)g(x)h(x)$$

for $x \geq \max(x_1, x_3)$. □

Note in particular that the largest of the O-terms in a sum absorbs everything smaller. The notation $f(x) \ll g(x)$ due to I. M. Vinogradov is synonymous with $f(x) = O(g(x))$. The notation $f(x) \asymp g(x)$ of G. H. Hardy means that $f(x) \ll g(x)$ and $g(x) \ll f(x)$. There is also a small-o notation due to Landau. The statement $f(x) = o(g(x))$ means that $\lim_{x\to+\infty} f(x)/g(x) = 0$, or equivalently, for any $\varepsilon > 0$ there exists some $x_0(\varepsilon)$ so that $|f(x)| \leq \varepsilon g(x)$ for $x \geq x_0(\varepsilon)$.

If the function f in an estimate $f = O(g)$ depends on one or more parameters, say $f_a(x) = O_a(g(x))$, the question of uniformity of the estimate is often of considerable importance. If it is possible to choose both C and x_0 in the statement that

$$|f_a(x)| \leq Cg(x) \quad \text{for} \quad x \geq x_0$$

independently of the parameter a, then the estimate is said to be *uniform* in a. Neglect of uniformity is a recognized source of error in number theoretical arguments. The case where C is independent of a while x_0 depends on a in some (unobvious) way is especially insidious; the expected inequality may hold on *some* such interval for *each* value of a, but there may be *no* such interval on which the inequality holds for *all* values of a.

From the next section onward, and all through the book, we frequently encounter integrals that must be estimated. That is to say, be shown to be small in absolute value, or at least not too big. So some remarks on this topic are in order at this point.

Let $f : I \to \mathbb{C}$ be a continuous function on some interval $I \subseteq \mathbb{R}$. Consider the problem of bounding

$$\left| \int_I f(x)\, dx \right|$$

from above. The inequality

$$\left| \int_I f(x)\, dx \right| \le \int_I |f(x)|\, dx$$

is basic here, but we will briefly describe some techniques that go a little further. Note that

$$\left| \int_I f(x)\, dx \right| \le \int_I |f(x)|\, dx \le \int_I M\, dx = M\ell(I)$$

if $|f(x)| \le M$ for $x \in I$. Here $\ell(I)$ denotes the length of I. An extension of this argument yields

$$\left| \int_I f(x)g(x)\, dx \right| \le M \int_I |g(x)|\, dx$$

if $|f(x)| \le M$ for $x \in I$. For integrals where the integrand is the product of two functions, one of which is oscillatory, integration by parts is often useful. We have

$$\int_a^b f(x)g(x)\, dx = f(b)G(b) - \int_a^b f'(x)G(x)\, dx$$

assuming f continuously differentiable on $[a, b]$ and putting

$$G(x) = \int_a^x g(u)\, du.$$

If $g(x)$ oscillates on $[a, b]$, there is a good possibility that $G(x)$ will grow slowly on the interval. If in addition $f(x)$ changes fairly slowly on $[a, b]$, its derivative $f'(x)$ will be small in magnitude, and integration by parts may yield a very favorable estimate of the integral of $f(x)g(x)$ over $[a, b]$. This is a standard technique for estimating Fourier transforms.

Proposition 1.4 (Hölder inequality). *If I is an interval and f and g are continuous functions on I then*

$$\int_I |f(x)g(x)|\,dx \le \left(\int_I |f(x)|^p\,dx\right)^{1/p} \left(\int_I |g(x)|^q\,dx\right)^{1/q},$$

where $1 < p, q < \infty$ with $1/p + 1/q = 1$.

Proof. The case where

$$\int_I |f(x)|^p\,dx = 0 \quad \text{or} \quad \int_I |g(x)|^q\,dx = 0$$

is trivial. Multiplying f and g by suitable positive real numbers, we may therefore assume that

$$\int_I |f(x)|^p\,dx = \int_I |g(x)|^q\,dx = 1$$

by homogeneity. Now

$$|f(x)g(x)| = (|f(x)|^p)^{1/p}\,(|g(x)|^q)^{1/q} = (|f(x)|^p)^{1/p}\,(|g(x)|^q)^{1-1/p},$$

so

$$\log(|f(x)g(x)|) = \frac{1}{p}\log(|f(x)|^p) + \left(1 - \frac{1}{p}\right)\log(|g(x)|^q).$$

But the logarithm function is strictly concave, that is to say, all the chords lie strictly below the graph except for their endpoints. Hence

$$\begin{aligned}\frac{1}{p}&\log(|f(x)|^p) + \left(1 - \frac{1}{p}\right)\log(|g(x)|^q)\\ &\le \log\left(\frac{1}{p}\log(|f(x)|^p) + \left(1 - \frac{1}{p}\right)\log(|g(x)|^q)\right),\end{aligned}$$

and so

$$|f(x)g(x)| \le \frac{1}{p}\log(|f(x)|^p) + \left(1 - \frac{1}{p}\right)\log(|g(x)|^q).$$

Integrating over I yields

$$\int_I |f(x)g(x)|\,dx \le \frac{1}{p}\cdot 1 + \left(1 - \frac{1}{p}\right)\cdot 1 = 1,$$

and this proves the inequality. □

The case $p = 2$ is especially important; this is the Cauchy-Schwarz inequality. The Hölder inequality extends by induction to estimate integrals of products of more than two functions. We have

$$\int_I |f_1(x)\cdots f_n(x)|\,dx \le \left(\int_I |f_1(x)|^{p_1}\,dx\right)^{1/p_1} \cdots \left(\int_I |f_n(x)|^{p_n}\,dx\right)^{1/p_n}$$

if $1 < p_1, \ldots, p_n < \infty$ with $1/p_1 + \cdots + 1/p_n = 1$.

1.4. The Mertens estimates

Sums

$$\sum_{p\le x} f(p)$$

over primes occur frequently in analytic number theory. In the simpler cases, f is a positive, continuous, monotone function of a real variable that does not change rapidly. If the sum diverges as $x \to +\infty$, its asymptotic behavior may be guessed from the heuristic

$$\sum_{p\le x} f(p) \sim \sum_{2\le n\le x} \frac{f(n)}{\log(n)} \sim \int_2^x \frac{f(u)\,du}{\log(u)},$$

which is inspired by the observation that the density of the primes near n is close to $1/\log(n)$. The latter statement is a formulation of the Prime Number Theorem, and indeed the PNT with a good estimate for the error term is a natural tool with which to estimate such sums. As an example of the heuristic in action, consider the sum of $\log(p)/p$ over primes $p \le x$. The guess for the asymptotic behavior is

$$\sum_{p\le x} \frac{\log(p)}{p} \sim \sum_{2\le n\le x} \frac{\log(n)/n}{\log(n)} \sim \int_2^x \frac{du}{u} \sim \log(x),$$

and this is actually correct. Indeed, F. C. J. Mertens proved in 1874 that the absolute error in the asymptotic approximation is bounded. This may be expressed as

$$\sum_{p\le x} \frac{\log(p)}{p} = \log(x) + O(1)$$

by means of the big-O notation for the error term.

The heuristic for guessing the asymptotic behavior of sums over primes will not perform satisfactorily if f does not have the nice properties assumed. If, for example, f changes sign, there will be cancellation in the sum, and the underlying rationale for the heuristic does not take account of this.

Partial summation is perhaps the tool most frequently applied in analytic number theory. The basic version is the identity

$$\sum_{m=1}^{n} a_m b_m = b_n \sum_{m=1}^{n} a_m - \sum_{m=1}^{n-1} (b_{m+1} - b_m) \sum_{k=1}^{m} a_k.$$

This is an analogue, for sums, of integration by parts. The partial summation identity is easily proved by observing that $b_j(a_1 + \cdots + a_j) = b_j(a_1 + \cdots + a_{j-1}) + a_j b_j$ and applying mathematical induction. The partial summation identity yields a formula that is very convenient for estimating weighted sums of arithmetic functions.

Proposition 1.5 (Partial summation). *Let f be an arithmetic function and g a continuous function with piecewise continuous derivative on $[1,\infty)$. Then*

$$\sum_{n\le x} f(n)g(n) = F(x)g(x) - \int_1^x F(u)g'(u)\,du,$$

where F is the summatory function of f.

Proof. Calculate

$$\begin{aligned}\sum_{n\le x} f(n)g(n) &= F(x)g([x]) - \sum_{n=1}^{[x]-1}(g(n+1)-g(n))F(n)\\ &= F(x)g([x]) - \sum_{n=1}^{[x]-1} F(n)\int_n^{n+1} g'(u)\,du\\ &= F(x)g([x]) - \sum_{n=1}^{[x]-1}\int_n^{n+1} F(u)g'(u)\,du\\ &= F(x)g([x]) - \int_1^{[x]} F(u)g'(u)\,du\end{aligned}$$

by the partial summation identity and the Fundamental Theorem of Calculus. Then replace $[x]$ by x in the last step, for the resulting changes cancel. ☐

The partial summation formula is best understood in terms of the Stieltjes integral, but we eschew this refinement. The following bound for integrals involving the sawtooth function $S(x) = x - [x] - 1/2$ is sometimes useful.

Proposition 1.6. *If $1 \le a \le b$ the estimate*

$$\left|\int_a^b \frac{S(x)}{x^s}\,dx\right| \le \left(\frac{1}{8} + \frac{|s|}{16\sigma}\right) a^{-\sigma}$$

holds for any complex number $s = \sigma + it$ with positive real part σ.

Proof. If

$$\beta(x) = \frac{1}{16} + \int_0^x S(u)\,du$$

then $|\beta(x)| \le 1/16$. Integration by parts gives

$$\begin{aligned}\left|\int_a^b \frac{S(x)}{x^s}\,dx\right| &= \left|\frac{\beta(x)}{x^s}\Big|_a^b - \int_a^b (-s)\frac{\beta(x)}{x^{s+1}}\,dx\right| \le 2\cdot\frac{1/16}{a^\sigma} + \int_a^b |s|\frac{1/16}{x^{\sigma+1}}\,dx\\ &= \frac{1}{8a^\sigma} - \frac{|s|}{16\sigma}\frac{1}{x^\sigma}\Big|_a^b \le \frac{1}{8a^\sigma} + \frac{|s|}{16\sigma}\frac{1}{a^\sigma},\end{aligned}$$

since $S(x)$ is piecewise continuous. ☐

The last inequality yields asymptotic estimates for the summatory functions of $\log(m)$ and $1/m$. The first of these is a weak version of Stirling's formula.

Proposition 1.7. *The estimates*

$$\sum_{m=1}^{n} \log(m) = n\log(n) - n + \frac{\log(n)}{2} + 1 + R_n$$

and

$$\sum_{m=1}^{n} \frac{1}{m} = \log(n) + \gamma + \frac{1}{2n} + S_n$$

hold for all positive integers n with $|R_n| \leq 3/16$ and $|S_n| \leq 3/(16n^2)$.

Proof. The partial summation formula of Proposition 1.5 gives

$$\begin{aligned}\sum_{m=1}^{n} \log(m) &= n\log(n) - \int_1^n [u]\frac{du}{u} \\ &= n\log(n) - n + 1 + \frac{1}{2}\log(n) + \int_1^n S(u)\frac{du}{u}\end{aligned}$$

when $f(n) \equiv 1$ and $g(x) = \log(x)$. The last integral is bounded by $3/16$ in absolute value by Proposition 1.6. Using the partial summation formula again yields

$$\begin{aligned}\sum_{m=1}^{n} \frac{1}{m} &= n\cdot\frac{1}{n} - \int_1^n [u]\left(-\frac{du}{u^2}\right) \\ &= 1 + \log(n) + \frac{1}{2n} - \frac{1}{2} - \int_1^\infty S(u)\frac{du}{u^2} + \int_n^\infty S(u)\frac{du}{u^2}.\end{aligned}$$

The last term is bounded by $3/(16n^2)$ by Proposition 1.6. □

The real number $\gamma = 0.5772\ldots$ is known as the Euler-Mascheroni constant. It is unknown whether this is irrational. Note that Proposition 1.7 yields the version $T(x) = x\log(x) - x + O(\log(x))$ of Stirling's formula that is most commonly applied in analytic number theory.

Proposition 1.8 (Euler-Maclaurin summation formula)**.** *If $A < B$ are integers and f a continuous function on the interval $[A, B]$ with f' piecewise continuous there, then*

$$\sum_{n=A}^{B} f(n) = \int_A^B f(u)\,du + \frac{f(A)+f(B)}{2} + \int_A^B S(u)f'(u)\,du$$

with $S(u) = u - [u] - 1/2$ the sawtooth function.

Proof. Partial summation yields

$$\sum_{n=1}^{B} f(n) = Bf(B) - \int_1^B [u]f'(u)\,du$$

and

$$\sum_{n=1}^{A} f(n) = Af(A) - \int_1^A [u]f'(u)\,du.$$

Then

$$\int_A^B [u]f'(u)\,du = uf(u)\Big|_A^B - \sum_{n=A+1}^{B} f(n)$$

after taking the difference. Now

$$\int_A^B \left(u - \frac{1}{2}\right) f'(u)\,dt = \left(u - \frac{1}{2}\right) f(u)\Big|_A^B - \int_A^B f(u)\,du$$

by integration by parts. Subtracting the next to last formula from the last formula yields the Euler-Maclaurin summation formula. □

Despite the fact that anything obtainable from the Euler-Maclaurin summation formula may also be obtained by partial summation, resort to the former is sometimes more convenient. Moreover, repeated integration by parts in the Euler-Maclaurin summation formula yields a technique for obtaining precise approximations to sums. We make no use of this, so it is not covered here.

The next two results are due to Mertens. These depend on the Chebyshev bound $\psi(x) = O(x)$ in an essential way.

Proposition 1.9. *The estimates*

$$\sum_{p \le x} \frac{\log(p)}{p} = \sum_{n \le x} \frac{\Lambda(n)}{n} + O(1) = \log(x) + O(1)$$

hold.

Proof. First

$$x \sum_{n \le x} \frac{\Lambda(n)}{n} = \sum_{n \le x} \Lambda(n) \left[\frac{x}{n}\right] + O\left(\sum_{n \le x} \Lambda(n)\right)$$

because $0 \le x - [x] < 1$. Now

$$T(x) = \sum_{m \le x} \Lambda(m) \left[\frac{x}{m}\right]$$

yields

$$\sum_{n\le x}\frac{\Lambda(n)}{n}=\frac{T(x)}{x}+O\left(\frac{\psi(x)}{x}\right)=\log(x)+O(1)$$

by Stirling's formula and Proposition 1.1. The series

$$\sum_{p}\sum_{k=2}^{\infty}\frac{\log(p)}{p^k}$$

converges, and $\Lambda(n)$ is zero off the prime powers. □

Sometimes it is necessary to remove a factor from the terms of a sum. This is an important application of the partial summation formula, and is illustrated in the proof of the next result.

Proposition 1.10. *The estimate*

$$\sum_{p\le x}\frac{1}{p}=\log\log(x)+a+O\left(\frac{1}{\log(x)}\right)$$

holds with some constant a.

Proof. First

$$\begin{aligned}\sum_{p\le x}\frac{1}{p}&=\sum_{p\le x}\frac{\log(p)}{p}\frac{1}{\log(p)}\\&=\left(\sum_{p\le x}\frac{\log(p)}{p}\right)\frac{1}{\log(x)}-\int_2^x\left(\sum_{p\le u}\frac{\log(p)}{p}\right)\frac{(-1)\,du}{u\log^2(u)}\end{aligned}$$

by partial summation. Then

$$\begin{aligned}\sum_{p\le x}\frac{1}{p}&=1+O\left(\frac{1}{\log(x)}\right)+\int_2^x\frac{du}{u\log(u)}\\&\quad+\int_2^x\left(\sum_{p\le u}\frac{\log(p)}{p}-\log(u)\right)\frac{du}{u\log^2(u)}\\&=1+O\left(\frac{1}{\log(x)}\right)+\log\log(x)-\log\log(2)\\&\quad+\int_2^\infty\left(\sum_{p\le u}\frac{\log(p)}{p}-\log(u)\right)\frac{du}{u\log^2(u)}+\int_x^\infty\frac{O(1)}{u\log^2(u)}\,du\\&=\log\log(x)+a+O\left(\frac{1}{\log(x)}\right)\end{aligned}$$

by Proposition 1.9 and integration by parts. □

The next result is of considerable significance in prime number theory for various considerations of a probabilistic nature. The fact that the constant b in the formula is positive is important in such contexts. Actually b equals the Euler-Mascheroni constant γ, though we won't prove this.

Proposition 1.11 (Mertens' formula). *The estimate*

$$\prod_{p\leq x}\left(1-\frac{1}{p}\right)\sim\frac{e^{-b}}{\log(x)}$$

holds with some constant b.

Proof. First

$$\log\left(\prod_{p\leq x}\left(1-\frac{1}{p}\right)\right)=\sum_{p\leq x}\log\left(1-\frac{1}{p}\right)=-\sum_{p\leq x}\sum_{k=1}^{\infty}\frac{1}{kp^k}$$

where

$$-\sum_{p\leq x}\sum_{k=1}^{\infty}\frac{1}{kp^k}=-\sum_{p\leq x}\frac{1}{p}-\sum_{p}\sum_{k=2}^{\infty}\frac{1}{kp^k}+\sum_{p>x}\sum_{k=2}^{\infty}\frac{1}{kp^k}.$$

The first term on the right-hand side may be estimated by means of Proposition 1.10, the second term is a convergent infinite series, and the third term tends to zero as $x\to+\infty$. Thus

$$\prod_{p\leq x}\left(1-\frac{1}{p}\right)\sim\exp(-\log\log(x)-b)=\frac{e^{-b}}{\log(x)}$$

by exponentiating. □

About half of all the integers n with $y<n\leq x$ for x and $x-y$ large are even, one third are divisible by three, and so forth. A suggestive way of phrasing this observation is to say that the chance of a randomly chosen large integer n being divisible by a prime p is $1/p$. An integer $n\geq 2$ that is not divisible by any prime $p\leq\sqrt{n}$ is itself prime. So if for $\sqrt{x}\leq y<n\leq x$ the events $p|n$ and $q|n$ for distinct primes $p,q\leq\sqrt{x}$ are independent in the sense of probability theory, the chance of n being prime should be

$$\prod_{p\leq\sqrt{x}}\left(1-\frac{1}{p}\right)\sim\frac{e^{-\gamma}}{\log(\sqrt{x})}=\frac{2e^{-\gamma}}{\log(x)}>\frac{1.12}{\log(x)}.$$

But the density of the primes near x is close to $1/\log(x)$ by the Prime Number Theorem. We conclude that the events $p|n$ and $q|n$ are not independent. It is easy to persuade oneself that independence must hold for pairs of distinct primes that are very small compared with n. Thus Mertens' formula reveals an aspect of divisibility of integers by comparatively large primes.

1.5. Sums over divisors

We remind the reader that an arithmetic function f is a mapping $f : \mathbb{N} \to \mathbb{C}$. Some arithmetic functions such as log arise by restricting functions of a real variable to the positive integers. But in most cases of interest $f(n)$ is determined by arithmetical information about the integer n. The *divisor function* $d(n)$ given as the number of positive divisors of the positive integer n is an example. Each positive integer n has a unique factorization

$$n = p_1^{\alpha_1} \cdots p_r^{\alpha_r}$$

into primes and the divisors d of n are the integers of the form

$$d = p_1^{\beta_1} \cdots p_r^{\beta_r}$$

where the β_j are integers satisfying $0 \le \beta_j \le \alpha_j$ for $j = 1, 2, \ldots, r$. Hence $d(n) = (\alpha_1 + 1)(\alpha_2 + 1) \cdots (\alpha_r + 1)$ is a formula for the divisor function $d(n)$ given in terms of the prime factorization of n.

An arithmetic function f is *additive* if $f(mn) = f(m) + f(n)$ whenever $\gcd(m, n) = 1$. It is *multiplicative* if $f \not\equiv 0$ and $f(mn) = f(m)f(n)$ whenever $\gcd(m, n) = 1$. It is *totally additive* or *totally multiplicative* if the corresponding property holds without requiring the condition $\gcd(m, n) = 1$. A multiplicative or additive function can be unambiguously prescribed by giving its values on the prime powers, and a totally multiplicative or totally additive function by giving its values on the primes. Note also that $f(1) = 1$ if f is multiplicative.

The function $\log(n)$ and the function $\Omega(n)$ that counts the prime divisors of n with multiplicity are totally additive. The function $\omega(n)$ that counts the distinct prime divisors of n is additive, but not totally additive. The identity function id given by $n \mapsto n$ is totally multiplicative. So is the *Liouville function*

$$\lambda(n) = (-1)^{\Omega(n)}.$$

The divisor function $d(n)$ is multiplicative, but not totally multiplicative. The Euler phi-function $\phi(n)$ is also multiplicative. Another multiplicative arithmetic function is the *radical*

$$\operatorname{rad}(n) = \prod_{p|n} p.$$

It is also called the *squarefree kernel.*

The von Mangoldt function $\Lambda(n)$ is an important arithmetic function that is neither additive nor multiplicative.

The product of two multiplicative functions is multiplicative and the sum of two additive functions is additive. But a more important algebraic operation on arithmetic functions is the *Dirichlet convolution.* If f and g are arithmetic functions then

$$(f * g)(n) = \sum_{d|n} f(d)g\Big(\frac{n}{d}\Big) = \sum_{km=n} f(k)g(m)$$

is their Dirichlet convolution. It is a straightforward exercise to see that under addition and Dirichlet convolution, the arithmetic functions form a commutative ring with multiplicative neutral element e where $\mathrm{e}(1) = 1$ and $\mathrm{e}(n) = 0$ for $n \geq 2$. This is called the *Dirichlet ring.* Denoting the constant function equal to 1 by 1 we note $d = 1 * 1$ as an example of Dirichlet convolution. Another convolution identity is $1 * \Lambda = \log$. This is easily proved by observing that

$$(1 * \Lambda)(n) = \sum_{d|n} \Lambda(d) = \sum_{p^k|n} \log(p) = \log(n),$$

since Λ is zero off the prime powers. This can replace the Legendre identity as the point of entry for the method of Chebyshev.

Proposition 1.12. *If f and g are multiplicative, so is $f * g$.*

Proof. If $\gcd(m, n) = 1$, then the divisors $d|mn$ are precisely those positive integers of the form $d = bc$ where $b|m$ and $c|n$. Hence

$$\begin{aligned}(f * g)(mn) &= \sum_{d|mn} f(d)g\Big(\frac{mn}{d}\Big) = \sum_{b|m,c|n} f(bc)g\Big(\frac{mn}{bc}\Big)\\ &= \sum_{b|m,c|n} f(b)f(c)g\Big(\frac{m}{b}\Big)g\Big(\frac{n}{c}\Big)\\ &= \sum_{b|m} f(b)g\Big(\frac{m}{b}\Big)\sum_{c|n} f(c)g\Big(\frac{n}{c}\Big)\\ &= (f * g)(m)(f * g)(n)\end{aligned}$$

by the multiplicativity of f and g. □

Since 1 is multiplicative, Proposition 1.12 shows that d is also multiplicative. The *sum-of-divisors function*

$$\sigma(n) = \sum_{d|n} d$$

is given by the Dirichlet convolution $\sigma = 1 * \mathrm{id}$, so σ is multiplicative, because id is.

Part of the significance of Proposition 1.12 is that for Dirichlet convolutions of multiplicative functions it affords a straightforward means of calculation; it is enough to calculate their values on prime powers. Let us, for example, calculate the Dirichlet convolution $1 * \phi$. Both factors are multiplicative, so the calculation

$$(1 * \phi)(p^\alpha) = \sum_{p^\beta | p^\alpha} 1 \cdot \phi(p^\beta) = 1 + \sum_{\beta=1}^{\alpha} (p-1)p^{\beta-1} = p^\alpha$$

yields the convolution identity $1 * \phi = \mathrm{id}$ of Gauss.

The *Möbius mu-function* μ is the unique multiplicative arithmetic function with values $\mu(p) = -1$ on the primes p, and values $\mu(p^k) = 0$ on the prime powers p^k with $k \geq 2$. The Möbius function has a strong combinatorial flavor. It is closely connected to the principle of inclusion and exclusion, and to the fact that the integers form a partially ordered set under the relation of divisibility. The importance of the Möbius function is due to the convolution identity

$$\sum_{d|n} \mu(d) = \mathrm{e}(n).$$

There are many ways to establish that $1 * \mu = \mathrm{e}$, but the quickest is to recall that μ is multiplicative. Then so is $1 * \mu$, and thus $(1 * \mu)(p^\alpha) = 1 + (-1) + 0 + 0 + \cdots = 0$ yields $(1 * \mu)(n) = 0$ for all $n \geq 2$.

Proposition 1.13 (First Möbius inversion formula). *If $g = 1 * f$ then $f = \mu * g$ and conversely.*

Proof. If $g = 1 * f$, then $\mu * g = \mu * (1 * f) = (\mu * 1) * f = \mathrm{e} * f = f$, and if $f = \mu * g$ then $1 * f = 1 * (\mu * g) = (1 * \mu) * g = \mathrm{e} * g = g$. □

This shows that 1 is a unit in the Dirichlet ring, and μ is its multiplicative inverse. An arithmetic function f is a unit if and only if $f(1) \neq 0$. Under this condition a Dirichlet inverse g for f may be constructed incrementally from

$$g(n) = \mathrm{e}(n) - \sum_{n \neq d | n} f\left(\frac{n}{d}\right) g(d).$$

The relation $f(1)g(1) = \mathrm{e}(1) = 1$ shows that the condition $f(1) \neq 0$ is necessary. Constructing an explicit Dirichlet inverse is usually infeasible, except in the very important case when f is multiplicative. The first Möbius inversion formula is often formulated as the statement

$$\text{“}f(n) = \sum_{d|n} \mu(d) g\left(\frac{n}{d}\right) \quad \text{if and only if} \quad g(n) = \sum_{d|n} f(d)\text{”}$$

about divisor sums.

Proposition 1.14 (Second Möbius inversion formula). *Suppose that F is a function on the interval $[1, \infty)$. If*

$$G(x) = \sum_{n \le x} F\Big(\frac{x}{n}\Big)$$

then

$$F(x) = \sum_{n \le x} \mu(n) G\Big(\frac{x}{n}\Big)$$

and conversely on this interval.

Proof. First

$$\begin{aligned}\sum_{n \le x} \mu(n) G\Big(\frac{x}{n}\Big) &= \sum_{n \le x} \mu(n) \sum_{m \le x/n} F\Big(\frac{x/n}{m}\Big) = \sum_{mn \le x} \mu(n) F\Big(\frac{x}{mn}\Big)\\ &= \sum_{N \le x} F\Big(\frac{x}{N}\Big) \sum_{n \mid N} \mu(n) = F(x)\end{aligned}$$

and then

$$\begin{aligned}\sum_{n \le x} F\Big(\frac{x}{n}\Big) &= \sum_{n \le x} \sum_{m \le x/n} \mu(m) G\Big(\frac{x/n}{m}\Big) = \sum_{mn \le x} \mu(m) G\Big(\frac{x}{mn}\Big)\\ &= \sum_{N \le x} G\Big(\frac{x}{N}\Big) \sum_{m \mid N} \mu(m) = G(x)\end{aligned}$$

since $1 * \mu = e$. □

The second Möbius inversion formula throws light on the method of Chebyshev. The relation

$$\psi(x) = \sum_{n \le x} \mu(n) T\Big(\frac{x}{n}\Big)$$

holds by Möbius inversion. Since $T(x)$ is quite precisely known, it might seem possible to estimate $\psi(x)$ fairly accurately by means of this formula. The problem here is that there is a great deal of cancellation in the sum, due to the oscillation of sign of $\mu(n)$. Too little is known about the behavior of $\mu(n)$ for this approach to promise much success. But the estimates of Chebyshev may be obtained by replacing $\mu(n)$ by an approximation of a particular kind. The approximation associated with the proof of Proposition 1.1 is $\mu(n) \approx \mathrm{e}_1(n) - 2\mathrm{e}_2(n)$ where $\mathrm{e}_k(n)$ is the arithmetic function that equals

1 for $n = k$ and is zero otherwise. The calculation

$$\sum_{n\le x}(\mathrm{e}_1(n) - 2\mathrm{e}_2(n))T\Big(\frac{x}{n}\Big) = \sum_{n\le x}(\mathrm{e}_1(n) - 2\mathrm{e}_2(n)) \sum_{m\le x/n} \psi\Big(\frac{x/n}{m}\Big)$$
$$= \sum_{k\le x}\psi\Big(\frac{x}{k}\Big)\sum_{d|k}(\mathrm{e}_1(d) - 2\mathrm{e}_2(d))$$
$$= \sum_{k\le x}(-1)^{k-1}\psi\Big(\frac{x}{k}\Big)$$

may be taken as the framework of the proof of Proposition 1.1. It can be generalized by replacing $\mathrm{e}_1 - 2\mathrm{e}_2$ with a more complicated linear combination f of $\mathrm{e}_1, \mathrm{e}_2, \ldots, \mathrm{e}_N$. Then $1 * f$ is required to take only the values $0, \pm 1$, and the nonzero values of this function should start with $(1 * f)(1) = 1$ and alternate. Chebyshev chose the linear combination $f = \mathrm{e}_1 - \mathrm{e}_2 - \mathrm{e}_3 - \mathrm{e}_5 + \mathrm{e}_{30}$ to obtain his better estimates.

We exhibit an example of the use of Möbius inversion to establish an arithmetically significant estimate. We find a quite precise bound for the error term in an asymptotic estimate for the summatory function

$$\Phi(x) \stackrel{\text{def}}{=} \sum_{n\le x}\phi(n)$$

of the Euler totient. The convolution identity of Gauss yields

$$\sum_{n\le x}\Phi\Big(\frac{x}{n}\Big) = \sum_{n\le x}\sum_{d\le x/n}\phi(d) = \sum_{nd\le x}\phi(d) = \sum_{m\le x}\sum_{d|m}\phi(d)$$
$$= \sum_{m\le x}(1*\phi)(m) = \sum_{m\le x} m = \frac{1}{2}[x]([x]+1),$$

and then

$$\Phi(x) = \sum_{n\le x}\mu(n)\frac{1}{2}\Big[\frac{x}{n}\Big]\Big(\Big[\frac{x}{n}\Big]+1\Big) = \frac{1}{2}\sum_{n\le x}\mu(n)\Big(\frac{x}{n}+O(1)\Big)\Big(\frac{x}{n}+O(1)\Big)$$
$$= \frac{x^2}{2}\sum_{n\le x}\frac{\mu(n)}{n^2} + O\Bigg(x\sum_{n\le x}\frac{1}{n}\Bigg) = \frac{x^2}{2}\sum_{n=1}^{\infty}\frac{\mu(n)}{n^2} + O\Bigg(x^2\sum_{n>x}\frac{1}{n^2}\Bigg)$$
$$+ O(x\log(x)) = \frac{x^2}{2}\sum_{n=1}^{\infty}\frac{\mu(n)}{n^2} + O(x\log(x))$$

by the second Möbius inversion formula. Absolutely convergent series may be multiplied together to yield absolutely convergent series, and so

$$\Bigg(\sum_{m=1}^{\infty}\frac{1}{m^2}\Bigg)\Bigg(\sum_{n=1}^{\infty}\frac{\mu(n)}{n^2}\Bigg) = \sum_{N=1}^{\infty}\frac{1}{N^2}\sum_{mn=N}1\cdot\mu(n) = \sum_{N=1}^{\infty}\frac{\mathrm{e}(N)}{N^2} = 1.$$

Then

$$\Phi(x) = \frac{3}{\pi^2}x^2 + O(x\log(x))$$

by the famous formula

$$\sum_{m=1}^{\infty} \frac{1}{m^2} = \frac{\pi^2}{6}$$

of Euler.

An Introduction to the Theory of Numbers by G. H. Hardy and E. M. Wright contains interesting material on arithmetic functions and applications of elementary techniques in number theory. Another good source for such material is *Introduction to the Theory of Numbers* by H. N. Shapiro.

1.6. The hyperbola method

Many arithmetic functions fluctuate rapidly and substantially, but we may still want precise information about their growth. There are various ways to approach such questions, differing not just in the methods used, but more fundamentally in the kind of statement at which one aims. A positive function g is a *maximal order* for an arithmetic function f if for any $\varepsilon > 0$ the inequality $|f(n)| \leq (1+\varepsilon)g(n)$ holds for all n sufficiently large, while the inequality $|f(n)| \leq (1-\varepsilon)g(n)$ fails for infinitely many n no matter the choice of $\varepsilon > 0$. For the concept to be useful, the maximal order should be some simple function that grows reasonably evenly. Otherwise we could just choose $g \equiv f$ and be done. Naturally there is also an analogous concept of minimal orders for arithmetic functions, though for functions that fluctuate greatly, minimal orders are often of little interest. An easy example showing the strengths and weaknesses of this approach is the function

$$\Omega(n) = \sum_{p^\alpha | n} 1$$

that counts the prime divisors of n with multiplicity. To see how large $\Omega(n)$ could be for given n, it is natural to look at integers that have many prime factors for their size. Because repeated prime factors are counted, this leads us to the powers of 2. Indeed the inequality

$$n = p_1^{\alpha_1} \cdots p_r^{\alpha_r} \geq 2^{\alpha_1 + \cdots + \alpha_r} = 2^{\Omega(n)}$$

is strict unless n is a power of 2, and it shows that $\Omega(n) \leq k$ for $2^k \leq n < 2^{k+1}$. So $\log(n)/\log(2)$ is a maximal order for $\Omega(n)$. The advantage here is that the statement is valid for all n individually; the disadvantage is that $\Omega(n)$ is actually very much smaller than the maximal order $\log(n)/\log(2)$ for most n; see Proposition 2.7. The analogous question for the divisor function $d(n)$ lies a little deeper. This function does not itself have a tractable maximal order, but its logarithm does.

Proposition 1.15. $\log(d(n))$ *has maximal order* $\log(2)\log(n)/\log\log(n)$.

Proof. Write the prime factorization of an arbitrary positive integer n in logarithmic form

$$\log(n) = \alpha_1 \log(p_1) + \cdots + \alpha_r \log(p_r).$$

The inequality $\alpha_k \log(2) \le \alpha_k \log(p_k) \le \log(n)$ is an immediate consequence, so $\alpha_k \le \log(n)/\log(2)$. Furthermore

$$\log(d(n)) = \log(\alpha_1 + 1) + \cdots + \log(\alpha_r + 1),$$

and the inequality $\log(\alpha_k + 1) \le \alpha_k \log(2)$ also holds. Apply the first inequality when $\log(p_k)$ is comparatively small, and the second inequality otherwise. Suppose $\log(p_k) \le c$ precisely when $1 \le k \le m$, where c is a parameter. Then

$$\begin{aligned}
\log(d(n)) &= \sum_{k=1}^{m} \log(\alpha_k + 1) + \sum_{k=m+1}^{r} \log(\alpha_k + 1) \\
&\le e^c \log\left(\frac{\log(n)}{\log(2)} + 1\right) + \sum_{k=m+1}^{r} \alpha_k \log(2) \\
&\le e^c \log\left(\frac{\log(n)}{\log(2)} + 1\right) + \frac{\log(2)}{c} \sum_{k=m+1}^{r} \alpha_k \log(p_k) \\
&\le e^c \log\left(\frac{\log(n)}{\log(2)} + 1\right) + \frac{\log(2)\log(n)}{c}
\end{aligned}$$

since $m \le \exp(c)$. Now choose $c = (1-\delta)\log\log(n)$ with $0 < \delta < 1$. Now

$$\log(d(n)) \le \log^{1-\delta}(n) \log\left(\frac{\log(n)}{\log(2)} + 1\right) + (1-\delta)^{-1} \frac{\log(2)\log(n)}{\log\log(n)}.$$

Since δ may be chosen arbitrarily close to 0, for every $\varepsilon > 0$ there is some $n(\varepsilon)$ so that $\log(d(n)) < (1+\varepsilon)\log(2)\log(n)/\log\log(n)$ for $n \ge n(\varepsilon)$.

Let p be any prime so large that $\vartheta(p) \ge p/e$ and let $n = 2\cdot 3 \cdots p$ be the product of all the primes up to and including p. Then

$$\begin{aligned}
\frac{\log(d(n))}{\frac{\log(2)\log(n)}{\log\log(n)}} &= \frac{\pi(p)\log(2)}{\frac{\log(2)\vartheta(p)}{\log(\vartheta(p))}} \ge \frac{\pi(p)\log(2)}{\frac{\log(2)\pi(p)\log(p)}{\log(\vartheta(p))}} \\
&= \frac{\log(\vartheta(p))}{\log(p)} \ge \frac{\log(p/e)}{\log(p)} = 1 - \frac{1}{\log(p)}.
\end{aligned}$$

But p can be taken arbitrarily large. □

The above result immediately yields the weaker bound $d(n) \ll_\varepsilon n^\varepsilon$, valid for any $\varepsilon > 0$. This bound is usually more convenient in applications.

Another approach to study the growth of the rapidly fluctuating arithmetic function $d(n)$ is to consider a *local average* such as

$$\frac{1}{h}\sum_{x-h<n\le x} d(n).$$

The fluctuations of $d(n)$ are smoothed out by the process of averaging. Then

$$\frac{1}{h}\sum_{x-h<n\le x} d(n) = \frac{D(x)-D(x-h)}{h}$$

where $D(x)$ is the summatory function of the divisor function. The use of local averaging to study the growth of arithmetic functions can be traced back to article 301 in the *Disquisitiones Arithmeticae.* Gauss was interested in the growth of the class number and the number of genera for binary quadratic forms, as (irregularly fluctuating) arithmetic functions of the discriminant. He quoted results on the rate of growth of their local averages, but judged the proofs to be too difficult to include in the *Disquisitiones.*

To calculate a local average by differencing an estimate for the associated summatory function is not always efficient. When h is small, one is apt to run into the same kind of problem as one does in numerics when subtracting floating-point numbers that are nearly equal.

The bijection $d \mapsto n/d$ on the set of divisors d of an integer n is called the *Dirichlet interchange.* Since $d < \sqrt{n}$ is equivalent to $n/d > \sqrt{n}$ it is clear that

$$d(n) = 2\sum_{\sqrt{n}>d|n} 1$$

unless n is a square. In the latter case the divisor $\sqrt{n}$ is missing and it is necessary to add 1 on the right-hand side. Applying this formula to the definition of $D(x)$ gives

$$\begin{aligned} D(x) &= [\sqrt{x}] + \sum_{n\le x} 2\sum_{\sqrt{n}>d|n} 1 = [\sqrt{x}] + 2\sum_{d\le\sqrt{x}}\sum_{d^2<kd\le x} 1 \\ &= [\sqrt{x}] + 2\sum_{d\le\sqrt{x}}\left(\left[\frac{x}{d}\right]-d\right) = 2\sum_{m\le\sqrt{x}}\left[\frac{x}{m}\right] - [\sqrt{x}]^2. \end{aligned}$$

The latter formula is due to D. F. E. Meissel.

Proposition 1.16. *The estimate*

$$D(x) = x\log(x) + (2\gamma-1)x + O(x^{1/2})$$

holds.

Proof. We have

$$\begin{aligned}
D(x) &= 2\sum_{n\le\sqrt{x}}\left[\frac{x}{n}\right]-[\sqrt{x}]^2\\
&= 2\sum_{n\le\sqrt{x}}\left(\frac{x}{n}+O(1)\right)-(\sqrt{x}+O(1))^2\\
&= 2x(\log(\sqrt{x})+\gamma+O(1/\sqrt{x}))-x+O(x^{1/2})\\
&= x\log(x)+(2\gamma-1)x+O(x^{1/2})
\end{aligned}$$

by Proposition 1.7. □

This estimate is due to J. P. G. Lejeune Dirichlet. It implies that the arithmetic average of $d(n)$ over the range $1\le n\le x$ is asymptotic to $\log(x)$ as $x\to+\infty$. One says that $d(n)$ has *average order* $\log(x)$. From Proposition 1.15 it is easy to see that $d(n)$ is sometimes larger than any fixed power of $\log(n)$. But Proposition 1.16 implies that $d(n)$ is only rarely so large. The notation $\Delta(x)=D(x)-x\log(x)-(2\gamma-1)x$ is traditional for the error term in the estimate in Proposition 1.16. The problem of bounding $\Delta(x)$ is known as the *Dirichlet divisor problem.* More precisely, the divisor problem is to find the least ϑ for which an estimate $\Delta(x)=O(x^{\vartheta+\varepsilon})$ holds for all $\varepsilon>0$. The result just proved shows that $\vartheta\le 1/2$. For x large and h rather smaller than x, say $h<x/2$, one obtains

$$\begin{aligned}
\frac{1}{h}\sum_{x-h<n\le x}d(n) &= \frac{D(x)-D(x-h)}{h}\\
&= \frac{x\log(x)+(2\gamma-1)x+\Delta(x)}{h}\\
&\quad -\frac{(x-h)\log(x-h)+(2\gamma-1)(x-h)+\Delta(x-h)}{h}\\
&= \log(x)+2\gamma+O\left(\frac{h}{x}\right)+O\left(\frac{x^{1/2}}{h}\right)
\end{aligned}$$

by the estimate $\Delta(x)=O(x^{1/2})$. The error is a sum of two terms, one of which dominates when h is large and the other when h is small. In such situations one would usually try to choose the parameter optimally to obtain a small error term overall. Minimizing $h/x+x^{1/2}/h$ over h for x fixed, one sees that $h=x^{3/4}$ is an optimal choice. Thus

$$\frac{1}{h}\sum_{x-h<n\le x}d(n)=\log(x)+2\gamma+O(x^{-1/4}),\qquad h=x^{3/4}.$$

The asymptotic law of growth $\log(x)+2\gamma$ for the local average of $d(n)$ was Dirichlet's main application in his 1849 paper on the divisor problem. The asymptotic estimate for the local average may be improved in two different ways: By shortening the interval or reducing the error term. Modifying the estimate as it stands by shortening the interval as much as possible, we lose the error term and obtain the asymptotic estimate

$$\frac{1}{h}\sum_{x-h<n\leq x} d(n) \sim \log(x)+2\gamma, \qquad h=o(x^{1/2}/\log(x)).$$

Using a better estimate in the divisor problem, the estimate for the local average may be improved both by reducing the error term and shortening the interval.

One may also prove Proposition 1.16 from Dirichlet's formula

$$D(x)=\sum_{dk\leq x}1=\sum_{n\leq x}\left[\frac{x}{n}\right].$$

That $D(x)=x\log(x)+O(x)$ is immediate from this formula. The better estimate for the error term may be obtained by observing that the sum equals the number of integer lattice points in the region of the uv-plane given by the inequalities $u\geq 1$, $v\geq 1$ and $uv\leq x$. One can then recover the formula of Meissel by observing that the union of the two subregions obtained by imposing the inequalities $u\leq\sqrt{x}$ and $v\leq\sqrt{x}$ equals the original region, while their intersection equals the square given by $1\leq u\leq\sqrt{x}$ and $1\leq v\leq\sqrt{x}$. The interpretation of $D(x)$ in terms of the number of lattice points under a hyperbola is very important for more advanced work on the divisor problem. See Figure 2 on page 26 for an illustration.

The Dirichlet interchange and the approach to the Meissel formula based on counting lattice points under a hyperbola are closely connected. The technique is usually called the *Dirichlet hyperbola method.* It has other applications and so we exhibit a more general formulation due to H. G. Diamond.

Proposition 1.17 (Dirichlet hyperbola method). *If f is an arithmetic function with summatory function F and g an arithmetic function with summatory function G then*

$$\sum_{n\leq x}(f*g)(n)=\sum_{k\leq y}f(k)G\Big(\frac{x}{k}\Big)+\sum_{m\leq x/y}g(m)F\Big(\frac{x}{m}\Big)-F(y)G\Big(\frac{x}{y}\Big)$$

for $1\leq y\leq x$.

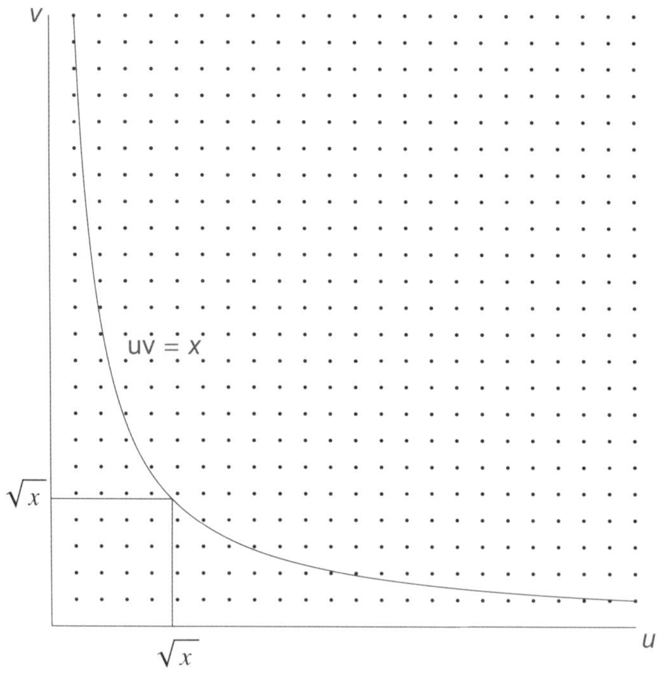

Figure 2. Lattice points in the divisor problem

Proof. We have

$$\begin{aligned}\sum_{n\le x}(f*g)(n) &= \sum_{n\le x}\sum_{km=n} f(k)g(m)\\ &= \sum_{k\le y} f(k)\sum_{km\le x} g(m) + \sum_{k>y}\sum_{km\le x} f(k)g(m)\\ &= \sum_{k\le y} f(k)G\Big(\frac{x}{k}\Big) + \sum_{m\le x/y} g(m)\sum_{y<k\le x/m} f(k)\\ &= \sum_{k\le y} f(k)G\Big(\frac{x}{k}\Big) + \sum_{m\le x/y} g(m)\left(\sum_{k\le x/m} f(k) - \sum_{k\le y} f(k)\right)\\ &= \sum_{k\le y} f(k)G\Big(\frac{x}{k}\Big) + \sum_{m\le x/y} g(m)F\Big(\frac{x}{m}\Big) - F(y)G\Big(\frac{x}{y}\Big)\end{aligned}$$

for any y with $1 \le y \le x$. □

Questions about divisors more delicate than their gross count have also been studied. The monograph *Divisors* by R. R. Hall and G. Tenenbaum is a good source for information of this kind.

Elementary Methods in the Analytic Theory of Numbers by A. O. Gelfond and Y. V. Linnik is a classic that covers some of the material of this chapter in greater depth, and other topics as well. *Elementary methods in number theory* by M. B. Nathanson and *Not Always Buried Deep* by P. Pollack also treat elementary techniques in analytic number theory. Material of this kind may also be found in *Introduction to the Theory of Numbers* by G. H. Hardy and E. M. Wright, *Lectures on Elementary Number Theory* by Hans Rademacher, and *Introduction to the Theory of Numbers* by H. N. Shapiro.

1.7. Notes

Gauss never published his empirical investigations on the distribution of the primes, but these are known from a letter that he wrote on Christmas Eve of 1849 to a former student of his, the astronomer J. F. F. Encke, and also from cryptic jottings in his research diary and on a flyleaf of a logarithm table. See pages 444–447 of volume II and pages 11–18 of volume X of his collected works [**Gau33**].

Legendre published the prime factorization of the factorial in the 1808 edition of his treatise *Essai sur la Théorie des Nombres* [**Leg08**]. There he also proposed

$$\pi(x) \approx \frac{x}{\log(x) - 1.08366}$$

as an excellent approximation. Today it is known that $\mathrm{li}(x)$ is a much better approximation for very large x, and that the asymptotically best approximation to $\pi(x)$ of the form $x/(\log(x)-A)$ is obtained for $A = 1$. But Legendre's approximation is better than Gauss' approximation in the interval between $x = 10^2$ and $x = 4{\cdot}10^6$, which stretches beyond the range of the tables of primes available in the early nineteenth century. Gauss makes a comment in his letter to Encke on the approximation of Legendre, to the effect that he does not care to commit himself as to what limit $A(x)$ in

$$\pi(x) = \frac{x}{\log(x) - A(x)},$$

may tend to as $x \to +\infty$.

Legendre made yet a third discovery of great importance to the development of prime number theory. Since antiquity an algorithm had been known for efficiently constructing tables of primes. The algorithm is called the Sieve of Eratosthenes, after the Hellenistic scholar Eratosthenes of Cyrene. We will explain how his algorithm may be used to construct a table of primes up to 30. Start with a list of the integers n with $2 \leq n \leq 30$. Keep the integer 2 but strike out all its proper multiples. Then keep the next integer 3, but strike out all its proper multiples. Next keep 5, but strike out all its proper multiples. The integers left in the list are the primes $2 \leq p \leq 30$. For every composite integer $n \leq 30$ has some prime divisor $p \leq \sqrt{30} < 5.5$. Note that we obtain the primes in the interval $[6, 30]$ by removing from the set of integers in that interval those that lie in the three arithmetic progressions $2\mathbb{Z}$, $3\mathbb{Z}$ and $5\mathbb{Z}$. Legendre reformulated the Sieve of Eratosthenes in terms of the principle of exclusion and inclusion from combinatorics, potentially making it available to count primes analytically. A modern version of Legendre's sieve formula is

$$\pi(x) = \pi(\sqrt{x}) - 1 + \sum_{d|P} \mu(d) \left[\frac{x}{d}\right], \qquad P = \prod_{p \le \sqrt{x}} p.$$

Though the Legendre sieve formula gives $\pi(x)$ exactly, it is difficult to extract strong information about the distribution of the primes from it and many years passed before the idea of sieving had any impact on number theory beyond the preparation of tables and other computational work. The main objective of modern sieve theory is to find upper and lower bounds on the number of elements of a finite set of integers remaining after those elements that lie in some prescribed arithmetic progressions have been removed. This is such a flexible framework that a wide variety of arithmetical problems are amenable to sieve methods to some extent.

The young French mathematician J. Merlin began making progress on sieve theory around 1911, but he fell in the Great War. From 1915 V. Brun obtained results on the primes by means of sieving, that were not accessible by any other method. Since then, a vast amount of work has been done to improve sieve methods and develop new and more powerful ones. They have had a very broad impact on analytic number theory, leading to proofs of many important results, often in combination with ideas from outside sieve theory. Some of the most striking results obtained by means of sieves are:

- The series of reciprocals of primes p for which $p+2$ is prime, is finite or convergent (Brun [**Bru15**] in 1915.)
- There are infinitely many primes p for which $p+2$ has at most two prime factors (J. R. Chen [**Che73**] in 1966.)
- For every sufficiently large even integer n there is some prime p and some integer m with at most two prime factors so that $n = p + m$ (Chen [**Che73**] in 1966.)
- There are infinitely many integers m for which $m^2 + 1$ has at most two prime factors (H. Iwaniec [**Iwa78**] in 1978.)
- The polynomial $m^2 + n^4$ takes infinitely many prime values (J. B. Friedlander and H. Iwaniec [**FI98**] in 1998.)
- The polynomial $m^3 + 2n^3$ takes infinitely many prime values (D. R. Heath-Brown [**HB01**] in 2001.)

The first two results are related to the twin prime problem, to prove that there are infinitely many pairs of primes that differ by 2. The third is related to the binary Goldbach problem, asserting that every even integer $n \ge 4$ is the sum of two primes, and the next two are related to the conjecture that the polynomial $m^2 + 1$ takes infinitely many prime values. All three problems are old, and all are unsolved. The first of the six results above is nowadays not very difficult to prove. Indeed M. Ram Murty and N. Saradha [**MS87**] discovered that even the sieve of Eratosthenes suffices, when combined with an elementary device due to R. A. Rankin. There is an exposition of their proof in *An Introduction to Sieve Methods and their Applications* by A. C. Cojocaru and M. Ram Murty [**CM06**]. The proofs of the other five results are much more difficult.

Chebyshev was the first mathematician to actually prove any results on the distribution of the primes of the kind envisioned by Gauss and Legendre. He published two papers on this topic, in 1848 and in 1850. In the first paper [**Che48**] Chebyshev shows that the limit

$$\lim_{s\to 1^+} \sum_{n=2}^{\infty} \left(\pi(n+1) - \pi(n) - \frac{1}{\log(n)} \right) \frac{\log^m(n)}{n^s}$$

exists and is finite for any choice of the integer m. This is a weak formulation of the statement that the density of the primes is $1/\log(x)$. To prove this result, he uses the Euler product formula that we shall consider in Chapter 3. He then deduces that for any $\varepsilon > 0$ and any integer m each of the inequalities $\pi(x) > \mathrm{li}(x) - \varepsilon x/\log^m(x)$ and $\pi(x) < \mathrm{li}(x) + \varepsilon x/\log^m(x)$ has arbitrarily large solutions x. Now it is immediately clear that *if* the ratio $\pi(x)/\mathrm{li}(x)$ tends to a limit, that limit must be 1, so that then the relative error in the approximation of Gauss tends to zero as $x \to \infty$. Chebyshev goes on to conclude that if we put $\pi(x) = x/(\log(x) - A(x))$ and $A(x)$ *has* a limit as $x \to +\infty$, then the limit must be 1. So if there is an asymptotically best approximation of the form $\pi(x) \approx x/(\log(x) - A)$ at all, that must be the approximation with $A = 1$ rather than with $A = 1.08366$ as proposed by Legendre. In his 1850 paper [**Che50**] Chebyshev proved an unconditional estimate that is equivalent to $0.921\cdot\mathrm{li}(x) \le \pi(x) \le 1.106\cdot\mathrm{li}(x)$ for all x sufficiently large, by a rather more elaborate version of the method that we used to prove Proposition 1.16. Even better upper and lower bounds were obtained by J. J. Sylvester [**Syl81**] using Chebyshev's method. Long after this work Diamond and Erdös [**DE80**] showed by means of the Prime Number Theorem that with sufficient calculation the method will yield estimates $c_1\cdot\mathrm{li}(x) \le \pi(x) \le c_2\cdot\mathrm{li}(x)$ with c_1 and c_2 arbitrarily close to 1.

The formula relating ψ and T was discovered independently of Chebyshev by A. de Polignac [**dP51**].

Chebyshev's proof of Bertrand's postulate in his 1850 paper [**Che50**] inaugurated the study of the local distribution of primes. Any nontrivial bound for the error term in the Prime Number Theorem implies existence of primes in short intervals. How short one can take the intervals by this approach depends on the quality of the bound on the error term in the PNT. However, in 1930 G. Hoheisel [**Hoh30**] by means of a different, analytic kind of argument succeeded in proving that the interval $(x - x^\theta, x]$ contains a prime for all x sufficiently large, with the exponent $\theta = 1 - 1/33000$. Even today, a bound for the error term in the Prime Number Theorem strong enough to allow this conclusion is not known. The exponent θ was gradually reduced over the years, by H. A. Heilbronn [**Hei33**] ($\theta = 0.996$ in 1933), N. G. Chudakov [**Chu36**] ($\theta = 3/4 + \varepsilon$ in 1936), A. E. Ingham [**Ing37**] ($\theta = 5/8 + \varepsilon$ in 1937), H. L. Montgomery [**Mon71**] ($\theta = 3/5 + \varepsilon$ in 1971), M. N. Huxley [**Hux72**] ($\theta = 7/12 + \varepsilon$ in 1972), H. Iwaniec and M. Jutila [**IJ79**] ($\theta = 13/23$ in 1979), Heath-Brown and Iwaniec [**HBI79**] ($\theta = 11/20$ in 1979), J. Pintz [**Pin84**] ($\theta = 17/31 - c$ for some small computable $c > 0$ in 1984), Iwaniec and Pintz [**IP84**] ($\theta = 23/42$ in 1984), C. J. Mozzochi [**Moz86**] ($\theta = 11/20 - 1/384$ in 1986), S. T. Lou and Q. Yao [**LY92, LY93**] ($\theta = 6/11$ in 1992 and $\theta = 6/11$ in 1993), R. C. Baker and G. Harman [**BH96**] ($\theta = .535$ in 1996), and Baker, Harman and Pintz [**BHP01**] ($\theta = 0.525$ in 2001.) There are heuristic arguments in favor of stronger conclusions. See page 422 of *Multiplicative Number Theory I. Classical Theory* by

H L. Montgomery and R. C. Vaughan [**MV07**], where the possibility that $\theta = \varepsilon$ with $\varepsilon > 0$ arbitrary is discussed. Also see the paper [**Gon93**] by S. M. Gonek. An even stronger conclusion follows if one accepts a probabilistic model of the distribution of the primes originated by H. Cramér [**Cra36**], and modified by A. Granville [**Gra95**] after work of H. Maier [**Mai85**]. This indicates that there should be some constant $C > 2e^{-\gamma}$ such that the interval $(x - C\log^2(x), x]$ contains a prime for all x sufficiently large, and that possibly any $C > 2e^{-\gamma}$ will do.

Defining $d_k = p_{k+1} - p_k$ one may, in view of the above considerations, ask how large d_k can become, say in terms of p_k. It is an immediate consequence of the Prime Number Theorem that $\limsup_{k\to+\infty} d_k/\log(p_k) \geq c$ with $c = 1$. R. J. Backlund [**Bac29**] showed in 1929 that one can take $c = 2$, and A. T. Brauer and H. Zeitz [**BZ30**] achieved $c = 4$ the year after. E. Westzynthius [**Wes31**] proved in 1933 that

$$\limsup_{k\to+\infty} \frac{d_k}{\frac{\log(p_k)\log_3(p_k)}{\log_4(p_k)}} \geq 2e^{\gamma},$$

where $\log_m(x)$ denotes the m times iterated logarithm. This was improved to

$$\limsup_{k\to+\infty} \frac{d_k}{\log(p_k)\log_3(p_k)} > 0$$

by G. Ricci [**Ric34**]. Further progress was made by Erdös [**Erd35**], who showed that

$$\limsup_{k\to+\infty} \frac{d_k}{\frac{\log(p_k)\log_2(p_k)}{(\log_3(p_k))^2}} > 0,$$

and by Rankin [**Ran38**], who showed in 1938 that

$$\limsup_{k\to+\infty} \frac{d_k}{\frac{\log(p_k)\log_2(p_k)\log_4(p_k)}{(\log_3(p_k))^2}} \geq c$$

with $c = 1/3$. Since then, only improvements in the constant c have been obtained, by A. Schönhage [**Sch63**], ($c = e^{\gamma_0/2}$ in 1963), Rankin [**Ran63**], ($c = e^{\gamma_0}$ in 1963), H. Maier and C. B. Pomerance [**MP90**], ($c = 1.312\ldots e^{\gamma_0}$ in 1990), and Pintz [**Pin97**], ($c = 2e^{\gamma_0}$ in 1997.)

The question of how small d_k can be in the long run is also of interest. If there are infinitely many twin primes, then $d_k = 2$ infinitely often. Defining $E = \liminf_{n\to\infty} d_k/\log(p_k)$, it is again an immediate consequence of the Prime Number Theorem that $E \leq 1$. Upper bounds for E were obtained by Erdös [**Erd40**] ($E < 1 - c$ for some small computable $c > 0$ in 1940), Rankin [**Ran50**] ($E \leq 42/43$ in 1950), Ricci [**Ric54a**] ($E \leq 15/16$ in 1954), E. Bombieri and H. Davenport [**BD66**] ($E \leq (2+\sqrt{3})/8$ in 1966), G. Z. Pil'tjaĭ [**Pil72**] ($E \leq (2\sqrt{2}-1)/4$ in 1972), Huxley [**Hux73, Hux77, Hux84**] ($E \leq 1/4 + \pi/16$ in 1973, $E \leq 0.4425\ldots$ in 1977 and $E \leq 0.4393\ldots$ in 1984), É. Fouvry and F. Grupp [**FG86**] ($E \leq 0.4342\ldots$ in 1986), H. Maier [**Mai88**] ($E \leq 0.2484\ldots$ in 1988) and finally D. A. Goldston, J. Pintz and C. Y. Yıldırım [**GPY09**] ($E = 0$ in 2005.) Recently Y. Zhang [**Zha14**] proved that $d_k \leq 70{\cdot}10^6$ infinitely often, which was a great advance. In work to appear in Annals of Mathematics, J. Maynard has shown that $p_{k+m} - p_k \leq c_m$ infinitely often for each positive integer m, with $c_1 = 600$ admissible.

The proof of Proposition 1.2 is modeled on the one given by Ramanujan [**Ram19**].

Landau gave sufficient conditions for the heuristic mentioned at the beginning of Section 1.4 to hold true; see page 201 of his treatise [**Lan74**] on prime number theory, or the paper [**Lan00**]. Partial summation goes back to a paper of N. H. Abel [**Abe26**] on power series. The Euler-Maclaurin summation formula [**Eul38, Mac42**] dates to the first half of the eighteenth century. The work of Mertens is in [**Mer74a, Mer74b**].

The formula for $d(n)$ dates to 1673 and is due to J. Kersey [**Ker73**]. Obscure today, to his contemporaries he was known as an author of well-regarded textbooks.

It seems that R. Descartes [**Des79**] was the first to explicitly note an arithmetic function multiplicative; in a posthumous manuscript he stated that the sum-of-divisors function $\sigma(n)$ has this property. From a letter [**Des98**] to M. Mersenne it seems likely that he knew this by 1638.

The von Mangoldt function, though not the notation for it used today, and the convolution identity $1 * \Lambda = \log$, are due to N. W. Bugaev [**Bug73**] and E. Césaro [**Cés88**].

The convolution identity $1 * \phi = \mathrm{id}$ was proved by Gauss in article 39 of the *Disquisitiones*, and this may well be the first instance of a divisor sum of an arithmetic function.

E. T. Bell [**Bel15**] and M. Cipolla [**Cip15**] independently in 1915 considered the set of arithmetic functions as an algebraic structure and gave Proposition 1.12. But a good many particular convolutions of multiplicative functions were known in the nineteenth century, so this result may well have been appreciated earlier.

A. F. Möbius [**Mőb31**] introduced the function named after him in 1831. But already Euler had considered infinite series whose terms involved values of the Möbius function.

Proposition 1.13 is due to J. W. R. Dedekind [**Ded57**] and J. Liouville [**Lio57**] independently in 1857, and Proposition 1.14 to Möbius [**Mőb31**].

The asymptotic estimate for $\Phi(x)$ is due to Mertens [**Mer74b**]. Dirichlet had obtained a similar estimate with x^{ε} in place of $\log(x)$ in [**Dir49**].

The maximal order of $\log(d(n))$ was found by S. Wigert [**Wig07**] around 1906 using the Prime Number Theorem. The dependence on the PNT was removed some years later by Ramanujan [**Ram15**]. The proof of Proposition 1.15 is a modified version of the one given by Wigert.

In two papers [**Dir38a, Dir38b**] of 1838 Dirichlet considers the question of how one can study arithmetic functions that fluctuate irregularly. It is not unreasonable to consider these papers as founding the theory of arithmetic functions, but Dirichlet himself refers to earlier work, and in particular to remarks of Gauss in article 301 of the *Disquisitiones*. What Dirichlet set out to do in the first of these papers was to find 'das asymptotische Gesetz' in the sense of Gauss, of the divisor function $d(n)$. He indicated an argument, based on the Lambert series expansion

$$\sum_{n=1}^{\infty} d(n)x^n = \sum_{m=1}^{\infty} \frac{x^m}{1-x^m},$$

that $\log(x) + 2\gamma$ is the asymptotic law of growth of $d(n)$ in the sense that the local average approaches $\log(x)+2\gamma$ as $x \to +\infty$ and $y \to +\infty$ in a suitable way. Dirichlet [**Dir49**] returned to the divisor problem in 1849 and gave his classical bound $O(\sqrt{x})$ for the error term. Using Dirichlet's result L. Kronecker in his lectures [**Kro01**] optimized the length of the interval to minimize the error bound for the local average of $d(n)$.

The Dirichlet divisor problem has a rich history, with a couple of obscure turns in the early stages. The first of these is a letter from Dirichlet to Kronecker dated July 23, 1858. Dirichlet had visited Kronecker for a few days in Ilsenburg, a resort by the Harz mountains, where the latter was spending his summer vacation. From the letter it is clear that they had discussed the divisor problem, and Dirichlet writes that he has now managed to substantially improve on his result of 1849 (*'.... die Summe ... bedeutend in die Enge zu treiben.'*) Dirichlet died on May 5, 1859 and his improvement never appeared. Apparently his Nachlass did not contain any material bearing on the problem, and nothing is known of the nature of the new method of which Dirichlet hints in the letter to Kronecker, nor of the extent of the improvement.

In 1903 G. F. Voronoi [**Vor03**] showed that $\vartheta \leq 1/3$ in the Dirichlet divisor problem. The proof is nearly forty pages long, and is based on the interpretation of $D(x)$ in terms of lattice points under a hyperbola. By means of Farey fractions Voronoi closely approximated the hyperbola by a polygon, and then he used the Euler summation formula to obtain the estimate $\Delta(x) = O(x^{1/3}\log(x))$. The following year he gave a different and much more analytic proof [**Vor04**].

As late as 1917 Voronoi's result was judged *'one of the deepest in the analytic theory of numbers'* by Hardy and Ramanujan [**HR17a**] in their paper on the normal number of prime factors of an integer. But in the same year I. M. Vinogradov [**Vin18a**] found a much easier proof. In 1922 J. G. van der Corput [**Cor22**] proved $\vartheta \leq 33/100$. His proof required estimates for exponential sums, and since that time further progress has depended on better estimates for such sums and on related techniques of counting lattice points.

The exponent in the Dirichlet divisor problem was slowly reduced over a long period, by van der Corput ($\vartheta = 27/82$ in 1928), T.-T. Chih [**Chi50**] ($\vartheta = 15/46$ in 1950), G. A. Kolesnik [**Kol69, Kol74, Kol82, Kol85**] ($\vartheta = 12/37$ in 1969, $\vartheta = 346/1067$ in 1973, $\vartheta = 35/108$ in 1982 and $\vartheta = 139/429$ in 1985), Iwaniec and Mozzochi [**IM88**] ($\vartheta = 7/22$ in 1988), and Huxley [**Hux93, Hux02, Hux03**] ($\vartheta = 23/73$ in 1993 and $\vartheta = 131/416$ in 2000.) In the other direction, Hardy [**Har15b, Har15a**] proved in 1914 that $\vartheta \geq 1/4$, and it is generally believed that $\vartheta = 1/4$ holds.

The formula of Meissel is in [**Mei54**]. Diamond's version of the hyperbola method is in his survey paper [**Dia82**] on elementary methods in prime number theory. The special case $y = x^{1/2}$ was noted by J. Franel [**Fra99**] in 1899.

Exercises

(1) a) Let $C(M, N)$ denote the binomial coefficient M choose N. Find the prime factorization of $C(2N, N)$.
b) Find a good upper bound for $C(2N, N)$.
c) Use a) and b) to show that $\vartheta(x) = O(x)$.

(2) Calculate $\mathrm{lcm}[1, 2, \ldots, N]$ in terms of $\psi(N)$.

(3) You are offered to make bets in favor of integers being squares. Integers n are drawn at random from the interval $x - x^{3/4} < n \leq x$ for some fixed very large x. Each time n is a square, you win $1.5\,x^{1/2}$ dollars. Each time it is not, you lose one dollar. Should you accept these bets?

(4) Prove the infinitude of primes by first proving the infinitude of squarefree numbers (J. Perrott.)

(5) Use the formula

$$\sum_{n \leq x} \psi\left(\frac{x}{n}\right) = T(x)$$

to prove that $\mathbf{a} \leq 1 \leq \mathbf{A}$ and thus show that if $\psi(x)/x$ has a limit as $x \to \infty$, that limit must equal 1 (Chebyshev.)

(6) Let $A \subseteq \mathbb{Z}$ and denote by $N_A(x)$ the number of elements $a \in A$ with $|a| \leq x$. Show that if $N_A(x)$ grows faster than any power of $\log(x)$, the total set of prime divisors of the elements of A is infinite. Show that if c is any positive constant, the total set of prime divisors of the sequence $[\exp(\log^c(n))]$ for $n = 1, 2, 3, \ldots$ is infinite. (Hint: The Fundamental Theorem of Arithmetic in logarithmic form.)

(7) † Let $N(a, b, c)$ denote the number of solutions of the inequality $ak + bm \leq c$ in nonnegative integers k and m. Express $N(a, b, c)$ in terms of sums involving the integer part function and establish the estimate

$$\left| N(a, b, c) - \frac{1}{2}\left(\frac{c}{a} + 1\right)\left(\frac{c}{b} + 1\right) \right| \leq \frac{1}{2} + \frac{3c}{4a} + \frac{3c}{4b}$$

when a, b and c are positive real numbers. About how many positive integers less than or equal to x are there of the form $2^k \cdot 3^m$ when x is large?

(8) Show that $n!$ is never a square for $n \geq 2$.

(9) Let n be an arbitrary positive integer. Show that counted with multiplicity every prime occurs at least as often in total as a factor of the integers in the interval $[n+1, 2n]$ as it does of the integers in the interval $[1, n]$.

(10) Counting prime factors without multiplicity, show that for every integer $n \geq 2$ there is some integer n' with $n < n' < 2n$ so that n and n' have the same number of prime factors.

(11) Use divisibility properties of the factorial to show that the sequence of primes has unbounded gaps. Then use a Chebyshev-type estimate to find a constant $c > 0$ so that there exist arbitrarily large pairs p_k, p_{k+1} of consecutive primes with $p_{k+1} - p_k \geq c\log(p_k)$.

(12) † a) Compute the normalized differences $(p_{k+1} - p_k)/\log(p_k)$ of pairs p_k, p_{k+1} of successive primes in intervals

$$(1000, 2000],\ (10000, 11000],\ (100000, 101000],\ (1000000, 1001000].$$

Sort and plot the normalized differences on each interval. Comment on: (i) the family resemblance of the various plots disregarding scale, (ii) the shape of the plots with special attention to the prevalence of small and large differences, and (iii) lack of "smoothness" of the plots.
b) Compute ratios $(p_{k+1} - p_k)/\log^2(p_k)$ to investigate an old conjecture of C. H. Cramér to the effect that this ratio is bounded. Since extreme values of the difference $p_{k+1} - p_k$ are sought, the computational strategy should be different from part a). Determine approximately the largest prime for which the time involved in computing the ratio is $1/10$ of a second, and calculate a thousand ratios for randomly chosen and well scattered primes of about the same order of magnitude. Sort and plot the ratios as in a). (Here a computer algebra system is needed, or else some programming. In the latter case, *Prime Numbers. A computational perspective* by Richard Crandall and Carl Pomerance, or *Prime numbers and computer methods for factorization* by Hans Riesel, may prove helpful.)

(13) Show that

$$\prod_{p \leq n} p \leq 5^n$$

for all $n \geq 1$. According to P. Erdös and L. Kalmár one can take 4 in the inequality, and according to D. Hanson one can take 3.

(14) a) Let $A_x = (x/2, x] \cup (x/4, x/3] \cup (x/6, x/5] \cup \cdots$. Show that

$$\sum_{n \in A_x} \Lambda(n) = \log(2)x + O(\log(x)),$$

and that the error term cannot be improved.
b) Deduce that the interval $(x/2, x]$ contains at least $x/(5\log(x))$ primes for all x sufficiently large.

(15) Show that

$$\int_{x_0}^{x} f(u)\,du = O\Big(\int_{x_0}^{x} g(u)\,du\Big)$$

if $f(x) = O(g(x))$ and f and g are integrable on bounded intervals.

(16) Prove that if f and g are real-valued continuous functions on a closed and bounded interval $[a,b]$ and g is positive there, then

$$\int_a^b f(x)g(x)\,dx = f(c)\int_a^b g(x)\,dx$$

for some $c \in (a,b)$. This is called the *first mean value theorem for integrals.*

(17) a) Prove that if f and g are real-valued continuous functions on a closed and bounded interval $[a,b]$ and f is monotone there, then

$$\int_a^b f(x)g(x)\,dx = f(a)\int_a^c g(x)\,dx + f(b)\int_c^b g(x)\,dx$$

for some $c \in (a,b)$. This is called the *second mean value theorem for integrals.* The second mean value theorem is useful for the estimation of integrals with an oscillatory factor in the integrand.

b) Show that

$$\left|\int_0^{2\pi} f(x)\cos(x)\,dx\right| \le |f(0) - f(2\pi)|$$

if f is a continuous monotone function.

(18) Show that

$$\lim_{h\to 0}\int_a^b \frac{f(x+h) - f(x)}{h}\,dx = f(b) - f(a)$$

if f is a continuous function on an open interval containing the closed interval $[a,b]$. Establish an integration by parts formula for two continuous functions, one of which is monotone.

(19) Make a guess for the asymptotic behavior of the sum

$$\sum_{p\le x}\frac{1}{\sqrt{p}}$$

and use partial summation and estimates of Chebyshev to show that your guess has the right order of growth.

(20) Show that

$$\sum_{pq\le x}\log(p)\log(q) \asymp x\log(x)$$

where p and q denote primes.

(21) a) Use integration by parts to show that

$$\int_n^{n+1} f(u)\,du = \frac{f(n)}{2} + \frac{f(n+1)}{2} - \int_n^{n+1} \left(u - n - \frac{1}{2}\right) f'(u)\,du.$$

Here n is an integer and f a continuous function on the interval $[n, n+1]$ with f' piecewise continuous there.

b) Use part a) to establish the Euler-Maclaurin summation formula.

(22) Prove the estimate

$$\sum_{n=-\infty}^{\infty} e^{-\pi n^2 u} = \frac{1}{\sqrt{u}} + O(1)$$

as $u \to 0^+$. This series comes from the theory of elliptic theta functions, and satisfies a functional equation that yields a far more precise estimate.

(23) † Establish the estimate

$$\sum_{n \le x} \log^m\left(\frac{x}{n}\right) = m!x + O(m! \log^m(x))$$

uniformly in positive integers m.

(24) Show without integration that $T(n) = n\log(n) + O(n)$ by subdividing the interval $[1, n]$ between successive powers of 2.

(25) Show that

$$\prod_{p \le x}\left(1 + \frac{1}{p}\right) \sim c\log(x)$$

as $x \to +\infty$, for some positive constant c.

(26) Show that

$$\sum_{p \le x} \frac{\log^2(p)}{p} = \frac{1}{2}\log^2(x) + O(\log(x))$$

as $x \to +\infty$.

(27) Show that the real number 1 is a point of accumulation of the sequence of ratios $(p_{k+1}/p_k)_{k=1}^{\infty}$ of successive primes.

(28) Show that the series

$$\sum_p \frac{1}{p(\log\log(p))^2}$$

converges.

(29) A divisor d of an integer n is called a *block divisor* if it is coprime to its complementary divisor n/d. In group theory such divisors are called *Hall divisors*. Count the number of block divisors of n.

(30) Show that the multiplicative arithmetic functions under Dirichlet convolution constitute a subgroup of the group of all arithmetic functions that have a Dirichlet inverse.

(31) Calculate $1 * \lambda$, $1 * \varrho$ and $d * \lambda$ where ϱ is the indicator function of the squares.

(32) Calculate $\mathrm{id} * \mathrm{id}$, $\phi * \sigma$ and $d * \phi$.

(33) Calculate $f * f$ for f totally multiplicative.

(34) Use the Binomial Theorem to show that $1 * \mu = \mathrm{e}$.

(35) Show that the sum of the primitive n-th roots of unity equals $\mu(n)$.

(36) Deduce the formula for T in terms of ψ by means of $1 * \Lambda = \log$.

(37) Show that $1*(\mu\cdot \log) = -\Lambda$.

(38) Show that

$$\sum_{n\leq x} \log(\mathrm{rad}(n)) = x\log(x) + O(x)$$

where rad is the radical.

(39) a) Show that the number of rationals in the interval $[0,1]$ written in lowest terms and with denominator $\leq x$ is asymptotic to $3x^2/\pi^2$.
b) Show that the chance that two randomly chosen large integers be coprime is $6/\pi^2$.
c) A lattice point (j,k) *occults* the lattice point (m,n) if both lie on the same ray from the origin and (j,k) lies closer to the origin. Show that the proportion of lattice points in $\mathbb{Z}\times\mathbb{Z}$ visible from the origin is $6/\pi^2$.

(40) Show that

$$\sum_{d|n} 2^{\omega(d)} = d(n^2),$$

and generalize (Dirichlet).

(41) Establish the estimate

$$\left| \sum_{n\leq x} d(\gcd(m,n)) - \frac{\sigma(m)}{m}x \right| \leq d(m)$$

for any fixed positive integer m.

(42) Show that

$$\sum_{n\leq x} \frac{\sigma(n)}{n} = \frac{\pi^2}{6}x + O(\log(x))$$

as $x \to +\infty$.

(43) Let $P^+(n)$ denote the largest prime factor of n. Show that the infinite series

$$\sum_{P^+(n)\leq x} \frac{\mu(n)}{n}$$

converges absolutely for every $x \geq 1$ and that the sum is always positive. Find the limit of the sum as $x \to +\infty$.

(44) An arithmetic function h is given, which is never zero. It is known that $h = fg$ where f is multiplicative and g is additive. Determine f and g from h.

(45) a) The sieve of Eratosthenes-Legendre: Show that

$$\sum_{d|P} \mu(d)\left[\frac{x}{d}\right] = \pi(x) - \pi(\sqrt{x}) + 1$$

if $P_{\sqrt{x}}$ denotes the product of the primes $p \leq \sqrt{x}$.
b) Show that

$$\pi(x) \leq \pi(z) - 1 + \sum_{d|P_z} \mu(d)\left[\frac{x}{d}\right]$$

if P_z denotes the product of the primes $p \leq z$.
c) Show that $\pi(x) \ll x/\log\log(x)$ by a choice of z in terms of x in part b). This is weaker than what we already have from Section 1.1, but see the next part.
d) In the sieve of Eratosthenes-Legendre, we calculated the number of primes in the interval $\sqrt{x} < n \leq x$ by removing those integers that lie in arithmetic progressions $p\mathbb{Z}$ with $p \leq \sqrt{x}$. Apply the same idea on the interval $x - h < n \leq x$ to show that a bound $\pi(x) - \pi(x - h) \ll h/\log\log(h)$ holds uniformly in x. Does this follow from the Chebyshev theory of Section 1.1?

(46) Find a maximal order and a minimal order of $\log(\phi(n))$.

(47) Use the Dirichlet interchange to show that $1 * \lambda = \varrho$, where ϱ is the characteristic function of the squares.

(48) Use the Dirichlet interchange to calculate $1 * \log$.

(49) A natural number n is called *perfect* if $\sigma(n) = 2n$. Use the Dirichlet interchange and congruences modulo 3 to show that $n \not\equiv 2 \pmod 3$ if n is perfect (J. Touchard).

(50) Find the arithmetic average of the fractional part $\{x/n\} = x/n - [x/n]$ of x/n over the interval $1 \leq n \leq x$ as $x \to +\infty$ (Dirichlet).

(51) Show that for every $\varepsilon > 0$ the equation $xu - yv = 1$ has $O_\varepsilon(R^{2+\varepsilon})$ solutions in integers in the ball $x^2 + y^2 + u^2 + v^2 \leq R^2$.

(52) a) Show that

$$\sum_{n\leq x} f(n)G\Big(\frac{x}{n}\Big) = \sum_{m\leq x} (G(m) - G(m-1))F\Big(\frac{x}{m}\Big)$$

if $f(n)$ is an arithmetic function with summatory function $F(x)$ and $G(x)$ is the summatory function of some arithmetic function (Dirichlet).

b) Find a constant $c(\theta)$ such that

$$\sum_{n\le x}[x/n]^{-\theta} = c(\theta)x + O(1)$$

holds uniformly for $\theta \ge 0$. Uniformity means that the constants C and x_0 that are implicit in the O-term do not depend on the parameter θ.

(53) Show that

$$\sum_{n\le x}\log(\phi(n)) = x\log(x) + (c-1)x + O(\log(x))$$

where

$$c = \sum_p \frac{1}{p}\log\left(1-\frac{1}{p}\right).$$

Find the average order of $\log(\phi(n))$. Calculate the local average and choose the length h of the interval in terms of x so as to minimize the error term.

(54) a) Use the identity

$$\sum_{n\le x}\left(\left[\frac{x}{n}\right] - \psi\left(\frac{x}{n}\right) - 2\gamma\right) = D(x) - T(x) - 2\gamma[x]$$

and the Dirichlet bound in the divisor problem to show that $\psi(x) = O(x)$ independently of the Chebyshev method. This idea is due to Nina Spears. Her approach may be refined to yield close upper and lower numerical bounds for $\psi(x)/x$. Though unfortunately only at the cost of extensive computation.

Note that this establishes the bound $\vartheta(x) = O(x)$ and Proposition 1.9 without reliance on the method of Chebyshev.

b) Show that

$$\sum_{x/K<p\le x}\frac{\log(p)}{p} = \log(K) + O(1)$$

where $K \ge 1$ is an arbitrary constant, and the error is uniform in K.

c) Apply b) to show that $\vartheta(x) \asymp x$ without reliance on the method of Chebyshev (H. N. Shapiro).

(55) † Find an asymptotic estimate for the sum

$$\sum_{n\le x}\frac{d(n)}{\log(n)+2\gamma},$$

with a bound for the error term.

(56) † Let $c \geq 1$ be a fixed real number, and let $f_c(n)$ be the number of divisors d of n satisfying the inequality $1/c \leq n/d^2 \leq c$. Show that

$$\sum_{n \leq x} f_c(n) = x \log(c) + O(\sqrt{cx})$$

where the constants implied in the O-term do not depend on c. (The estimate is *uniform* in c.) Then conclude that 50% of the divisors d of the integers n in the interval $1 \leq n \leq x$ satisfy the inequality

$$\frac{1}{\sqrt{x}} \leq \frac{n}{d^2} \leq \sqrt{x}$$

as $x \to +\infty$. An estimate that contains a parameter frequently becomes much more useful if we can establish that the estimate is uniform in the parameter.

Chapter 2

Topics on Arithmetic Functions

2.1. ⋆ The neighborhood method

An estimate for the summatory function F of an arithmetic function f sometimes yields an estimate for the summatory function

$$\sum_{n\le x}(f*g)(n)$$

of a Dirichlet convolution $f*g$ by what is called the *neighborhood method.* The scaffolding

$$\begin{aligned}\sum_{n\le x}(f*g)(n) &= \sum_{n\le x}\sum_{km=n} f(k)g(m)\\ &= \sum_{m\le x} g(m) \sum_{k\le x/m} f(k) = \sum_{m\le x} g(m)F\Big(\frac{x}{m}\Big)\end{aligned}$$

of the method is obtained by an interchange of the order of summation. We may also obtain it from Proposition 1.17 by choosing $y=1$. If g is in some sense a small perturbation of the multiplicative identity e in the Dirichlet ring, one could hope that the approximation

$$\sum_{n\le x}(f*g)(n) \approx \sum_{n\le x}(f*\mathrm{e})(n) = \sum_{n\le x} f(n) = F(x)$$

might have some merit. This is generally too optimistic. But the method can sometimes be made to work nonetheless if $|g(m)|$ is small on the average. A convolution $f*g$ with g close to the identity may be thought of as lying in a neighborhood of f, and this accounts for the name of the method.

The arithmetic function $\mu^2(n)$ is the indicator function of the set of squarefree numbers, and

$$Q(x) \stackrel{\text{def}}{=} \sum_{n\le x} \mu^2(n)$$

is the counting function of the squarefrees.

Proposition 2.1. *The estimate*

$$Q(x) = \frac{6}{\pi^2}x + O(\sqrt{x})$$

holds.

Proof. To apply the neighborhood method we must express μ^2 as a suitable convolution $f * g$. It is important that the summatory function of f be easy to estimate. We will try with $f = 1$ since μ^2 and 1 are both multiplicative, and both equal on the primes. Then we have to solve $\mu^2 = 1 * g$ for g. Möbius inversion yields $g = \mu^2 * \mu$ and since g is thus multiplicative, we see that $g(m) = 0$ if m is not a square, while $g(j^2) = \mu(j)$ otherwise. Hence

$$\begin{aligned} Q(x) &= \sum_{n\le x}(1 * g)(n) = \sum_{m\le x} g(m) \sum_{k\le x/m} 1 \\ &= \sum_{j^2\le x} \mu(j)\left[\frac{x}{j^2}\right] = x \sum_{j\le x^{1/2}} \frac{\mu(j)}{j^2} + O(x^{1/2}). \end{aligned}$$

Now

$$\sum_{j\le x^{1/2}} \frac{\mu(j)}{j^2} = \sum_{j=1}^{\infty} \frac{\mu(j)}{j^2} - \sum_{j>x^{1/2}} \frac{\mu(j)}{j^2} = \sum_{j=1}^{\infty} \frac{\mu(j)}{j^2} + O\left(\frac{1}{x}\right)$$

since $|\mu(j)| \le 1$. But we already know that

$$\sum_{k=1}^{\infty} \frac{\mu(m)}{m^2} = \frac{6}{\pi^2}$$

by the reasoning concerning $\Phi(x)$ in Section 1.5. □

Since $Q(x)/x \to 6/\pi^2$ as $x \to +\infty$, the proportion of squarefree integers among the positive integers is $6/\pi^2$ in an asymptotic sense, and the average order of the arithmetic function μ^2 is the constant $6/\pi^2$. We say that μ^2 possesses the *mean value* $6/\pi^2$. The mean value of an arithmetic function f is given by

$$\mathcal{M}(f) \stackrel{\text{def}}{=} \lim_{x\to+\infty} \frac{1}{x} \sum_{n\le x} f(n)$$

if the limit exists. We note that the Prime Number Theorem may be expressed as $\mathcal{M}(\Lambda) = 1$. Later we will see that the PNT may also be expressed as $\mathcal{M}(\mu) = 0$.

One may also interpret $\mathcal{M}(\mu^2) = 6/\pi^2$ to say that the chance that a randomly chosen large integer be squarefree is $6/\pi^2$. We have several times given "probabilistic" interpretations of arithmetical statements, but carefully avoided the words *probability* and *likelihood* that have precise meanings in probability theory. Instead we have preferred the comfortably vague word *chance,* for the phrase "a randomly chosen large integer" does not on its own convey any precise meaning, though it does convey a definite impression. The prescription to choose an element of $\mathbb{N}_n = \{1, 2, \ldots, n\}$ randomly is unproblematic in probability theory; just use a random variable X that is uniformly distributed on $\mathbb{N}_n$. But in analytic number theory one is mainly interested in what happens in the long run, and $\mathbb{N}$ cannot carry a uniformly distributed random variable.

This difficulty can be overcome, and the tools of probability theory brought to bear on arithmetic functions. The resulting theory is called probabilistic number theory. It was pioneered first by P. Turán, and then by P. Erdös, M. Kac, and A. F. Wintner, before the Second World War. Turán's proof of Proposition 2.6 is generally reckoned as the starting point of the theory. Accounts may be found in *Probabilistic Number Theory* by P. D. T. A. Elliott, *Probabilistic Methods in the Theory of Numbers* by J. Kubilius, and *Introduction to Analytic and Probabilistic Number Theory* by G. Tenenbaum.

Probabilistic number theory as such is outside our scope here. We should nonetheless equip ourselves with some precise terminology to discuss the distribution of integer sequences, and therefore introduce a concept of density. Any sequence A of natural numbers has a counting function

$$A(x) = \sum_{A \ni n \leq x} 1$$

that is the summatory function of the indicator function

$$\mathrm{I}_A(n) = \begin{cases} 1 & \text{if } n \in A, \\ 0 & \text{if } n \notin A, \end{cases}$$

of the sequence. The *asymptotic density* $\mathrm{d}A$ of a sequence A of natural numbers is given by

$$\mathrm{d}A = \mathcal{M}(\mathrm{I}_A) = \lim_{x \to +\infty} \frac{A(x)}{x}$$

if the limit exists. Naturally there are sequences for which this limit does not exist, and these sequences lack an asymptotic density. All sufficiently sparse sequences, such as the squares and the primes, have asymptotic density zero. Proposition 2.1 shows that the sequence of squarefrees has asymptotic density $6/\pi^2$. If a sequence has asymptotic density then it also has *logarithmic density*

$$\delta A = \lim_{x\to+\infty}\left(\sum_{A\ni n\le x} n^{-1}\right)\Big/\left(\sum_{n\le x} n^{-1}\right),$$

and $\delta A = \mathrm{d}A$. This is just an exercise in partial summation. There are sequences that have logarithmic density while lacking asymptotic density, as well as sequences that lack both densities. *Almost all* positive integers have a property if the number of exceptions $n \le x$ is $o(x)$. That almost all natural numbers are composite gives an example. It is obvious that almost all positive integers has some property if and only if the sequence of exceptions has asymptotic density equal to zero, or equivalently the sequence of positive integers that have the property has asymptotic density equal to one.

In the sense of asymptotic density more than half of all natural numbers are squarefree. But the squarefrees are irregularly distributed, and there are arbitrarily long gaps without any. Differencing the estimate for $Q(x)$ from Proposition 2.1 gives

$$\begin{aligned} Q(x) - Q(x-h) &= \frac{6}{\pi^2}x + O(x^{1/2}) - \frac{6}{\pi^2}(x-h) - O((x-h)^{1/2}) \\ &= \frac{6}{\pi^2}h + O(x^{1/2}) > 0 \end{aligned}$$

for $h = Cx^{1/2}$ with $C > 0$ a suitable constant. So all intervals $(x - Cx^{1/2}, x]$ contain a squarefree integer.

Proposition 2.2. *The intervals $(x - 5x^{1/3}, x]$ contain a squarefree integer for all x sufficiently large.*

Proof. The formula

$$\begin{aligned} Q(x) - Q(x-h) &= \sum_{x-h<n\le x} \mu^2(n) \\ &= \sum_{x-h<n\le x}\sum_{j^2|n}\mu(j) = \sum_{x-h<j^2k\le x}\mu(j) \end{aligned}$$

holds by the same reasoning as in the beginning of the proof of Proposition 2.1. To treat the last sum, we apply a very useful device; split the range of summation into subranges and treat each subsum separately. Then

$$\sum_{x-h<j^2k\le x}\mu(j) = \sum_{j\le x^{1/3}}\mu(j)\sum_{x-h<j^2k\le x} 1 + \sum_{j>x^{1/3}}\mu(j)\sum_{x-h<j^2k\le x} 1.$$

Here

$$\sum_{j\le x^{1/3}} \mu(j) \sum_{x-h<j^2k\le x} 1 = \sum_{j\le x^{1/3}} \mu(j)\left(\left[\frac{x}{j^2}\right] - \left[\frac{x-h}{j^2}\right]\right)$$
$$= \frac{6}{\pi^2}h + O(hx^{-1/3}) + R_1(x)$$

with $|R_1(x)| \le 2x^{1/3}$, since

$$\frac{h}{j^2} - 1 \le \left[\frac{x}{j^2}\right] - \left[\frac{x-h}{j^2}\right] \le \frac{h}{j^2} + 1.$$

Up to this point the argument closely tracks the proof of Proposition 2.1. To treat the other sum, observe that the right-hand side of the double inequality $x - h < j^2k \le x$ forces $k < x^{1/3}$ when $j > x^{1/3}$. For any positive integer k the number of positive integers j with $x - h < j^2k \le x$ is bounded by $1 + \sqrt{x/k} - \sqrt{(x-h)/k} = 1 + h/(\sqrt{k}(\sqrt{x} + \sqrt{x-h})) \le 1 + h/\sqrt{x}$. So if $h < \sqrt{x}$ there is at most one value of j for each value of k. If

$$R_2(x) = \sum_{j>x^{1/3}} \mu(j) \sum_{x-h<j^2k\le x} 1$$

then $|R_2(x)| \le x^{1/3}$, and so

$$Q(x) - Q(x-h) = \frac{6}{\pi^2}h + O(hx^{-1/3}) + R_1(x) + R_2(x)$$

with $|R_1(x) + R_2(x)| \le |R_1(x)| + |R_2(x)| \le 3x^{1/3}$. Now choose $h = 5x^{1/3}$ and observe that $5{\cdot}6/\pi^2 > 3$ so

$$\frac{6}{\pi^2}h + R_1(x) + R_2(x) > \eta x^{1/3}$$

for some small $\eta > 0$. Since the error term $O(hx^{-1/3}) = O(1)$ for this choice of h it is clear that $Q(x) - Q(x-h) > 0$ when x is large enough, if the condition $h < \sqrt{x}$ can be satisfied. But this condition is also satisfied when x is sufficiently large. □

This proof indicates that differencing an estimate for the summatory function may not be the most efficient way to study an arithmetic function on short intervals.

Analytic Number Theory by Jean-Marie De Koninck and Florian Luca is an introduction to analytic number theory with ample material on arithmetic functions. The theory of arithmetic functions as it was briefly presented in Sections 1.6 and 2.1 is much broadened and deepened in the treatise *Arithmetical Functions* by W. Schwarz and J. Spilker. *Introduction to Analytic and Probabilistic Number Theory* by G. Tenenbaum has extensive coverage of the topics touched upon in those sections, and much more.

2.2. ⋆ The normal order method

A real-valued function g is a *normal order* for a real-valued arithmetic function f if for any $\varepsilon > 0$ the inequality

$$-\varepsilon g(n) \leq f(n) - g(n) \leq \varepsilon g(n)$$

holds for $n \leq x$ with at most $o_\varepsilon(x)$ exceptions. G. H. Hardy and S. A. Ramanujan showed that $\omega(n)$ and $\Omega(n)$ both have normal order $\log\log(n)$ and that $\log(d(n))$ has normal order $\log(2)\log\log(n)$. We shall prove these important results by the *normal order method* of P. Turán. This method has a probabilistic interpretation, and may be significantly broadened and deepened, especially for additive arithmetic functions. The field of study opened up by Turán's proof is called probabilistic number theory.

Our main tool will be a special case of an inequality of Chebyshev. His inequality plays an important role in probability theory, but from this theory we shall only take the convenient notations

$$\mathrm{E}[f] = \frac{1}{N}\sum_{n\leq N} f(n) \quad \text{and} \quad \mathrm{Var}[f] = \mathrm{E}[(f - E[f])^2].$$

Note that in our arguments N will be a fixed positive integer, dependence on which we suppress wherever convenient. The *variance* $\mathrm{Var}[f]$ measures the deviation of f from its mean value $\mathrm{E}[f]$ over the interval $1 \leq n \leq N$. The mean value $\mathrm{E}[f]$ is called the *expectation* in probability theory. The formula

$$\mathrm{Var}[f] = \mathrm{E}[(f - \mathrm{E}[f])^2] = \mathrm{E}[f^2 - 2\mathrm{E}[f]f + \mathrm{E}[f]^2] = \mathrm{E}[f^2] - \mathrm{E}[f]^2$$

is a consequence of the linearity of expectation, and is useful in calculating variances.

Proposition 2.3 (Chebyshev's inequality). *The bound*

$$\frac{1}{N}\sum_{\substack{1\leq n\leq N \\ |f(n)-\mathrm{E}[f]|\geq\alpha}} 1 \leq \frac{\mathrm{Var}[f]}{\alpha^2}$$

holds for any real-valued arithmetic function f.

Proof. The inequality follows from

$$\alpha^2 \sum_{\substack{1\leq n\leq N \\ |f(n)-\mathrm{E}[f]|\geq\alpha}} 1 \leq \sum_{1\leq n\leq N} (f(n) - \mathrm{E}[f])^2 = \mathrm{Var}[f]$$

on dividing through by $N\alpha^2$. □

The basic result to be established by the Chebyshev inequality is that $\omega(n)$ has normal order $\log\log(n)$. The other statements follow from this one with a little more work. To apply the Chebyshev inequality to the arithmetic function ω it is necessary to calculate its expectation and variance.

Proposition 2.4. $\mathrm{E}[\omega] = \log\log(N) + O(1)$.

Proof. We have

$$\begin{aligned}\sum_{n\le N}\omega(n) &= \sum_{n\le N}\sum_{p|n} 1 = \sum_{p\le N}\sum_{\substack{n\le N\\ p|n}} 1 = \sum_{p\le N}\frac{N}{p} + O(N)\\ &= N\left(\log\log(N) + a + O\left(\frac{1}{\log(N)}\right)\right) + O(N)\\ &= N\log\log(N) + O(N)\end{aligned}$$

by Proposition 1.10. □

This proof yields slightly more than what is stated. But what is stated is what we need.

Proposition 2.5. $\mathrm{Var}[\omega] = O(\log\log(N))$.

Proof. The calculation

$$\begin{aligned}\sum_{n\le N}\omega^2(n) &= \sum_{n\le N}\left(\sum_{p|n}1\right)\left(\sum_{q|n}1\right) = \sum_{p\le N}\sum_{q\le N}\sum_{\substack{n\le N\\ p|n,q|n}} 1\\ &= \sum_{pq\le N}\left(\frac{N}{pq} + O(1)\right) + \sum_{p\le N}\left(\frac{N}{p} - \frac{N}{p^2}\right)\\ &= N\sum_{p\le N}\frac{1}{p}\sum_{q\le N/p}\frac{1}{q} + O(N\log\log(N))\\ &= N\sum_{p\le N}\frac{1}{p}\sum_{q\le N}\frac{1}{q} - N\sum_{p\le N}\frac{1}{p}\sum_{N/p<q\le N}\frac{1}{q} + O(N\log\log(N))\\ &= N(\log\log(N))^2 - N\sum_{p\le N^{1/2}}\frac{1}{p}\sum_{n/p<q\le N}\frac{1}{q}\\ &\quad - N\sum_{N^{1/2}<p\le N}\frac{1}{p}\sum_{n/p<q\le N}\frac{1}{q} + O(N\log\log(N))\\ &= N(\log\log(N))^2 + O(N\log\log(N))\end{aligned}$$

establishes the value of $\mathrm{E}[\omega^2]$ by repeated use of Proposition 1.10. Then the claim follows from $\mathrm{Var}[\omega] = \mathrm{E}[\omega^2] - \mathrm{E}[\omega]^2$ and Proposition 2.4. □

Proposition 2.6. *$\omega(n)$ has normal order* $\log\log(n)$.

Proof. Chebyshev's inequality yields

$$\frac{1}{N}\sum_{|\omega(n)-\log\log(N)-O(1)|\geq\alpha} 1 = O\left(\frac{\log\log(N)}{\alpha^2}\right)$$

by Propositions 2.4 and 2.5. Choosing $\alpha = \varepsilon\log\log(N)/4$ shows that the number of solutions of

$$|\omega(n) - \log\log(N) - O(1)| \geq \frac{\varepsilon}{4}\log\log(N)$$

in the interval $1 \leq n \leq N$ is $O(16\varepsilon^{-2}N/\log\log(N)) = o_\varepsilon(N)$ for any ε in the interval $0 < \varepsilon < 4$. The inequality

$$|\log\log(N) - \log\log(n) - O(1)| \geq \frac{\varepsilon}{4}\log\log(N)$$

implies that

$$n \leq e^{k\log(N)^{1-\varepsilon/4}}$$

for some positive constant k and thus it has $o_\varepsilon(N)$ solutions. Then

$$|\omega(n) - \log\log(n)| \geq \frac{\varepsilon}{2}\log\log(N)$$

with at most $o_\varepsilon(x)$ exceptions. But the inequality

$$\frac{\varepsilon}{2}\log\log(N) \geq \varepsilon\log\log(n)$$

implies that

$$n \leq e^{\log(N)^{1/2}}$$

and thus it has $o_\varepsilon(N)$ solutions. So finally $|\omega(n) - \log\log(n)| \geq \varepsilon\log\log(n)$ has $o_\varepsilon(N)$ solutions. □

This argument actually yields a much more precise result. There exists some positive constant K such that

$$|\omega(n) - \log\log(n)| < C(\log\log(n))^{1/2}$$

holds on the interval $1 \leq n \leq N$ with at most KN/C^2 exceptions for all C sufficiently large.

The Hardy-Ramanujan result on the normal order of $\omega(n)$ was deepened by P. Erdös and M. Kac, who proved that in the limit $\omega(n)$ is distributed like a Gaussian random variable with mean $\log\log(n)$ and variance $\sqrt{\log\log(n)}$. This so-called Erdös-Kac theorem was an early result in probabilistic number theory and initiated the study of the distribution properties of additive arithmetic functions. A precise statement and proof may be found in Chapter III.4 of *Introduction to Analytic and Probabilistic Number Theory* by Gérald Tenenbaum.

Proposition 2.7. $\Omega(n)$ *has normal order* $\log\log(n)$.

Proof. On the one hand $\omega(n) \leq \Omega(n)$, while on the other hand

$$\sum_{n\leq N}(\Omega(n)-\omega(n)) = \sum_{n\leq N}\sum_{\substack{p^\alpha|n\\ \alpha\geq 1}} 1 - \sum_{n\leq N}\sum_{p|n} 1 = \sum_{n\leq N}\sum_{\substack{p^\alpha|n\\ \alpha\geq 2}} 1$$
$$= \sum_{\substack{p^\alpha\leq N\\ \alpha\geq 2}}\sum_{\substack{n\leq N\\ p^\alpha|n}} 1 = \sum_{\substack{p^\alpha\leq N\\ \alpha\geq 2}}\left(\frac{N}{p^\alpha}+O(1)\right) = O(N).$$

But this implies that there are only $o_\varepsilon(N)$ integers n in the interval $1 \leq n \leq N$ for which $|\Omega(n)-\omega(n)| > \varepsilon\log\log(n)/2$, and there are only $o_\varepsilon(N)$ integers in the same interval for which $|\omega(n) - \log\log(n)| \geq \varepsilon\log\log(n)/2$. Thus the number of integers n in $1 \leq n \leq N$ for which $|\Omega(n)-\log\log(n)| \geq \varepsilon\log\log(n)$ is $O_\varepsilon(N)$. ☐

We could also prove this result directly by the normal order method without making use of the analogous result for $\omega(n)$.

Proposition 2.8. $\log(d(n))$ *has normal order* $\log(2)\log\log(n)$.

Proof. The inequality

$$2^{\omega(n)} = 2^r \leq (\alpha_1+1)\cdots(\alpha_r+1) = d(n) \leq 2^{\alpha_1+\cdots+\alpha_r} = 2^{\Omega(n)}$$

holds for any positive integer $n = p_1^{\alpha_1}\cdots p_r^{\alpha_r}$. Hence $\log(2)\omega(n) \leq \log(d(n)) \leq \log(2)\Omega(n)$, and since $\log(2)\omega(n)$ and $\log(2)\Omega(n)$ have the common normal order $\log(2)\log\log(n)$ by Propositions 2.6 and 2.7, so does $\log(d(n))$. ☐

The last result implies that $d(n)$ is not much larger than $\log^{\log(2)+\varepsilon}(n)$ for most n. The average order of $d(n)$ is $\log(n)$, thus a minority of exceptionally large values of $d(n)$ make the bulk of the contribution to its average order.

2.3. ⋆ The Mertens function

The summatory function

$$M(x) = \sum_{n\leq x}\mu(n)$$

of the Möbius function is important in the theory of the distribution of primes, and is named after Mertens because he was the first to publish extensive computations on it. It may be interpreted as the excess of the number of squarefree integers in the interval $[1, x]$ with an even number of prime factors over those with an odd number of prime factors. It also has another interpretation as an alternating sum

$$M(x) = \sum_{k=0}^{\infty} (-1)^k \pi_k(x)$$

where $\pi_k(x)$ denotes the number of integers $n \leq x$ that are products of exactly k distinct prime factors.

The behavior of $M(x)$ is poorly understood, so it is perhaps surprising that it satisfies a very simple identity, which actually determines $M(x)$ uniquely.

Proposition 2.9. *The identities*

$$\sum_{n \leq x} \mu(n) \left[\frac{x}{n}\right] = 1$$

and

$$\sum_{n \leq x} M\left(\frac{x}{n}\right) = 1$$

hold for $x \geq 1$.

Proof. The identities follow from the second Möbius inversion formula with

$$\sum_{n \leq x} 1 = [x]$$

or

$$M(x) = \sum_{n \leq x} \mu(n) \cdot 1,$$

respectively. □

The infinite series

$$\sum_{n=1}^{\infty} \frac{\mu(n)}{n}$$

converges. This is as difficult to prove as the Prime Number Theorem. But it is easy to show that the partial sums stay bounded.

Proposition 2.10. *The bound*

$$\left| \sum_{n \leq x} \frac{\mu(n)}{n} \right| \leq 1$$

holds for all $x \geq 1$.

Proof. Proposition 2.9 yields

$$\sum_{n\le x}\mu(n)\frac{x}{n}=1+\sum_{n\le x}\mu(n)\left(\frac{x}{n}-\left[\frac{x}{n}\right]\right).$$

Assume without loss of generality that x is an integer. Then

$$\left|\sum_{n\le x}\mu(n)\frac{x}{n}\right|\le 1+\sum_{n\le x}\left|\mu(n)\left(\frac{x}{n}-\left[\frac{x}{n}\right]\right)\right|\le 1+(x-1)=x,$$

from which the desired bound follows by dividing through by x. □

The next result implies that if $M(x)$ ultimately keeps a constant sign, then it cannot grow linearly. Though this is useful to us, it is quite weak information, for it is known that $M(x)=o(x)$ and that $M(x)$ changes sign infinitely often.

Proposition 2.11. *The improper integral*

$$\int_1^\infty \frac{M(t)}{t^2}\,dt$$

converges.

Proof. First

$$\begin{aligned}\int_1^N \frac{M(t)}{t^2}\,dt &= \int_N^1 M\left(\frac{N}{u}\right)\left(\frac{N}{u}\right)^{-2}\left(-\frac{N}{u^2}\right)du\\ &=\frac{1}{N}\int_1^N M\left(\frac{N}{u}\right)du\end{aligned}$$

by the change of variable $t=N/u$. Then

$$\begin{aligned}\int_1^N \frac{M(t)}{t^2}\,dt &= \frac{1}{N}\sum_{n=1}^{N-1}\int_n^{n+1} M\left(\frac{N}{u}\right)du\\ &=\frac{1}{N}\sum_{n=1}^{N-1}\left(\int_n^{n+1} M\left(\frac{N}{u}\right)du - M\left(\frac{N}{n}\right)\right)\\ &=\frac{1}{N}\sum_{n=1}^{N-1}\int_n^{n+1}\left(M\left(\frac{N}{u}\right)-M\left(\frac{N}{n}\right)\right)du\\ &=\frac{1}{N}\sum_{n=1}^{N-1}\int_n^{n+1} O\left(\frac{N}{u}-\frac{N}{n}\right)du=\sum_{n=1}^{N-1}O(n^{-2}),\end{aligned}$$

so the integral converges. □

From the convergence of the series

$$\sum_{n=1}^{\infty} \frac{\mu(n)}{n}$$

it is quite straightforward to show that the sum must be zero. The assumption that

$$\sum_{n=1}^{\infty} \frac{\mu(n)}{n} = c \geq 0$$

implies that there is some x_0 so that

$$F(x) = \sum_{n \leq x} \mu(n) \left(\frac{x}{n} - \left[\frac{x}{n} \right] \right) = x \sum_{n \leq x} \frac{\mu(n)}{n} - 1 \geq \frac{cx}{2}$$

for $x \geq x_0$. But then

$$\begin{aligned} x - [x] = \sum_{n \leq x} F\left(\frac{x}{n}\right) &= \sum_{x/n \leq x_0} F\left(\frac{x}{n}\right) + \sum_{x/n > x_0} F\left(\frac{x}{n}\right) \\ &\geq \frac{cx \log(x)}{2} + O(x), \end{aligned}$$

which forces $c = 0$, and similarly if $c \leq 0$.

Partial summation yields

$$\sum_{n \leq x} \frac{\mu(n)}{n} = \frac{M(x)}{x} + \int_1^x \frac{M(t)}{t^2}\, dt.$$

Thus the partial sums of the series

$$\sum_{n=1}^{\infty} \frac{\mu(n)}{n}$$

approach a limit if and only if $M(x)/x$ approaches a limit as $x \to +\infty$, by Proposition 2.11. If the latter limit were different from zero, the improper integral in Proposition 2.11 would diverge. So $M(x) = o(x)$ easily follows from convergence of the series and vice versa.

We are now going to show that the bound $M(x) = o(x)$ may be deduced from the Prime Number Theorem, and vice versa, by relatively simple arguments.

Assume the PNT in the form $\psi(x) \sim x$; thus for arbitrary $\varepsilon > 0$ there exists an $x_0(\varepsilon)$ such that $|\psi(x) - x| \leq \varepsilon x$ for $x \geq x_0(\varepsilon)$. Then $M(x) = o(x)$ is proved by estimating the sum

$$\sum_{n \leq x} \mu(n) \psi\left(\frac{x}{n}\right)$$

in two different ways. On the one hand

$$\left|\sum_{n\le x}\mu(n)\psi\Big(\frac{x}{n}\Big) - x\sum_{n\le x}\frac{\mu(n)}{n}\right| = \left|\sum_{n\le x}\mu(n)\Big(\psi\Big(\frac{x}{n}\Big)-\frac{x}{n}\Big)\right|$$
$$= \left|\sum_{x/n<x_0(\varepsilon)}\mu(n)\Big(\psi\Big(\frac{x}{n}\Big)-\frac{x}{n}\Big) + \sum_{x/n\ge x_0(\varepsilon)}\mu(n)\Big(\psi\Big(\frac{x}{n}\Big)-\frac{x}{n}\Big)\right|$$
$$\le \sum_{n>x/x_0(\varepsilon)}|\psi(x_0(\varepsilon)) + x_0(\varepsilon)| + \sum_{n\le x/x_0(\varepsilon)}\varepsilon\frac{x}{n} = \varepsilon x\log(x) + O(x).$$

On the other hand

$$\sum_{n\le x}\mu(n)\psi\Big(\frac{x}{n}\Big) = \sum_{mn\le x}\mu(n)\Lambda(m) = \sum_{N\le x}\sum_{m|N}\mu\Big(\frac{N}{m}\Big)\Lambda(m)$$
$$= \sum_{N\le x}\sum_{p|N}\sum_{p|p^k|N}\mu\Big(\frac{N}{p^k}\Big)\log(p) = \sum_{N\le x}\sum_{p|N}\log(p)\mu\Big(\frac{N}{p}\Big)$$
$$= -\sum_{N\le x}\mu(N)\sum_{p|N}\log(p) = -\sum_{N\le x}\mu(N)\log(N)$$
$$= -M(x)\log(x) + \int_1^x M(u)\frac{du}{u} = -M(x)\log(x) + O(x)$$

by partial summation. The inequality

$$\left|-M(x)\log(x) + O(x) - x\sum_{n\le x}\frac{\mu(n)}{n}\right| \le \varepsilon x\log(x) + O(x)$$

follows. Dividing through by $x\log(x)$ and using

$$\left|\sum_{n\le x}\frac{\mu(n)}{n}\right| \le 1$$

yields

$$\left|\frac{M(x)}{x} + O\Big(\frac{1}{\log(x)}\Big)\right| \le \varepsilon + O\Big(\frac{1}{\log(x)}\Big).$$

Thus $\limsup_{x\to+\infty}|M(x)/x| \le \varepsilon$, and since $\varepsilon > 0$ was arbitrary, $M(x) = o(x)$ follows.

The Prime Number Theorem may be deduced from the bound $M(x) = o(x)$ by means of the Dirichlet hyperbola method. The identity

$$[x] - \psi(x) - 2\gamma = \sum_{n\le x}(\mu * (d - \log - 2\gamma))(n)$$

holds because $\mu * (d - \log - 2\gamma) = 1 - \Lambda - 2\gamma e$. Then the Dirichlet hyperbola method yields

$$[x] - \psi(x) - 2\gamma$$
$$= \sum_{n \le y} \mu(n) F\Big(\frac{x}{n}\Big) + \sum_{n \le x/y} (d(n) - \log(n) - 2\gamma) M\Big(\frac{x}{n}\Big) - M(y) F\Big(\frac{x}{y}\Big)$$

with

$$F(x) = \sum_{n \le x} (d(n) - \log(n) - 2\gamma) = D(x) - T(x) - 2\gamma[x] = O(x^{1/2}).$$

Clearly $M(y)F(x/y) = o(x)$ no matter how y is chosen, and

$$\sum_{n \le y} \mu(n) F\Big(\frac{x}{n}\Big) = O(x^{1/2} y^{1/2}) = o(x)$$

if y is chosen so that $x/y \to \infty$ as $x \to \infty$. Putting

$$\eta(y) = \sup_{z \ge y} \frac{|M(z)|}{z}$$

it is clear that $\eta(y)$ is monotone decreasing with $\eta(y) \to 0$ as $y \to +\infty$. Then

$$\left| \sum_{n \le x/y} (d(n) - \log(n) - 2\gamma) M\Big(\frac{x}{n}\Big) \right| \le x\eta(y) \sum_{n \le x/y} \frac{|d(n) - \log(n) - 2\gamma|}{n}.$$

For every x sufficiently large it is possible to find an $y_1 = y_1(x)$ so that

$$\sum_{n \le x/y_1} \frac{|d(n) - \log(n) - 2\gamma|}{n} \le \eta(x^{1/2})^{-1/2}$$

while $x/y_1(x) \to +\infty$ as $x \to +\infty$. Putting $y = y(x) = \max(y_1(x), x^{1/2})$ we see that

$$x - \psi(x) = o(x) + O(x\eta(y)\eta(x^{1/2})^{-1/2}) = o(x) + O(x\eta(y)^{1/2}) = o(x)$$

so $\psi(x) \sim x$.

The above argument suggests that a viable route to the Prime Number Theorem would be to prove $M(x) = o(x)$ first, and indeed several such proofs are known. There is an exposition of one of these proofs, due to H. Daboussi, in *The Prime Numbers and Their Distribution* by Gérald Tenenbaum and Michel Mendès France.

2.4. Notes

L. B. Gegenbauer [**Geg85**] proved Proposition 2.1 in 1885. The best estimate currently known is

$$Q(x) = \frac{6}{\pi^2}x + O\left(x^{1/2}e^{-c\log^{3/5}(x)(\log\log(x))^{-1/5}}\right),$$

proved by A. Z. Walfisz [**Wal63**] by means of the same method of estimating exponential sums, due to I. M. Vinogradov, that has yielded the best currently known error term in the Prime Number Theorem.

Proposition 2.2 is due to K. F. Roth [**Rot51**] and the elementary formulation of his proof given here was pointed out by T. Estermann. One may ask for the best exponent θ for which $(x - x^{\theta+\varepsilon}, x]$ contains a squarefree number for all x sufficiently large, for all $\varepsilon > 0$. Successively smaller exponents were obtained by E. Fogels [**Fog41**] ($\theta = 2/5$ in 1941), Roth [**Rot51**] ($\theta = 3/13$ in 1951), H.-E. Richert [**Ric54b**] ($\theta = 2/9$ in 1954), Rankin [**Ran55**] ($\theta = 0.221982\ldots$ in 1955), P. G. Schmidt [**Sch64**] ($\theta = 109556/494419$ in 1964), S. W. Graham and G. Kolesnik [**GK88**] ($\theta = 1057/4785$ in 1988), O. Trifonov [**Tri88, Tri89**] ($\theta = 17/77$ in 1988), M. A. Filaseta [**Fil90**] ($\theta = 47/217$ in 1990), and Filaseta and Trifonov jointly [**FT90, FT92**] ($\theta = 3/14$ in 1990 and $\theta = 1/5$ in 1992.)

For the squarefree integers in short intervals, there is an interesting link with Diophantine analysis. The *ABC Conjecture* states that for every $\varepsilon > 0$ there is some $C(\varepsilon) > 0$ such that $c \le C(\varepsilon)\mathrm{rad}(abc)^{1+\varepsilon}$ for all coprime triples a, b, c of positive integers with $a + b = c$. The ABC Conjecture is a central problem of Diophantine analysis. It implies several of the deepest known results in that field, and a proof of the conjecture would resolve a bewildering variety of Diophantine problems. Granville [**Gra98**] has shown that the ABC Conjecture implies that for any $\varepsilon > 0$ the interval $(x - x^{\varepsilon}, x]$ contains a squarefree integer for all x sufficiently large.

The asymptotic estimate for the summatory function of $\omega(n)$ is in the 1917 paper [**HR17a**] of Hardy and Ramanujan on the normal number of prime factors of integers.

Normal orders of arithmetic functions were introduced by Hardy and Ramanujan [**HR17a**], and the determination of the normal order of $\omega(n)$, $\Omega(n)$ and $\log(d(n))$ is also due to them. The proof given here is much simpler, and is due to Turán [**Tur34**].

Proposition 2.9 is due to N. W. Bugaev [**Bug73**] and E. Meissel [**Mei54**], and Proposition 2.10 is due to J. P. Gram [**Gra86**]. Proposition 2.11 is in a paper [**Lan06**] of Landau, and the reasoning immediately following is also due to Landau [**Lan01**].

That the Prime Number Theorem implies $M(x) = o(x)$ was discovered by H. C. F. von Mangoldt [**Man97**] and that $M(x) = o(x)$ implies the PNT is due to Landau [**Lan11**].

Exercises

(1) Show that every positive integer n has a unique factorization $n = s_1 s_2 \cdots s_r$ into squarefree integers $s_k \geq 2$ with $s_1 | s_2 | \cdots | s_r$.

(2) What is the asymptotic density of an arithmetic progression?

(3) Show that any positive real number can be arbitrarily closely approximated by ratios s_1/s_2 of squarefree integers s_1 and s_2.

(4) Show that a sequence with logarithmic density zero has arbitrarily long gaps.

(5) Show that every sufficiently large natural number is a sum of two squarefrees.

(6) Calculate the mean value $\mathcal{M}(f)$ of the arithmetic function $f(n) = \phi(n)/n$.

(7) Find the asymptotic density of the integers n divisible by a prime $p > \sqrt{n}$.

(8) Use the Chinese remainder theorem and a Chebyshev-type estimate to find a constant $c > 0$ so that there exist arbitrarily large pairs s_k, s_{k+1} of consecutive squarefrees with $s_{k+1} - s_k \geq c \log(s_k)/\log\log(s_k)$.

(9) a) Show that if a sequence A has asymptotic density $\mathrm{d}A$ then it also has logarithmic density δA, and that $\delta A = \mathrm{d}A$.
b) Show that

$$\sum_{n \leq x} \frac{\mu^2(n)}{n} \sim \frac{6}{\pi^2} \log(x)$$

as $x \to +\infty$.

(10) Show that

$$\sum_{n \leq x} 2^{\omega(n)} = \frac{6}{\pi^2} x \log(x) + O(x)$$

as $x \to +\infty$ (Dirichlet).

(11) Asymptotically determine the average number of representations of integers $n \leq x$ as sums of a squarefree and a power of 2.

(12) Construct an infinite sequence in $\mathbb{N}$ whose logarithmic density does not exist. Construct an infinite sequence in $\mathbb{N}$ that has logarithmic density but whose asymptotic density does not exist.

(13) Compare the average order and the normal order of $\omega(n)$. How large can $\omega(n)$ be in terms of n? Do you think that $\omega(n)$ can take a significant number of abnormally large values?

(14) † Show that

$$\sum_{n\le x} d(n)\Omega(n) = 2x\log(x)\log\log(x) + cx\log(x) + O(x\log\log(x))$$

where c is some constant.

(15) Determine a normal order for the additive arithmetic function $\log(\phi(n))$. (Don't make a song and dance out of it; this is easy.)

(16) Obtain a more precise version of Proposition 2.4.

(17) Bound the variance $\mathrm{Var}[\Omega]$.

(18) Show that for any $\varepsilon > 0$ the inequality $|\omega(n) - \log\log(n)| < \varepsilon\log\log(n)$ has no more than $O(\varepsilon^{-2}N/\log\log(N))$ solutions in the interval $1 \le n \le N$ as $N \to +\infty$.

(19) Show that

$$\pi(x) = \sum_{n\le x} \omega(n) M\left(\frac{x}{n}\right),$$

where $\omega(n)$ is the number of distinct prime factors of n. Let $\omega_A(n)$ denote the number of distinct divisors of n from a fixed set A of positive integers, and generalize.

(20) Show that the sequence of squarefrees with an odd number of prime factors has logarithmic density equal to $3/\pi^2$. Interpret the analogous statement with asymptotic density in terms of a famous theorem.

(21) a) Let $(\mathbf{e}_1, \mathbf{e}_2, \ldots, \mathbf{e}_N)$ be the standard basis of $\mathbb{R}^N$. Define a vector space

$$\mathcal{L}_N = \mathrm{linspan}\{\mathbf{e}_n \mid 1 \le n \le N \text{ and } \mu(n) \ne 0\},$$

a codimension one subspace

$$\mathcal{M}_N = \mathrm{linspan}\{\mathbf{e}_n \mid 2 \le n \le N \text{ and } \mu(n) \ne 0\},$$

and a vector

$$\mathbf{v}_N = \sum_{\substack{1\le n\le N\\ \mu(n)\ne 0}} \mathbf{e}_n \in \mathcal{L}_N.$$

Show that

$$\frac{|M(N)|}{\sqrt{Q(N)}} = \mathrm{dist}(\mathbf{v}_N, \mathcal{M}_N),$$

where dist denotes Euclidean distance in $\mathcal{L}_N$. What bound for $M(N)$ do you get by bounding the distance from $\mathbf{v}_N$ to $\mathcal{M}_N$ by the distance from $\mathbf{v}_N$ to the origin?

b) Consider the distance problem in part a) as a least squares problem, and determine the normal equations for the unique minimizer.

(22) a) Establish the inequality

$$k\,\pi_{k+1}(x) \le \sum_{p \le \sqrt{x}} \pi_k\left(\frac{x}{p}\right)$$

for any positive integer k (Landau.) Here $\pi_k(x)$ denotes the number of squarefree integers $k \le x$ with precisely k prime factors.

† b) Show that the bound

$$\pi_{k+1}(x) \le \frac{Kx}{\log(x)} \frac{(\log\log(x) + C)^k}{k!}$$

holds for $x \ge 2$ with some positive constants C and K, by induction or otherwise (Hardy and Ramanujan.)

c) That $\pi_k(x)$ is unimodal (single-peaked) in k for each fixed x was one of three closely related unimodality conjectures stated by Erdös in 1948. These conjectures were proved by M. Balazard in the late 1980s. Deduce that

$$M(x) = O\left(\frac{x}{\sqrt{\log\log(x)}}\right)$$

from the unimodality of $\pi_k(x)$. Balazard's proof of unimodality is more difficult than the usual proofs of the Prime Number Theorem. But the possibility remains that an easy proof of his result might be discovered, and this would then yield an easy proof of the PNT. Unfortunately the unimodality does not reduce to combinatorics, but depends in some way on the distribution of the primes, as one can see by considering an example.

Chapter 3

Characters and Euler Products

3.1. The Euler product formula

An unordered sum

$$\sum_{j \in J} a_j$$

of terms $a_j \in [0, +\infty)$ always has a value $v \in [0, +\infty]$. If the set

$$S = \left\{ \sum_{j \in I} a_j \,\middle|\, I \subseteq J \text{ is finite} \right\}$$

has a finite upper bound, then put $v = \sup(S)$, otherwise $v = +\infty$. And

$$\prod_{j \in J} a_j$$

with $a_j \in [1, +\infty)$ always has a value $v \in [1, +\infty]$. If the set

$$P = \left\{ \prod_{j \in I} a_j \,\middle|\, I \subseteq J \text{ is finite} \right\}$$

has a finite upper bound, then put $v = \sup(P)$, otherwise $v = +\infty$.

Proposition 3.1. *The identity*

$$\prod_p \sum_{k=0}^{\infty} f(p^k) = \sum_{n=1}^{\infty} f(n)$$

holds for any nonnegative multiplicative arithmetic function f.

Proof. Let $n_1, \ldots, n_h$ be arbitrary distinct positive integers, and $p_1, \ldots, p_r$ all the primes dividing these integers. Then

$$\sum_{i=1}^{h} f(n_i) \leq \prod_{j=1}^{r} \sum_{k=0}^{m} f(p_j^k)$$

where m is the maximal exponent of all the prime powers dividing the integers $n_1, \ldots, n_h$. This is a consequence of the existence of prime factorizations of integers. Now

$$\sum_{i=1}^{h} f(n_i) \leq \prod_{p} \sum_{k=0}^{\infty} f(p^k)$$

and thus

$$\sum_{n=1}^{\infty} f(n) \leq \prod_{p} \sum_{k=0}^{\infty} f(p^k).$$

Let $p_1, \ldots, p_r$ be be arbitrary distinct primes and m an arbitrary positive integer. Let $n_1, \ldots, n_h$ be all the integers that are products of powers of these primes, with exponents less than or equal to m. Then

$$\prod_{j=1}^{r} \sum_{k=0}^{m} f(p_j^k) = \sum_{i=1}^{h} f(n_i).$$

This is a consequence of the uniqueness of prime factorizations of integers. Now

$$\prod_{j=1}^{r} \sum_{k=0}^{m} f(p_j^k) \leq \sum_{n=1}^{\infty} f(n)$$

and thus

$$\prod_{p} \sum_{k=0}^{\infty} f(p^k) \leq \sum_{n=1}^{\infty} f(n).$$

Combining the two inequalities yields the desired formula. □

This result is a version of the famous Euler product formula. Applying it to the nonnegative multiplicative function $f(n) = 1/n$ yields

$$\prod_{p} \sum_{k=0}^{\infty} \frac{1}{p^k} = \sum_{n=1}^{\infty} \frac{1}{n}$$

and then

$$\prod_{p} \frac{1}{1 - \frac{1}{p}} = \sum_{n=1}^{\infty} \frac{1}{n}$$

after summing the geometric series. This is an identity of Euler that is sometimes stated to be invalid. To the contrary, the identity is perfectly valid, and shows that there are infinitely many primes. The right-hand side is the harmonic series, which diverges to $+\infty$. Moreover it is well known that if a series

$$\sum_{j=1}^{\infty} a_j$$

converges absolutely, then the associated infinite product

$$\prod_{j=1}^{\infty}(1+a_j)$$

converges. Hence

$$\sum_p \frac{1}{p} = +\infty,$$

otherwise the left-hand side of the Euler identity would be a convergent infinite product.

Proposition 3.2. *If f is a multiplicative arithmetic function and the series*

$$\sum_{n=1}^{\infty} |f(n)| \quad \textit{or the product} \quad \prod_p \sum_{k=0}^{\infty} |f(p^k)|$$

converges, then

$$\prod_p \sum_{k=0}^{\infty} f(p^k) = \sum_{n=1}^{\infty} f(n),$$

where both the series and product are absolutely convergent.

Proof. First assume that the series converges. The inequality

$$\prod_p \left(1 + \left|\sum_{k=1}^{\infty} f(p^k)\right|\right) \leq \prod_p \sum_{k=0}^{\infty} |f(p^k)| = \sum_{n=1}^{\infty} |f(n)|$$

holds by applying Proposition 3.1 to $|f|$, so the infinite product is absolutely convergent. Furthermore

$$\left|\prod_{p\leq x} \sum_{k=0}^{\infty} f(p^k) - \sum_{n\leq x} f(n)\right| = \left|\sum_{p_1,\ldots,p_r\leq x} f(p_1^{k_1})\cdots f(p_r^{k_r}) - \sum_{n\leq x} f(n)\right|$$

$$= \left|\sum_{p_1,\ldots,p_r\leq x} f(p_1^{k_1}\cdots p_r^{k_r}) - \sum_{n\leq x} f(n)\right| \leq \sum_{n>x} |f(n)| \to 0$$

as $x \to +\infty$, by absolute convergence and unique factorization into primes.

Next assume that the product converges. Then the series also converges, again by Proposition 3.1 applied to $|f|$. $\square$

Most of the interesting arithmetic functions grow no faster than polynomially. If f is such a function, the function $f(n)n^{-s}$ will tend to zero rapidly enough for the series

$$\sum_{n=1}^{\infty} f(n)n^{-s}$$

to converge absolutely, for suitable choices of s. Here s may be taken as a complex number. The function n^{-s} is multiplicative for each fixed s, so $f(n)n^{-s}$ is multiplicative if f is. Under the assumption that f is multiplicative and that the above series converges absolutely, Proposition 3.2 yields the version

$$\prod_{p}\sum_{k=0}^{\infty} f(p^k)p^{-ks} = \sum_{n=1}^{\infty} f(n)n^{-s}$$

of the Euler product formula. If f is totally multiplicative, the Euler product formula takes the form

$$\prod_{p}\frac{1}{1-f(p)p^{-s}} = \sum_{n=1}^{\infty} f(n)n^{-s}$$

because the sums under the product sign are geometric series.

The *Riemann zeta function* is defined by the Euler product

$$\zeta(s) \stackrel{\text{def}}{=} \prod_{p}(1-p^{-s})^{-1}$$

for $\operatorname{Re}(s) > 1$. It is standard notation in analytic number theory, going back to papers of Dirichlet and Riemann, to write the complex variable as $s = \sigma + it$ where $\sigma = \operatorname{Re}(s)$ and $t = \operatorname{Im}(s)$. The estimate

$$\sum_{p}|p^{-s}| \le \sum_{n=1}^{\infty} n^{-\sigma} \le 1 + \int_{1}^{\infty} x^{-\sigma}\,dx = \frac{\sigma}{\sigma-1}$$

shows that the Euler product for $\zeta(s)$ converges absolutely in the half plane $\sigma > 1$, and uniformly on any compact set in this half plane. Thus $\zeta(s)$ is holomorphic in $\sigma > 1$, and the Euler product shows that $\zeta(s) \neq 0$ in the same half plane. Moreover

$$\zeta(s) = \sum_{n=1}^{\infty} n^{-s}$$

for $\sigma > 1$ by Proposition 3.2. We could equally well have defined the Riemann zeta function by this series expansion, but towards the end of the book we shall see good reasons to regard the Euler product as more fundamental, though both are important. The series expansion is used to analytically continue $\zeta(s)$ to the whole complex plane as a meromorphic function, and to obtain auxiliary information of an analytic nature. The Euler product is clearly of high significance from an arithmetic viewpoint.

A series of the form

$$\sum_{n=1}^{\infty} a_n n^{-s}$$

with coefficients $a_n \in \mathbb{C}$ and variable $s \in \mathbb{C}$ is called a *Dirichlet series.* The Riemann zeta function is the best known Dirichlet series. The Dirichlet series

$$\sum_{n=1}^{\infty} f(n) n^{-s}$$

of an arithmetic function f can be regarded as a type of *generating function* for f, that is to say, as a representation of the sequence $(f(n))_{n=1}^{\infty}$ by a convenient analytical expression, in the form of an infinite series. In this context no questions of convergence arise, and we speak of *formal Dirichlet series.* Generating functions are devices applied in several subjects, in particular combinatorics, number theory and probability theory. These generating functions are usually the more familiar formal power series

$$\sum_{n=0}^{\infty} g(n) z^n$$

that are used to investigate additive problems in number theory. Formal power series are appropriate for additive problems because of the property $z^m z^n = z^{m+n}$ of the monomials. In the same way formal Dirichlet series are appropriate for multiplicative problems because of the property $m^{-s}n^{-s} = (mn)^{-s}$ of the Dirichlet monomials.

Multiplication of formal Dirichlet series has an interesting arithmetical interpretation. Rearrangement of a double series yields

$$\left(\sum_{k=1}^{\infty} f(k)k^{-s}\right)\left(\sum_{m=1}^{\infty} g(m)m^{-s}\right) = \sum_{k}\sum_{m} f(k)k^{-s}g(m)m^{-s}$$

$$= \sum_{n} n^{-s} \sum_{km=n} f(k)g(m) = \sum_{n=1}^{\infty} (f * g)(n) n^{-s}$$

where $f * g$ denotes the Dirichlet convolution of f and g. As an example, the formal Dirichlet series expansion

$$\sum_{n=1}^{\infty} \phi(n) n^{-s} = \frac{\zeta(s-1)}{\zeta(s)}$$

follows by the convolution identity $1 * \phi = id$ of Gauss.

The Dirichlet ring is isomorphic to the ring of formal Dirichlet series, and the latter is isomorphic to the ring $\mathbb{C}[[X_1, X_2, \ldots]]$ of power series in infinitely many variables. This fact was used by E. D. Cashwell and C. J. Everett to show that unique factorization into irreducibles holds in the Dirichlet ring.

By interpreting the Euler product formula

$$\prod_p \sum_{k=0}^{\infty} f(p^k)p^{-ks} = \sum_{n=1}^{\infty} f(n)n^{-s}$$

as an identity between formal Dirichlet series, we can free it of all considerations of convergence. In this framework the operation of multiplying out the infinitely many factors on the left-hand side is well defined. The formal identitities

$$\sum_{n=1}^{\infty} \mu(n)n^{-s} = \prod_p (1-p^{-s}) = \zeta(s)^{-1},$$

$$\sum_{n=1}^{\infty} d(n)n^{-s} = \prod_p \sum_{k=0}^{\infty} (k+1)p^{-ks} = \prod_p \frac{1}{(1-p^{-s})^2} = \zeta^2(s),$$

$$\sum_{n=1}^{\infty} \Lambda(n)n^{-s} = -\sum_p \frac{-\log(p)p^{-s}}{1-p^{-s}} = -\sum_p \frac{(1-p^{-s})'}{1-p^{-s}} = -\frac{\zeta'(s)}{\zeta(s)},$$

are examples of Dirichlet series identities obtained by means of the Euler product formula, and there are many others.

There is an illuminating discussion of generating functions with special emphasis on number theory in *Introduction to Analytic Number Theory* by D. J. Newman, and also an example not to be missed.

3.2. Convergence of Dirichlet series

Proposition 3.3. *If a Dirichlet series*

$$\sum_{n=1}^{\infty} a_n n^{-s}$$

converges at a point $s_0 \in \mathbb{C}$, *then it converges uniformly in the angular sector* $|s-s_0| \le C(\sigma-\sigma_0)$ *for any fixed* $C>0$.

Proof. Suppose, by translation and without loss of generality, that $s_0 = 0$. Putting $r_n = a_{n+1} + a_{n+2} + \cdots$, we obtain

$$\sum_{n=M+1}^{N} a_n n^{-s} = \sum_{n=M+1}^{N} (r_{n-1} - r_n)n^{-s}$$

$$= r_M(M+1)^{-s} + \sum_{n=M+1}^{N-1} r_n((n+1)^{-s} - n^{-s}) - r_N N^{-s}$$

for $M < N$ by partial summation. Now

$$\begin{aligned}|(n+1)^{-s} - n^{-s}| &\le \left|\int_n^{n+1} (-s)u^{-s-1}\,du\right| \\ &\le \int_n^{n+1} |s|u^{-\sigma-1}\,du = \frac{|s|}{\sigma}(n^{-\sigma} - (n+1)^{-\sigma}).\end{aligned}$$

For $\varepsilon > 0$ arbitrary, choose $n_0(\varepsilon)$ so that $|r_n| < \varepsilon/(2C)$ for $n \ge n_0(\varepsilon)$. Then

$$\begin{aligned}&\left|\sum_{n=M+1}^{N} a_n n^{-s}\right| \\ &\le \frac{|s|}{\sigma}\left(|r_M|(M+1)^{-\sigma} + \sum_{n=M+1}^{N-1} |r_n|(n^{-\sigma} - (n+1)^{-\sigma}) + |r_N|N^{-\sigma}\right) \\ &< \frac{|s|}{\sigma}\frac{\varepsilon}{C} \le \varepsilon\end{aligned}$$

for $N > M \ge n_0(\varepsilon)$, since $|s| \le C\sigma$. □

If a Dirichlet series converges at s_0, then it converges in each point s with $\sigma > \sigma_0$. For $|s - s_0| \le C(\sigma - \sigma_0)$ for some sufficiently large $C > 0$. So if the Dirichlet series converges at s_0, then it converges in the half plane $\sigma > \sigma_0$ and if it diverges at s_0, then it diverges in the half plane $\sigma < \sigma_0$. Hence every Dirichlet series has an *abscissa of convergence* σ_c such that it converges to the right of the line $\sigma = \sigma_c$ and diverges to the left of this line, which is called the *line of convergence* for the series. It is possible that a Dirichlet series may converge nowhere, in which case $\sigma_c = +\infty$, or that it may converge everywhere, in which case $\sigma_c = -\infty$. If $\sigma_c < \infty$, we say that the Dirichlet series is convergent.

Proposition 3.4. *Suppose that $\alpha \ge 0$ and that*

$$F(x) = \sum_{n \le x} a_n = O(x^{\alpha+\varepsilon})$$

for any $\varepsilon > 0$. Then the Dirichlet series

$$\sum_{n=1}^{\infty} a_n n^{-s}$$

has abscissa of convergence $\sigma_c \le \alpha$.

Proof. We have

$$\sum_{n=1}^{\infty} a_n n^{-s} = \lim_{x \to +\infty} F(x)x^{-s} + s\int_1^{\infty} F(t)t^{-s-1}\,dt$$

by partial summation. Now

$$\int_1^\infty |F(t)t^{-s-1}|\,dt \le \int_1^\infty Ct^{\alpha+\varepsilon}t^{-\sigma-1}\,dt < \infty,$$

and $F(x)x^{-s} \to 0$, for $\sigma > \alpha + \varepsilon$. □

The abscissa of absolute convergence σ_a of the Dirichlet series

$$\sum_{n=1}^{\infty} a_n n^{-s}$$

is defined as the abscissa of convergence of

$$\sum_{n=1}^{\infty} |a_n| n^{-s}.$$

The corresponding vertical line is called the *line of absolute convergence.* The abscissa of convergence and the abscissa of absolute convergence of a Dirichlet series may be different. As an example, consider the Dirichlet series

$$A(s) = \sum_{n=1}^{\infty} (-1)^{n-1} n^{-s} = (1 - 2^{1-s})\zeta(s)$$

where the summatory function

$$F(x) = \sum_{n \le x} (-1)^{n-1}$$

of the coefficients is bounded. Clearly $\sigma_c \le 0$ by Proposition 3.4, while the series diverges for $s = 0$. Hence $\sigma_c = 0$ for this Dirichlet series, but

$$\sum_{n=1}^{\infty} |(-1)^{n-1}| n^{-s} = \sum_{n=1}^{\infty} n^{-s},$$

and so $\sigma_a = 1$.

Proposition 3.5. *The inequality $\sigma_a - \sigma_c \le 1$ holds for any Dirichlet series.*

Proof. Suppose, by translation and without loss of generality, that $\sigma_c = 0$. Then $a_n n^{-s}$ tends to zero as $n \to +\infty$ if $\sigma > 0$. In particular $|a_n| n^{-\varepsilon/2}$ is bounded for any fixed $\varepsilon > 0$, but then

$$\sum_{n=1}^{\infty} |a_n| n^{-\varepsilon/2} n^{-1-\varepsilon/2} = \sum_{n=1}^{\infty} |a_n| n^{-1-\varepsilon}$$

converges, and so $\sigma_a \le 1 + \varepsilon$, for any $\varepsilon > 0$. □

The next result implies that every convergent Dirichlet series is uniquely determined by its sum. This allows us to specify the abscissa of convergence and of absolute convergence of

$$A(s) = \sum_{n=1}^{\infty} a_n n^{-s}$$

unambiguously by the notations $\sigma_c(A)$ and $\sigma_a(A)$ respectively.

Proposition 3.6. *Suppose that*

$$A(s) = \sum_{n=1}^{\infty} a_n n^{-s} \quad \text{and} \quad B(s) = \sum_{n=1}^{\infty} b_n n^{-s}$$

are convergent Dirichlet series, and $(s_k)_{k=1}^{\infty} \subseteq \mathbb{C}$ *a sequence of points with* $\sigma_k \to +\infty$ *as* $k \to +\infty$, *for which* $A(s_k) = B(s_k)$ *for* $k \in \mathbb{N}$. *Then* $a_n = b_n$ *for* $n \in \mathbb{N}$.

Proof. Each of the two Dirichlet series converges absolutely in a half plane to the right of some vertical line, by Proposition 3.5. Hence they both converge absolutely on $\sigma > \alpha$, say. Then they converge uniformly on $\sigma > \alpha + 1$, by the Weierstrass M-test. Now

$$a_1 = \lim_{k \to +\infty} A(s_k) = \lim_{k \to +\infty} B(s_k) = b_1$$

by uniform convergence, since $\sigma_k \to +\infty$. Subtracting the common constant term from both series, and multiplying by 2^s, and repeating this argument, we obtain $a_2 = b_2$. Continuing this way, we see that $a_n = b_n$ for all $n \in \mathbb{N}$. □

Most of the theory of Dirichlet series useful in analytic number theory may be found in *The General Theory of Dirichlet's Series* by G. H. Hardy and M. Riesz and in Chapter IX of *The Theory of Functions* by E. C. Titchmarsh. Beyond this material, there are important estimates for finite Dirichlet series in *Analytic Number Theory* by H. Iwaniec and E. Kowalski and in *Ten Lectures on the Interface Between Analytic Number Theory and Harmonic Analysis* by H. L. Montgomery.

3.3. Harmonics

Ideas from harmonic analysis have played an important role in analytic number theory ever since Dirichlet founded the subject in the 1830s. In harmonic analysis one studies the approximation, or even the exact representation, of functions by finite linear combinations

$$f(x) = c_1 b_1(x) + c_2 b_2(x) + \cdots + c_N b_N(x),$$

or by analogous infinite series or integrals, where all the *basis functions* $b_n(x)$ share a common symmetry. This common symmetry is usually determined by an action of some group on the space on which the functions are defined. Basis functions that share a common symmetry are called *harmonics.*

The best known example from this circle of ideas is the classical Fourier series

$$f(x) \sim \sum_{n=-\infty}^{\infty} \hat{f}_n \, e(nx)$$

where the harmonics $e(nx)$ given by

$$e(x) \stackrel{\text{def}}{=} e^{2\pi i x}$$

share the common symmetry of being periodic with period 1. That is, the harmonics are invariant under the action of the additive group $\mathbb{Z}$ of integers on the real line $\mathbb{R}$ given by $x \mapsto k + x$.

The simplest kind of harmonic analysis important in number theory pertains to arithmetic functions $f : \mathbb{Z} \to \mathbb{C}$ that are periodic with period q. There are various arithmetic functions that are most naturally defined on $\mathbb{Z}$ rather than $\mathbb{N}$ and that are periodic. An example is the Legendre symbol $(n|p)$ modulo an odd prime p, which becomes a totally multiplicative arithmetic function when extended by $(n|p) = 0$ for $n \equiv 0 \pmod{p}$. It is clearly periodic with period p.

The basis functions

$$e(mn/q) = e^{2\pi i mn/q}, \qquad m = 1, 2, \ldots, q$$

are periodic with period q. Supposing that f is an arithmetic function with period q, we shall determine a function $\hat{f}$ so that f has a *finite Fourier expansion*

$$f(n) = q^{-1/2} \sum_{m=1}^{q} \hat{f}(m) e(mn/q)$$

valid for for all n. Multiplying through by $e(-kn/q)$ and summing over $n = 1, 2, \ldots, q$ yields

$$\begin{aligned} q^{-1/2} \sum_{n=1}^{q} f(n) e(-kn/q) &= q^{-1/2} \sum_{n=1}^{q} \left(q^{-1/2} \sum_{m=1}^{q} \hat{f}(m) e(mn/q) \right) e(-kn/q) \\ &= q^{-1} \sum_{m=1}^{q} \hat{f}(m) \sum_{n=1}^{q} e((m-k)n/q) = \hat{f}(k) \end{aligned}$$

by the summation

$$\sum_{n=1}^{q} e\left(\frac{(m-k)n}{q} \right) = \begin{cases} q & \text{if } m \equiv k \pmod{q}, \\ 0 & \text{otherwise,} \end{cases}$$

of a finite geometric series. Thus the coefficients in the finite Fourier expansion are given by

$$\hat{f}(m) = q^{-1/2} \sum_{n=1}^{q} f(n) e(-mn/q).$$

Considered as a function, $\hat{f}$ is called the *finite Fourier transform* of f. These coefficients were found on the unproved assumption that f actually has a finite Fourier expansion, but now that we have discovered a formula for them, the calculation

$$\begin{aligned} q^{-1/2} \sum_{m=1}^{q} \hat{f}(m) e(mn/q) &= q^{-1/2} \sum_{m=1}^{q} \left(q^{-1/2} \sum_{k=1}^{q} f(k) e(-mk) \right) e(mn/q) \\ &= \frac{1}{q} \sum_{k=1}^{q} f(k) \sum_{m=1}^{q} e(m(n-k)/q) = f(n) \end{aligned}$$

shows that f really has such an expansion. As the coefficients are unique, the harmonics $e(mn/q)$ with $m = 1, 2, \ldots, q$ are linearly independent. Since there are q different harmonics $e(mn/q)$, this checks with the observation that the linear space of arithmetic functions of period q has dimension q.

The space of arithmetic functions of period q is isomorphic with the space of complex-valued functions on the cyclic group $\mathbb{Z}/q\mathbb{Z}$. The harmonics $e(mn/q)$ are themselves periodic of period q, and may be considered as complex-valued functions on this group. The above expansion is called the finite Fourier expansion on $\mathbb{Z}/q\mathbb{Z}$. Later we shall consider Fourier expansions on more complicated finite groups.

The calculation

$$\begin{aligned} \sum_{m=1}^{q} |\hat{f}(m)|^2 &= \sum_{m=1}^{q} q^{-1/2} \sum_{k=1}^{q} \overline{f(k)} e(mkq) q^{-1/2} \sum_{n=1}^{q} f(n) e(-mn/q) \\ &= q^{-1} \sum_{k=1}^{q} \sum_{n=1}^{q} \overline{f(k)} f(n) \sum_{m=1}^{q} e((k-n)m/q) = \sum_{n=1}^{q} |f(n)|^2 \end{aligned}$$

establishes the Plancherel formula for $\mathbb{Z}/q\mathbb{Z}$. Introducing the Hermitian inner product

$$\langle f | g \rangle = \sum_{n=1}^{q} \overline{f(n)} g(n),$$

the Plancherel formula states that the Fourier transform is a unitary operator on the linear space of complex-valued functions on $\mathbb{Z}/q\mathbb{Z}$.

As an important example, we consider the finite Fourier expansion modulo an odd prime p whose coefficients are given by the Legendre symbol modulo p. The calculation

$$p^{-1/2}\sum_{k=1}^{p}\left(\frac{k}{p}\right)e(kn/p) = p^{-1/2}\sum_{m=1}^{p}\left(\frac{m\overline{n}}{p}\right)e(m\overline{n}n/p)$$
$$= \left(\frac{n}{p}\right)p^{-1/2}\sum_{m=1}^{p}\left(\frac{m}{p}\right)e(m/p)$$

is valid if p does not divide n, for then n has a multiplicative inverse $\overline{n}$ modulo p. While if p does divide n, then the first and the third expression are equal because both are zero. The sum

$$\tau_p = \sum_{m=1}^{p}\left(\frac{m}{p}\right)e(m/p)$$

is the *classical Gauss sum.* Denoting an element of a complete collection of quadratic residues modulo p by m' and an element of a complete collection of quadratic nonresidues modulo p by m'' one sees that

$$\tau_p = \sum_{m'} e(m'/p) - \sum_{m''} e(m''/p)$$

while

$$\sum_{m'} e(m'/p) + \sum_{m''} e(m''/p) = -1$$

by summing a finite geometric series. For the Legendre symbol $(m|p)$ equals zero if $p|m$, otherwise it equals 1 if m is a quadratic residue modulo p and -1 if not. The congruence $x^2 \equiv m \pmod p$ has two solutions $x \equiv \pm k \pmod p$ for each quadratic residue m and the single solution $x \equiv 0$ for $m = 0$. Hence

$$\tau_p = 1 + \tau_p - 1 = 1 + 2\sum_{m'} e(m'/p) = \sum_{k=1}^{p} e(k^2/p)$$

expressed in the same notation. This formula yields an easy calculation of the absolute value of the classical Gauss sum. We have

$$|\tau_p|^2 = \tau_p\overline{\tau_p} = \sum_{k=1}^{p} e(k^2/p)\sum_{m=1}^{p} e(-m^2/p)$$
$$= \sum_{k=1}^{p}\sum_{m=1}^{p} e((k-m)(k+m)/p) = \sum_{j}\sum_{\ell} e(j\ell/p) = \sum_{\ell} e(0{\cdot}\ell/p) = p$$

where $j = k - m$ and $\ell = k + m$ range over a complete collection of residues modulo p. In particular $\tau_p \neq 0$ and thus

$$\left(\frac{n}{p}\right) = p^{-1/2}\sum_{k=1}^{p}\frac{\sqrt{p}}{\tau_p}\left(\frac{k}{p}\right)e(kn/p)$$

is the finite Fourier expansion of the Legendre symbol.

3.4. Group representations

The example in the previous section may be widely generalized by replacing $\mathbb{Z}/q\mathbb{Z}$ by a group G acting on some space X. The case where G is a finite group acting on itself by multiplication is particularly tractable because the expansions into harmonics are finite sums and thus no issues of convergence arise. Two elementary cases are especially important in analytic number theory: the cyclic group $\mathbb{Z}/q\mathbb{Z}$ of residue classes modulo a positive integer q under addition, which we already considered in the previous section, and the abelian group $(\mathbb{Z}/q\mathbb{Z})^\times$ of reduced residue classes modulo q under multiplication.

The *general linear group* $\mathrm{GL}(V)$ on a finite-dimensional vector space V consists of the automorphisms of V (the linear isomorphisms of V with itself) under composition. This is a group that is nonabelian as soon as the dimension is larger than one. From now on we consider vector spaces V over $\mathbb{C}$ exclusively. Introducing an ordered basis for V shows $\mathrm{GL}(V)$ to be isomorphic to the group of nonsingular complex matrices of size $\dim(V)$ by $\dim(V)$ under matrix multiplication. A (complex, finite-dimensional) *representation* ρ of G on the *representation space* $V = V_\rho$ is a homomorphism $\rho : G \to \mathrm{GL}(V)$. We might tend to think of G as an abstract group, or perhaps a group that we do not know much about, while $\mathrm{GL}(V)$ is a very concrete group; simply a matrix group. One algebraic structure is being represented by another that is more tangible. In this particular situation the procedure turns out to be very fruitful indeed, because it brings the resources of linear algebra to bear on the group theory. But we are not going to pursue applications of representation theory to group theory; the point is rather its potential to produce harmonics.

Though one speaks of finite-dimensional versus infinite-dimensional representations, the dimension $\dim(V_\rho) = \deg(\rho)$ is usually called the *degree* of the representation in the finite-dimensional case, rather than its dimension. It is, of course, the degree of the characteristic polynomial of the linear operator $\rho(g)$. Every group has a *trivial representation* ρ_0 of degree one defined by the constant map $g \mapsto \mathrm{id}_\mathbb{C}$ on G.

Suppose ρ and ρ' are two representations of the same group G on the representation spaces $V = V_\rho$ and $W = V_{\rho'}$ respectively. A linear transformation $\varphi : V \to W$ such that $\varphi\rho(g) = \rho'(g)\varphi$ for each $g \in G$ will be called a *linear map* between ρ and ρ'. If this is an isomorphism the two representations are said to be *equivalent*. For the most part one does not distinguish between equivalent representations.

As an example, we consider the group S_3 of permutations on three letters. As always, the group has a trivial representation ρ_0. Furthermore it contains an odd permutation, and any permutation group with an odd permutation has a representation of degree one that is not equivalent to the trivial representation, namely the map $\pi \mapsto \text{sgn}(\pi)$ given by the sign of the permutation. (If a permutation group contains no odd permutation, then the sign map just yields the trivial representation.) So now we have a representation ρ_1 of S_3 of degree one that is not equivalent to ρ_0. To find a third representation ρ_2 of S_3 we argue geometrically. Consider the equilateral triangle in the plane with corners at $(1,0), T(1,0), T^2(1,0)$ where the linear operator $T : \mathbb{R}^2 \to \mathbb{R}^2$ given by

$$T(x,y) = \begin{bmatrix} -\frac{1}{2} & -\frac{\sqrt{3}}{2} \\ \frac{\sqrt{3}}{2} & -\frac{1}{2} \end{bmatrix} \begin{bmatrix} x \\ y \end{bmatrix}$$

rotates the triangle 120 degrees counterclockwise around the origin, taking each corner into the corner to its left as seen from the origin. The rotation T and the reflection U around the real axis given by $U(x,y) = (x,-y)$ together generate a group isomorphic to S_3. The map that labels the corners $(1,0), T(1,0), T^2(1,0)$ by the letters a, b, c sends T to the permutation (abc) and U to $(a)(bc)$, and these two permutations generate S_3. Clearly we now have a representation ρ of S_3 on $\mathbb{C}^2$ by $\rho((abc)) = T$ and $\rho((a)(bc)) = U$. It is not equivalent to the others that we found, since it has degree 2. We note that all the representations we found are defined over $\mathbb{R}$, though generally a group may be expected to have representations that are not equivalent to any representation defined over $\mathbb{R}$.

A subspace W of V is called an *invariant subspace* for an operator T on V if $TW \subseteq W$. In this case the restriction $T|_W$ of T to W is itself a linear operator, and not merely a linear transformation. If $\rho : G \to \text{GL}(V)$ is a representation and $W \subseteq V$ an invariant subspace of $\rho(g)$ for each $g \in G$, we say that W is ρ-invariant. Then $\rho(g)|_W$ is an automorphism of W for each $g \in G$, since the kernel of $\rho(g)|_W$ is zero and $\dim(W) < +\infty$. Thus $\rho|_W : G \to \text{GL}(W)$ is also a representation, called a *subrepresentation* of ρ. If we choose an ordered basis $\mathcal{B} = (e_1, \ldots, e_k)$ for W and extend it to an ordered basis $\mathcal{C} = (e_1, \ldots, e_k, e_{k+1}, \ldots, e_n)$ of V, then the representation ρ can be expressed as a block matrix

$$[\rho(g)]_{\mathcal{C},\mathcal{C}} = \begin{bmatrix} A(g) & B(g) \\ 0 & D(g) \end{bmatrix}$$

in terms of the basis of V, and the subrepresentation $\rho|_W$ is expressed as $[\rho(g)|_W]_{\mathcal{B},\mathcal{B}} = A(g)$ in terms of the basis for W. The *direct sum* $\rho \oplus \rho'$ of two representations $\rho : G \to \text{GL}(W)$ and $\rho' : G \to \text{GL}(W')$ of the same group G is the representation $\rho \oplus \rho' : G \to \text{GL}(W \oplus W')$ given by $(\rho \oplus \rho')(g)(w \oplus w') = \rho(g)w \oplus \rho'(g)w'$ for each $g \in G$. Extending a basis $\mathcal{B} = (e_1, \ldots, e_k)$ for W by a basis $\mathcal{B}' = (e_{k+1}, \ldots, e_n)$ for W' to a basis

$\mathcal{C} = (e_1, \ldots, e_k, e_{k+1}, \ldots, e_n)$ for $W \oplus W'$, we see that the direct sum of the two representations can be expressed as a block matrix

$$[\rho \otimes \rho'(g)]_{\mathcal{C},\mathcal{C}} = \begin{bmatrix} A(g) & 0 \\ 0 & D(g) \end{bmatrix}$$

in terms of the basis for $W \oplus W'$. Clearly a subrepresentation that is a direct summand is very different from a general subrepresentation; in the former case the invariant subspace W has a complemented subspace W' that is also ρ-invariant. A representation that has no proper nonzero subrepresentations is called *irreducible.* Every representation of degree one is irreducible. A representation is called *completely reducible* if it is the direct sum of irreducible ones. Completely reducible representations may be studied in terms of irreducibles, and the following result is thus very important.

Proposition 3.7 (Maschke complete reducibility theorem)**.** *Every complex finite-dimensional representation of a finite group is completely reducible.*

Proof. The representation space V carries a Hermitian inner product, for example,

$$\langle c_1e_1 + \cdots + c_ne_n | d_1e_1 + \cdots + d_ne_n \rangle = \overline{c_1}d_1 + \cdots + \overline{c_n}d_n$$

where $(e_1, \ldots, e_n)$ is some ordered basis for V. We then define a new Hermitian inner product

$$\langle v|w \rangle_\rho = \frac{1}{|G|} \sum_{g \in G} \langle \rho(g)v | \rho(g)w \rangle$$

by averaging. This inner product is visibly invariant under ρ. That is to say, $\rho(g)$ is a unitary operator with respect to this inner product, for each $g \in G$. If ρ is not itself irreducible, there is a proper nonzero ρ-invariant subspace W of V. But then the orthocomplement $W' = W^\perp$ of W with respect to $\langle u|v \rangle_\rho$ is also ρ-invariant. For if $w' \in W'$ then

$$\langle \rho(g)w | \rho(g)w' \rangle_\rho = \langle w|w' \rangle_\rho = 0$$

for all $w \in W$ and all $g \in G$. This implies that W' is also ρ-invariant, for $\rho(g)w' \in W'$ since $\rho(g)|_W$ is an automorphism of W for each $g \in G$. We have now obtained a decomposition of ρ as a direct sum of two representations on W and W' respectively with $0 < \dim(W), \dim(W') < \dim(V)$. This process can be continued with each direct summand, but must ultimately stop since dimension is strictly decreasing. Then we are left with a decomposition of the representation as a direct sum, each summand of which must be irreducible. □

The averaging idea in this proof of Maschke's theorem implies that every finite-dimensional complex representation of a finite group is equivalent to a unitary representation, whatever Hermitian inner product prescribes what it shall mean to be unitary. For suppose that ρ is a representation, V its representation space and $\langle \cdots | \cdots \rangle$ any Hermitian inner product on V. There exists some operator $\varphi : V \to V$ such that $\langle \varphi v | \varphi w \rangle = \langle v|w\rangle_\rho$ in the notation of the proof. The representation $\rho' = \varphi\rho\varphi^{-1}$ satisfies

$$\begin{aligned}\langle \rho' v | \rho' w \rangle &= \langle \varphi\rho\varphi^{-1} v | \varphi\rho\varphi^{-1} w\rangle = \langle \rho\varphi^{-1} v | \rho\varphi^{-1} w\rangle_\rho \\ &= \langle \varphi^{-1} v | \varphi^{-1} w\rangle_\rho = \langle \varphi\varphi^{-1} v | \varphi\varphi^{-1} w\rangle = \langle v|w\rangle.\end{aligned}$$

Clearly ρ' is equivalent to ρ, moreover, with the same representation space. The desired statement follows. We shall make use of this observation when discussing the Fourier transform on finite groups.

The direct sum decomposition into irreducible representations is unique in the sense that in any two such decompositions the same multiplicities of irreducible direct summands will appear.

Proposition 3.8 (Schur's Lemma). *A linear map between irreducible finite-dimensional complex representations is an isomorphism or else identically zero. If the two representations have the same representation space, the linear map must be some complex multiple of the identity map.*

Proof. Suppose φ is a linear map between irreducible complex representations ρ and ρ'. Then $\ker(\varphi)$ is a ρ-invariant subspace of V_ρ and $\operatorname{im}(\varphi)$ is a ρ'-invariant subspace of $V_{\rho'}$. If φ is not identically zero, then φ is both a monomorphism and an epimorphism.

If $V_\rho = V_{\rho'} = V$ then the operator φ on the finite-dimensional complex vector space V has an eigenvalue $\lambda \in \mathbb{C}$. In that case the ρ-invariant subspace $\ker(\varphi - \lambda \mathrm{id}_V)$ is different from zero and so $\varphi - \lambda \mathrm{id}_V \equiv 0$. □

By Schur's Lemma the identity map from one decomposition $m_1\rho_1 \oplus \cdots \oplus m_r\rho_r$ of a representation, with pairwise inequivalent irreducibles ρ_j, onto another such decomposition $m'_1\rho'_1 \oplus \cdots \oplus m'_s\rho'_s$ of the same representation, maps each direct summand $m_j\rho_j$ into some $m'_k\rho'_k$, and vice versa. Thus the $m_j\rho_j$ are equal to the $m'_k\rho'_k$ in some order, which is the precise form of the uniqueness statement mentioned above. There is no implication that the ρ_j will be equal to the ρ'_k in some order, and this does not hold in general. It does of course hold if all the multiplicities equal one. The unique blocks $m_j\rho_j$ obtained by collecting equivalent irreducible representations are called *isotypical components.*

Any finite-dimensional representation ρ may be expressed as a matrix function $[\rho(g)]_{\mathcal{B},\mathcal{B}} = [a_{ij}(g)]$ on G relative to a fixed ordered basis $\mathcal{B}$ of V_ρ. The scalar functions a_{ij} are called *matrix entries* of the representation. There is nothing canonical about particular matrix entries; choosing another ordered basis for V_ρ yields a different collection of matrix entries. But a system of linear relations

$$a_{ij}(gh) = \sum_m a_{im}(g)a_{mj}(h)$$

between matrix entries holds by the definition of a representation. So the matrix entries transform in a predictable way under the mappings $h \mapsto gh$ on G, and may be viewed as harmonics on G. The matrix entries have a kind of orthogonality property.

Proposition 3.9 (Schur orthogonality theorem). *Suppose that ρ and ρ' are inequivalent irreducible finite-dimensional complex representations. Let a_{ij} be the matrix entries of ρ with respect to an ordered basis $\mathcal{B}$ of V_ρ and b_{kl} the matrix entries of ρ' with respect to an ordered basis $\mathcal{B}'$ of $V_{\rho'}$. Then*

$$\sum_g a_{ij}(g^{-1})b_{kl}(g) = 0$$

when the sum is taken over all elements g of G. Furthermore

$$\frac{1}{|G|}\sum_g a_{ij}(g^{-1})a_{kl}(g) = \frac{\delta_{il}\delta_{jk}}{\deg(\rho)},$$

where δ_{il} is the Kronecker delta that is equal to 1 when $i = l$ and zero otherwise.

Proof. Let M be an arbitrary matrix of size $\deg(\rho)$ by $\deg(\rho')$ and define a linear transformation $\varphi_M : V_\rho \to V_{\rho'}$ in terms of ordered bases by

$$[\varphi_M]_{\mathcal{B},\mathcal{B}'} = \sum_g [a_{ij}(g^{-1})]M[b_{kl}(g)].$$

Then

$$\begin{aligned}
[\rho(f)]_{\mathcal{B},\mathcal{B}}[\varphi_M]_{\mathcal{B},\mathcal{B}'} &= \sum_g [\rho(f)]_{\mathcal{B},\mathcal{B}}[\rho(g^{-1})]_{\mathcal{B},\mathcal{B}}M[\rho'(g)]_{\mathcal{B}',\mathcal{B}'} \\
&= \sum_g [\rho(fg^{-1})]_{\mathcal{B},\mathcal{B}}M[\rho'(g)]_{\mathcal{B}',\mathcal{B}'} \\
&= \sum_h [\rho(h^{-1})]_{\mathcal{B},\mathcal{B}}M[\rho'(h)]_{\mathcal{B}',\mathcal{B}'}[\rho'(f)]_{\mathcal{B}',\mathcal{B}'} \\
&= [\varphi_M]_{\mathcal{B},\mathcal{B}'}[\rho'(f)]_{\mathcal{B}',\mathcal{B}'}
\end{aligned}$$

for each element f of G, by the change $h = fg^{-1}$ of summation variable; fg^{-1} runs through G once if g runs through G once. Clearly φ_M is a linear map between the representations ρ and ρ'.

If the two representations are assumed inequivalent then $\varphi_M \equiv 0$ by Schur's Lemma. Taking M to be the matrix whose entry in position (j,k) equals 1 and whose other entries are all zero, the first assertion follows.

If instead $\rho = \rho'$ then $\varphi_M = \lambda_M \mathrm{id}_V$ by Schur's Lemma. Hence

$$\sum_g [a_{ij}(g^{-1})] M [a_{kl}(g)] = \lambda_M \delta_{il} I$$

where I is the $\deg(\rho)$ by $\deg(\rho)$ identity matrix. Taking M as before we obtain

$$\sum_g a_{ij}(g^{-1}) a_{kl}(g) = \lambda_{jk} \delta_{il} \delta_{jk}.$$

Summing over g^{-1} instead of g yields

$$\sum_g a_{kl}(g^{-1}) a_{ij}(g) = \lambda_{jk} \delta_{il} \delta_{jk}.$$

Comparison with

$$\sum_g a_{kl}(g^{-1}) a_{ij}(g) = \lambda_{li} \delta_{kj} \delta_{li}$$

shows that $\lambda_{jk} = \lambda_{li}$ when $l = i$ and $k = j$. Thus $\lambda_{ii} = \lambda$ is independent of the index, and it remains only to find the value of λ. Now

$$\begin{aligned}\deg(\rho)^2 \lambda = \sum_{ij} \lambda = \sum_{ij} \sum_g a_{ij}(g^{-1}) a_{ji}(g) &= \sum_g \sum_{ij} a_{ij}(g^{-1}) a_{ji}(g) \\ = \sum_g \mathrm{tr}(\rho(g^{-1})\rho(g)) = \sum_g \mathrm{tr}(\mathrm{id}_V) &= |G| \deg(\rho)\end{aligned}$$

and so $\lambda = |G| / \deg(\rho)$ on canceling the common factor. □

3.5. Fourier analysis on finite groups

The *group character* χ_ρ of a finite-dimensional representation ρ of a group G is the trace $\chi_\rho(g) = \mathrm{tr}(\rho(g))$ of ρ as a linear operator on V_ρ. Characters for us will be characters of complex representations, and in particular they are complex-valued functions on G. Since $\chi_\rho(h^{-1}gh) = \mathrm{tr}(\rho(h^{-1}gh)) = \mathrm{tr}(\rho(h)^{-1}\rho(g)\rho(h)) = \mathrm{tr}(\rho(g)) = \chi_\rho(g)$, the characters are constant on the conjugacy classes of G. Such functions are called *class functions.* The natural domain of definition of the characters of a group is the set of conjugacy classes of the group. Note that every representation of degree one is just multiplication with its character. In particular every group has a *trivial character* $\chi_0 \equiv 1$ belonging to the trivial representation.

Proposition 3.10. *If χ is a character of a finite group G, then $\chi(g^{-1}) = \overline{\chi(g)}$ for every element $g \in G$.*

Proof. Since G is assumed finite, $g^{|G|} = e$ for every element $g \in G$. Hence the eigenvalues λ_j of $\rho(g)$ are $|G|$-th roots of unity, where $\chi = \chi_\rho$. But

$\zeta^{-1} = \overline{\zeta}$ for any complex number ζ with modulus equal to one. Now

$$\begin{aligned}\chi(g^{-1}) &= \mathrm{tr}(\rho(g^{-1})) = \mathrm{tr}(\rho(g)^{-1}) = \sum_j \lambda_j^{-1} \\ &= \sum_j \overline{\lambda_j} = \overline{\sum_j \lambda_j} = \overline{\mathrm{tr}(\rho(g))} = \overline{\chi(g)}\end{aligned}$$

since the trace of a linear operator on a complex vector space is equal to the sum of its eigenvalues. □

Thus if χ is a character of a finite group, so is $\overline{\chi}$. Characters with $\overline{\chi} = \chi$ are called *real characters.*

Define a Hermitian inner product

$$\langle \alpha | \beta \rangle_{L^2(G)} = \frac{1}{|G|} \sum_g \overline{\alpha(g)} \beta(g)$$

on the space of complex-valued functions on G.

Proposition 3.11 (First Frobenius orthogonality theorem). *If χ and χ' are irreducible characters of G, then $\langle \chi | \chi \rangle_{L^2(G)} = 1$ while $\langle \chi | \chi' \rangle_{L^2(G)} = 0$ if χ and χ' are different characters.*

Proof. We have

$$\langle \chi | \chi' \rangle_{L^2(G)} = \frac{1}{|G|} \sum_g \overline{\chi(g)} \chi'(g) = \frac{1}{|G|} \sum_g \mathrm{tr}(\rho(g^{-1}))\mathrm{tr}(\rho'(g))$$

where χ is the character of ρ and χ' is the character of ρ'. Then

$$\langle \chi | \chi' \rangle_{L^2(G)} = \frac{1}{|G|} \sum_g \sum_i a_{ii}(g^{-1}) \sum_k b_{kk}(g) = \sum_{ik} \frac{1}{|G|} \sum_g a_{ii}(g^{-1}) b_{kk}(g)$$

in terms of the matrix entries a_{ij} and b_{kl} of ρ and ρ' with respect to ordered bases of the associated representation spaces. If $\chi \neq \chi'$ then ρ and ρ' are inequivalent, and so

$$\frac{1}{|G|} \sum_g a_{ii}(g^{-1}) b_{kk}(g) = 0$$

by the Schur orthogonality theorem. If $\chi = \chi'$ then $b_{kk} = a_{kk}$ and

$$\sum_{ik} \frac{1}{|G|} \sum_g a_{ii}(g^{-1}) b_{kk}(g) = \sum_{ik} \frac{\delta_{ik}\delta_{ki}}{\deg(\rho)} = \sum_i \frac{1}{\deg(\rho)} = 1,$$

again by the Schur orthogonality theorem. □

That the irreducible characters form an orthonormal set has several striking consequences for the complex representations of a finite group. If ρ and ρ' are representations then $\chi_{\rho \oplus \rho'} = \mathrm{tr}(\rho \oplus \rho') = \mathrm{tr}(\rho) + \mathrm{tr}(\rho') = \chi_\rho + \chi_{\rho'}$ since the trace is linear over $\mathbb{C}$. The decomposition $\rho = m_1\rho_1 \oplus \cdots \oplus m_r\rho_r$

of an arbitrary representation as a direct sum of irreducible representations implies a decomposition $\chi = m_1\chi_1 + \cdots + m_r\chi_r$ of an arbitrary character as a sum of irreducible characters. But now $\langle \chi | \chi_j \rangle_{L^2(G)} = m_1 \langle \chi_1 | \chi_j \rangle_{L^2(G)} + \cdots + m_r \langle \chi_r | \chi_j \rangle_{L^2(G)} = m_j$ by orthonormality. So the multiplicity m with which an irreducible representation ρ' appears in the direct sum decomposition of an arbitrary representation ρ is given by the formula $m = \langle \chi_\rho | \chi_{\rho'} \rangle_{L^2(G)}$. This implies that the character χ_ρ of a representation ρ characterizes the representation up to equivalence. Moreover $\langle \chi_\rho | \chi_\rho \rangle_{L^2(G)} = m_1^2 + \cdots + m_r^2$, so $\langle \chi_\rho | \chi_\rho \rangle_{L^2(G)} = 1$ is a necessary and sufficient condition for ρ to be irreducible. The space of class functions on G is finite-dimensional. Thus G has only finitely many irreducible representations, since their characters constitute an orthonormal set and the characters determine the representations.

For any finite group G let V be the vector space of functions $f : G \to \mathbb{C}$, and define for each $g_1 \in G$ a linear operator $\rho_{\text{reg}}(g_1)$ on V by $(\rho_{\text{reg}}(g_1)f)(g) = f(gg_1)$. It is clear that $\rho_{\text{reg}} \in \text{GL}(V)$, and by

$$(\rho_{\text{reg}}(g_1 g_2)f)(g) = f(gg_1g_2) = (\rho_{\text{reg}}(g_2)f)(gg_1) = (\rho_{\text{reg}}(g_1)\rho_{\text{reg}}(g_2)f)(g)$$

it is a homomorphism of G into $\text{GL}(V)$. It is called the *(right) regular representation* of G. The degree of the regular representation of G equals $|G|$.

The character χ_{reg} of ρ_{reg} is called the *regular character.* Let $\mathcal{B}$ be an ordered basis for V consisting of functions f_i each supported on a single element of G and taking the value 1 there. Let a_{ij} be the matrix entries of ρ_{reg} relative to $\mathcal{B}$. Then $a_{ii}(e) = 1$ since $\rho_{\text{reg}}(e)f_i = f_i$. Thus $\chi_{\text{reg}}(e) = a_{11} + \cdots + a_{|G||G|} = |G|$. But if $g \neq e$ then $a_{ii} = 0$ since $\rho_{\text{reg}}(g)f_i = f_j$ with $i \neq j$, and so $\chi_{\text{reg}}(g) = 0$ in the same way. The calculation of the regular character allows us to determine the multiplicity with which an irreducible representation ρ appears in the direct sum decomposition of the regular representation. The degree of a character is defined as the degree of the associated representation, and so $\deg(\chi) = \deg(\rho) = \dim(V_\rho) = \text{tr}(\text{id}_{V_\rho}) = \text{tr}(\rho(e)) = \chi(e)$ where $\chi = \chi_\rho$ and e is the identity element in G. Since

$$\langle \chi_{\text{reg}} | \chi_\rho \rangle_{L^2(G)} = \frac{1}{|G|} \sum_g \overline{\chi_{\text{reg}}(g)} \chi_\rho(g) = \frac{1}{|G|} \overline{\chi_{\text{reg}}(e)} \chi_\rho(e) = \deg(\rho)$$

the multiplicity of the irreducible representation ρ in the regular representation equals $\deg(\rho)$. The dimension of a direct sum is the sum of the dimensions of the summands, whence

$$\sum_\rho \deg(\rho)^2 = |G|,$$

where the sum is taken over a complete collection of pairwise inequivalent irreducible representations of G. This important formula shows again that a finite group has only finitely many irreducible finite-dimensional complex representations up to equivalence. Moreover, it shows that the total number of matrix entries of inequivalent irreducible representations of a finite group G equals $|G|$. The matrix entries form a linearly independent set by the Schur orthogonality theorem and thus are a basis for the space of complex-valued functions on G.

We may now complete our example from the previous section. The representations ρ_0 and ρ_1 are irreducible since they are of degree one. Moreover $\langle \chi_{\rho_2} | \chi_{\rho_2} \rangle_{L^2(G)} = (2^2 + (-1)^2 + (-1)^2 + 0^2 + 0^2 + 0^2)/6 = 1$ so ρ_2 is also irreducible. Since $\deg(\rho_0)^2 + \deg(\rho_1)^2 + \deg(\rho_2)^2 = 1^1 + 1^2 + 2^2 = 6$ while $|S_3| = 6$, up to equivalence the group S_3 has no irreducible representations beyond the three we already found, and every representation of this group is built from ρ_0, ρ_1, ρ_2 by taking direct sums.

Recall that for each finite-dimensional complex representation of a finite group, we can find an equivalent unitary representation on the same representation space, with respect to whatever Hermitian inner product the representation space may carry. Thus any finite group G has a complete collection of pairwise inequivalent irreducible finite-dimensional complex unitary representations. Once the representation spaces carry Hermitian inner products, we may take adjoints of representations with respect to the inner products, and will denote these by the superscript ad.

We introduce an inner product on the space of functions on $\hat{G}$ that take values in the linear operators on the representation spaces, by

$$\langle \alpha | \beta \rangle_{L^2(\hat{G})} = \frac{1}{|G|} \sum_{\rho \in \hat{G}} \deg(\rho) \operatorname{tr}\Big(\alpha(\rho)^{\mathsf{ad}} \beta(\rho)\Big).$$

Here $\alpha(\rho), \beta(\rho)$ are linear operators on V_ρ for each $\rho \in \hat{G}$.

Let G be a finite group and choose a fixed complete collection $\hat{G}$ of pairwise inequivalent irreducible finite-dimensional complex unitary representations of G. For each complex-valued function f on G define the *Fourier transform* of f on $\hat{G}$ by

$$\hat{f}(\rho) = |G|^{-1/2} \sum_{g \in G} f(g) \rho(g^{-1}).$$

The Fourier transform evaluated at ρ is a linear operator on the representation space of ρ. For each such function φ on $\hat{G}$ define the *inverse Fourier transform* of φ on G by

$$\check{\varphi}(g) = |G|^{-1/2} \sum_{\rho \in \hat{G}} \deg(\rho) \operatorname{tr}(\rho(g) \varphi(\rho)).$$

The inverse Fourier transform evaluated at g is a complex number.

Proposition 3.12 (Fourier inversion). *The maps $f \mapsto \hat{f}$ and $\varphi \mapsto \check{\varphi}$ are inverses.*

Proof. We have

$$\begin{aligned}\check{\hat{f}}(g) &= |G|^{-1}\sum_{\rho}\deg(\rho)\mathrm{tr}(\rho(g)\sum_{h} f(h)\rho(h^{-1}))\\ &= \sum_{h} f(h)\frac{1}{|G|}\sum_{\rho}\deg(\rho)\mathrm{tr}(\rho(g)\rho(h^{-1}))\\ &= \sum_{h} f(h)\frac{1}{|G|}\sum_{\rho}\deg(\rho)\mathrm{tr}(\rho(gh^{-1}))\\ &= \sum_{h} f(h)\frac{1}{|G|}\sum_{\rho}\deg(\rho)\chi_\rho(gh^{-1}).\end{aligned}$$

But if $j \in G$ with $j \neq e$, then

$$\sum_{\rho}\deg(\rho)\chi_\rho(j) = \left(\sum_{\rho}\deg(\rho)\chi_\rho\right)(j) = \chi_{\mathrm{reg}}(j) = 0.$$

Thus

$$\check{\hat{f}}(g) = f(g)\frac{1}{|G|}\sum_{\rho}\deg(\rho)\chi_\rho(e) = f(g)\frac{1}{|G|}\sum_{\rho}\deg(\rho)^2 = \psi(g).$$

Hence the composition of $f \mapsto \hat{f}$ with $\varphi \mapsto \check{\varphi}$ equals the identity on G. The dimension of the space of complex-valued functions on G is $|G|$. The dimension of the space of functions φ on $\hat{G}$ such that $\varphi(\rho)$ is an operator in the representation space of ρ equals the sum of $\deg(\rho)^2$ over $\hat{G}$. So $f \mapsto \hat{f}$ and $\varphi \mapsto \check{\varphi}$ are inverses. □

Proposition 3.13. *The irreducible complex characters of a finite group G constitute a basis for the space of complex-valued class functions.*

Proof. By the first Frobenius orthogonality theorem it will be enough to show that the irreducible characters span the space of class functions. If f is a class function, then

$$\begin{aligned}\rho(h)^{-1}\hat{f}(\rho)\rho(h) &= |G|^{-1/2}\sum_{g\in G} f(g)\rho(h)^{-1}\rho(g^{-1})\rho(h)\\ &= |G|^{-1/2}\sum_{g\in G} f(g)\rho(h^{-1}g^{-1}h)\\ &= |G|^{-1/2}\sum_{t\in G} f(hth^{-1})\rho(t^{-1})\\ &= |G|^{-1/2}\sum_{t\in G} f(t)\rho(t^{-1}) = \hat{f}(\rho),\end{aligned}$$

for any $h \in G$. Thus the Fourier transform $\hat{f}(\rho)$ of f is a linear map of the irreducible representation ρ. Then $\hat{f}(\rho) = \lambda(\rho)\mathrm{id}_{V_\rho}$ by Schur's Lemma. Now

$$\begin{aligned} f(g) = \check{\hat{f}}(g) &= |G|^{-1/2} \sum_{\rho \in \hat{G}} \deg(\rho) \mathrm{tr}(\rho(g)\lambda(\rho)\mathrm{id}_{V_\rho}) \\ &= |G|^{-1/2} \sum_{\rho \in \hat{G}} \deg(\rho)\lambda(\rho)\mathrm{tr}(\rho(g)) = \sum_{\rho \in \hat{G}} |G|^{-1/2} \deg(\rho)\lambda(\rho)\chi_\rho(g) \end{aligned}$$

by Fourier inversion. □

Note that the number of inequivalent irreducible representations and the number of conjugacy classes both equal the dimension of the space of class functions, thus they are equal.

Proposition 3.14 (Plancherel formula). *$\|\hat{f}\|_{L^2(\hat{G})} = \|f\|_{L^2(G)}$ holds for all complex-valued functions f on G.*

Proof. Note that $\rho(g^{-1})^{\mathsf{ad}} = \rho(g)$ for all $g \in G$ and $\rho \in \hat{G}$. We have

$$\begin{aligned} \|\hat{f}\|^2_{L^2(\hat{G})} &= \frac{1}{|G|} \sum_{\rho \in \hat{G}} \deg(\rho) \mathrm{tr}(\hat{f}(\rho)^{\mathsf{ad}} \hat{f}(\rho)) \\ &= |G|^{-2} \sum_{\rho \in \hat{G}} \deg(\rho) \mathrm{tr}\left(\sum_{g \in G} \overline{f(g)} \rho(g^{-1})^{\mathsf{ad}} \sum_{h \in G} f(h)\rho(h^{-1}) \right) \\ &= |G|^{-2} \sum_{g \in G} \sum_{h \in G} \overline{f(g)} f(h) \sum_{\rho \in \hat{G}} \deg(\rho) \mathrm{tr}(\rho(g)\rho(h^{-1})) \\ &= \frac{1}{|G|} \sum_{g \in G} \sum_{h \in G} \overline{f(g)} f(h) \frac{1}{|G|} \sum_{\rho \in \hat{G}} \deg(\rho)\chi_\rho(gh^{-1}) \\ &= \frac{1}{|G|} \sum_{g \in G} |f(g)|^2 = \|f\|^2_{L(G)} \end{aligned}$$

where

$$\sum_{\rho \in \hat{G}} \deg(\rho)\chi_\rho(gh^{-1}) = \begin{cases} |G| & \text{if } gh^{-1} = e, \\ 0 & \text{if } gh^{-1} \neq e, \end{cases}$$

as in the proof of the Fourier inversion theorem. □

The Plancherel formula shows that the Fourier transform on a finite group is a unitary linear operator. Both the Fourier inversion theorem and the Plancherel formula are central results in Fourier analysis on more general groups, in particular in Fourier analysis on the real line. We shall encounter the Fourier transform on $\mathbb{R}$ in Chapter 8, but will not make much use of it.

Proposition 3.15. *The irreducible finite-dimensional complex representations of any abelian group A all have degree equal to one.*

Proof. Suppose that A is abelian and let b be some arbitrary fixed element of A. Then $\rho(a)\rho(b) = \rho(ab) = \rho(ba) = \rho(b)\rho(a)$ for all $a \in A$. Thus $\rho(b)$ is a linear map of ρ to itself, and so $\rho(b) = \lambda(b)\mathrm{id}_{V_\rho}$ by Schur's Lemma. But such a representation has one-dimensional ρ-invariant subspaces, so it must have degree equal to one in order to be irreducible. □

This result implies that the irreducible complex characters of an abelian group A coincide with the homomorphisms $f : A \to \mathbb{C}^\times$ from A into the multiplicative group $\mathbb{C}^\times$ of nonzero complex numbers. For finite abelian groups, the situation is even simpler.

Proposition 3.16. *A finite abelian group A has $|A|$ irreducible characters, and their values are $|A|$-th roots of unity.*

Proof. Since $\deg(\rho) = 1$ for the irreducible representations of A by Proposition 3.15, the formula

$$\sum_\rho \deg(\rho)^2 = |A|$$

shows that A has $|A|$ irreducible characters. And the computation $\chi(a)^{|A|} = \chi(a^{|A|}) = \chi(e) = 1$ for arbitrary $a \in A$ shows that the values of the irreducible characters of A are $|A|$-th roots of unity. □

It follows from the last two results that the irreducible complex representations of finite abelian groups are necessarily unitary. This makes it natural to consider the irreducible characters of such groups as homomorphisms $\chi : A \to \mathbb{T}$ where $\mathbb{T}$ is the multiplicative group of complex numbers of unit modulus, called the *circle group.* With this as the starting point, characters of finite abelian groups may be developed independently of representation theory, as is often done in number theory.

Specializing the Fourier inversion theorem 3.12 to a finite abelian group A and taking Proposition 3.15 into account yields the form

$$\hat{f}(\chi) = |A|^{-1/2} \sum_{a \in A} f(a)\overline{\chi(a)} \quad \Longleftrightarrow \quad f(a) = |A|^{-1/2} \sum_{\chi \in \hat{A}} \hat{f}(\chi)\chi(a)$$

of Fourier inversion. Here f is an arbitrary complex-valued function on A, since the conjugacy classes of an abelian group are singletons. The Plancherel formula takes the form

$$\sum_{\chi \in \hat{A}} |\hat{f}(\chi)|^2 = \sum_{a \in A} |f(a)|^2$$

after canceling the common factor $1/|A|$.

3.6. Primes in arithmetic progressions

Our main objective in this section is to prove Dirichlet's theorem on primes in arithmetic progressions.

Proposition 3.17 (Dirichlet's theorem). *If q and a are positive integers with* $\gcd(q, a) = 1$*, then the arithmetic progression $qm+a$ for $m = 0, 1, 2, \ldots$ contains infinitely many primes.*

Simple arithmetical arguments in the style of Euclid's proof of the infinitude of primes suffice to establish Dirichlet's result in some special cases of arithmetic progressions with small difference q. As an example, we treat the progression $4q+1$. No prime p congruent to 3 modulo 4 is ever a factor of the polynomial x^2+1. This is obvious if p divides x. If p is an odd prime that does not divide x, but does divide x^2+1, then

$$1 \equiv x^{p-1} \equiv (x^2)^{(p-1)/2} \equiv (-1)^{(p-1)/2} \pmod{p}$$

by the little theorem of Fermat. Thus p has to be congruent to 1 modulo 4. Now suppose that $p_1, \ldots, p_r$ are primes congruent to 1 modulo 4. The odd integer $P = (2p_1 \cdots p_r)^2 + 1$ is not divisible by any prime congruent to 3 modulo 4, so it must have a prime factor p congruent to 1 modulo 4 and different from the primes $p_1, \ldots, p_r$. Thus the arithmetic progression $4m+1$ for $m = 0, 1, 2, \ldots$ contains infinitely many primes.

Nobody has yet found an easy approach to prove Dirichlet's theorem on primes in arithmetic progressions. There are various proofs of special cases that are easier than the general case. For example the proofs for progressions of the form $qm \pm 1$ that use special polynomials. These proofs may be found in *Introduction to Number Theory* by T. Nagell.

We are going to prove the theorem by the same method that Dirichlet himself used, with two important modifications. He introduced certain arithmetic functions in his proof based on the theory of primitive roots. Today these are called Dirichlet characters, and they play a very important role in analytic number theory. We shall approach these arithmetic functions from representation theory rather than from the theory of primitive roots. At one point in his proof of the general case Dirichlet made use of the class number formula from the theory of binary quadratic forms. At that point we shall deviate from his course, and use an alternative argument due to Mertens that avoids the theory of binary quadratic forms.

The proof that Dirichlet found generalizes the analytic proof of the infinitude of primes based on the Euler product formula. Two of the necessary ingredients we already possess; the Euler product formula for multiplicative arithmetic functions, and a little convergence theory for Dirichlet series. In addition, we require the Dirichlet characters. To prove the infinitude of primes in arithmetic progressions, it clearly suffices to show that

$$\sum_{p \equiv a \bmod q} \frac{1}{p} = +\infty$$

if q and a are positive integers with $\gcd(q,a) = 1$. Now the indicator function $\mathrm{I}_{q,a}(n)$ of the arithmetic progression $n \equiv a \pmod q$ is not multiplicative when $q \geq 2$, so the Euler product formula cannot be applied directly. Dirichlet overcame this difficulty by a Fourier analysis style argument.

Let $(\mathbb{Z}/q\mathbb{Z})^\times$ be the abelian group of reduced residue classes modulo q under multiplication. It has $\phi(q)$ characters χ. These lift to arithmetic functions on $\mathbb{Z}$ that we also denote by χ, by

$$\chi(n) = \begin{cases} \chi(n+q\mathbb{Z}) & \text{if } \gcd(q,n) = 1, \\ 0 & \text{if } \gcd(q,n) \geq 2. \end{cases}$$

There should be no notational confusion between the two functions denoted by χ, since they are defined on entirely different sets. The arithmetic functions χ are called *Dirichlet characters*. If $\gcd(q,mn) \geq 2$ then $\chi(mn) = 0$ and $\gcd(q,m) \geq 2$ or $\gcd(q,n) \geq 2$, so $\chi(m) = 0$ or $\chi(n) = 0$. Thus $\chi(mn) = \chi(m)\chi(n)$ holds in this case. If $\gcd(q,mn) = 1$ then $\gcd(q,m) = \gcd(q,n) = 1$, and $\chi(mn) = \chi(mn+q\mathbb{Z}) = \chi((m+q\mathbb{Z})(n+q\mathbb{Z})) = \chi(m+q\mathbb{Z})\chi(n+q\mathbb{Z}) = \chi(m)\chi(n)$, since the characters of an abelian group are homomorphisms into $\mathbb{C}^\times$. Hence the Dirichlet characters are totally multiplicative arithmetic functions, and periodic with period q.

The Fourier inversion equivalence

$$\hat{f}(\chi) = \frac{1}{\sqrt{\phi(q)}} \sum_{1 \leq n \leq q} f(n)\overline{\chi(n)} \quad \Longleftrightarrow \quad f(n) = \frac{1}{\sqrt{\phi(q)}} \sum_{\chi \bmod q} \hat{f}(\chi)\chi(n)$$

holds for any complex-valued arithmetic function f, periodic with period q, with $f(n) = 0$ if $\gcd(q,n) \geq 2$. The sum is taken over all Dirichlet characters χ modulo q. Calculating the Fourier transform

$$\hat{\mathrm{I}}_{q,a}(\chi) = \phi(q)^{-1/2} \sum_{1 \leq n \leq q} \mathrm{I}_{q,a}(n)\overline{\chi(n)} = \phi(q)^{-1/2}\,\overline{\chi(a)}$$

we obtain the harmonic analysis

$$\mathrm{I}_{q,a}(n) = \frac{1}{\phi(q)} \sum_{\chi \bmod q} \overline{\chi(a)}\chi(n)$$

of $\mathrm{I}_{q,a}$ in terms of Dirichlet characters when $\gcd(q,a) = 1$. Then

$$\sum_{\chi} \chi(n) = \begin{cases} \phi(q) & \text{if } n \equiv 1 \pmod q, \\ 0 & \text{if } n \not\equiv 1 \pmod q, \end{cases}$$

by substituting $a = 1$.

The *principal* Dirichlet character χ_0 modulo q is given by

$$\chi_0(n) = \begin{cases} 1 & \text{if } \gcd(q,n) = 1, \\ 0 & \text{if } \gcd(q,n) \geq 2. \end{cases}$$

It corresponds to the trivial representation ρ_0 of $(\mathbb{Z}/q\mathbb{Z})^\times$. The others are called *nonprincipal* characters. Now we also see that

$$\sum_{n=1}^{q} \chi(n) = |(\mathbb{Z}/q\mathbb{Z})^\times| \langle \chi_0 | \chi \rangle_{L^2(G)} = \begin{cases} \phi(q) & \text{if } \chi = \chi_0, \\ 0 & \text{if } \chi \neq \chi_0, \end{cases}$$

by the first Frobenius orthogonality theorem.

The functions

$$L(s,\chi) = \sum_{n=1}^{\infty} \chi(n) n^{-s}$$

are called *Dirichlet L-functions.*

Proposition 3.18. *The Dirichlet series defining $L(s,\chi)$ has abscissa of convergence $\sigma_c = 1$ if $\chi = \chi_0$ and $\sigma_c = 0$ if $\chi \neq \chi_0$.*

Proof. Since χ is periodic with period q and

$$\sum_{n=1}^{q} \chi(n) = \begin{cases} \phi(q) & \text{if } \chi = \chi_0, \\ 0 & \text{if } \chi \neq \chi_0, \end{cases}$$

we see that

$$\sum_{n \leq x} \chi(n) = \begin{cases} O(x) & \text{if } \chi = \chi_0, \\ O(1) & \text{if } \chi \neq \chi_0. \end{cases}$$

So $\sigma_c \leq 1$ if $\chi = \chi_0$ by Proposition 3.4, while the comparison

$$L(\sigma, \chi_0) \geq \sum_{m=1}^{\infty} (qm+1)^{-\sigma}$$

for $\sigma > 1$ shows that $\sigma_c \geq 1$.

Proposition 3.4 yields $\sigma_c \leq 0$ if $\chi \neq \chi_0$, while the terms of the Dirichlet series do not tend to zero when $s = 0$, so $\sigma_c \geq 0$. □

Applying Proposition 3.4 and the fact that χ is totally multiplicative, we obtain the Euler product formula

$$L(s,\chi) = \prod_p (1 - \chi(p) p^{-s})^{-1}$$

for the L-functions, valid in the half plane $\sigma > 1$. From this it is clear that

$$L(s,\chi_0) = \zeta(s) \prod_{p|q} (1 - p^{-s})$$

and that $L(s,\chi) \neq 0$ in $\sigma > 1$.

Proposition 3.19. *$L(1,\eta)\neq 0$ holds if η is a nonprincipal character.*

Proof. The estimates

$$\sum_{n\le x} n^{-\theta} = [x]x^{-\theta} - \int_1^x [t](-\theta)t^{-\theta-1}\,dt = \frac{x^{1-\theta}}{1-\theta} + c(\theta) + O(x^{-\theta})$$

and

$$\sum_{n\le x} \eta(n)n^{-\theta} = \begin{cases} O(1) & \text{if } \theta = 0,\\ L(\theta,\eta) + O(x^{-\theta}) & \text{if } \theta > 0,\end{cases}$$

hold on the interval $0\le\theta\le 1/2$ with some constant $c(\theta)$, by partial summation and Proposition 1.6. If f,g,h are arithmetic functions with f totally multiplicative, then

$$\begin{aligned}(f\cdot(g*h))(n) &= \sum_{km=n} f(n)g(k)h(m)\\ &= \sum_{km=n} f(k)f(m)g(k)h(m) = ((f\cdot g)*(f\cdot h))(n).\end{aligned}$$

Consequently

$$\begin{aligned}\sum_{n\le x}(\eta*1)(n)n^{-\theta} &= \sum_{n\le x^{1/2}}\eta(n)n^{-\theta}\sum_{m\le x/n} m^{-\theta} + \sum_{n\le x^{1/2}} n^{-\theta}\sum_{m\le x/n}\eta(m)m^{-\theta}\\ &\quad - \sum_{m\le x^{1/2}}\eta(m)m^{-\theta}\sum_{n\le x^{1/2}} n^{-\theta} = L(1,\eta)\frac{x^{1-\theta}}{1-\theta} + O(x^{1/2-\theta})\end{aligned}$$

for $0\le\theta\le 1/2$ by the Dirichlet hyperbola method.

The claim will be proved by contradiction, assuming $L(1,\eta)=0$. Then

$$\sum_{n\le x}(\eta*1)(n) = O(x^{1/2})$$

on choosing $\theta=0$, and so the Dirichlet series

$$\sum_{n=1}^{\infty}(\eta*1)(n)n^{-s}$$

converges in the half plane $\sigma>1/2$. Thus $L(\sigma,\eta)L(\sigma,\chi_0)$ approaches a finite limit as $\sigma\to 1^+$. For

$$\begin{aligned}L(s,\eta)L(s,\chi_0) &= L(s,\eta)\zeta(s)\prod_{p\mid q}(1-p^{-s})\\ &= \left(\sum_{n=1}^{\infty}(\eta*\mathbf{1})(n)n^{-s}\right)\prod_{p\mid q}(1-p^{-s})\end{aligned}$$

with q the modulus of η and $\sigma > 1$. Taking the product over all characters χ modulo q, we see that

$$\prod_{\chi \bmod q} L(s,\chi) = \prod_{p} \prod_{\chi \bmod q} (1-\chi(p)p^{-s})^{-1}$$

for $\sigma > 1$. If $z_0, \ldots, z_m$ are real numbers with $0 \le z_0, \ldots, z_m < 1$, then

$$(1-z_0)\cdots(1-z_m)\sum_{k=0}^{\infty}\frac{1}{k!}\left(\sum_{j=1}^{\infty}\frac{1}{j}(z_0^j+\cdots+z_m^j)\right)^k = 1$$

holds by the power series expansions of the exponential and the logarithm as functions of a real variable. But then the same identity holds if $z_0, \ldots, z_m$ are complex numbers with $|z_0|, \cdots, |z_m| < 1$, by absolute convergence. Hence

$$\begin{aligned}
&\prod_{\chi \bmod q} (1-\chi(p)p^{-\sigma})^{-1} \\
&\quad = \sum_{k=0}^{\infty}\frac{1}{k!}\left(\sum_{j=1}^{\infty}\frac{1}{j}((\chi_0(p)p^{-\sigma})^j+\cdots+(\chi_m(p)p^{-\sigma})^j)\right)^k \\
&\quad = \exp\left(\sum_{j=1}^{\infty}\frac{1}{j}(\chi_0(p^j)+\cdots+\chi_m(p^j))p^{-j\sigma}\right) \\
&\quad = \exp\left(\phi(q)\sum_{p^j\equiv 1 \bmod q}\frac{1}{j}p^{-j\sigma}\right) \ge 1
\end{aligned}$$

for $\sigma > 0$ with $\chi_0, \chi_1, \ldots, \chi_m$ the characters modulo q. Thus $\prod_\chi L(\sigma,\chi) \ge 1$ for $\sigma > 1$, and because $L(\sigma,\eta)L(\sigma,\chi_0)$ has a finite limit as $\sigma \to 1^+$, and $L(\sigma,\chi)$ is continuous at $\sigma = 1$ when $\chi \ne \chi_0$, we conclude that η is the only character χ modulo q for which $L(1,\chi) = 0$. But if η is a nonreal character, then $\overline{\eta} \ne \eta$ is also a character, with $L(1,\overline{\eta}) = \overline{L(1,\eta)} = 0$, and so η must be real. Then

$$(\eta * 1)(n) = \prod_{p|n}\sum_{p^k|n}\eta(p^k) = \prod_{p|n}\sum_{p^k|n}\eta(p)^k \ge 0,$$

while

$$(\eta * 1)(n^2) = \prod_{p|n}\sum_{p^k|n^2}\eta(p)^k \ge 1.$$

Thus

$$\sum_{n\le x}(\eta * \mathbf{1})(n)n^{-1/2} \ge \sum_{m\le x^{1/2}} m^{-1} \to +\infty$$

as $x \to +\infty$. But this contradicts the estimate

$$\sum_{n \le x} (\eta * \mathbf{1})(n) n^{-1/2} = O(1)$$

obtained by choosing $\theta = 1/2$. □

The *Dirichlet density* $\delta = \delta_{A,B}$ of a subsequence A of natural numbers in a sequence B of natural numbers is defined by

$$\delta_{A,B} = \lim_{\sigma \to 1^+} \left(\sum_{n \in A} n^{-\sigma} \right) \Big/ \left(\sum_{n \in B} n^{-\sigma} \right)$$

if the limit exists.

Proposition 3.20. *The Dirichlet density of the sequence of primes* $p \equiv a \pmod q$, *in the sequence of all primes, equals* $1/\phi(q)$ *if* $\gcd(q, a) = 1$.

Proof. An identity obtained in the course of the proof of Proposition 3.19 yields

$$\prod_{p \le x} (1 - \chi(p) p^{-\sigma})^{-1} = \exp\left(\sum_{p \le x} \sum_{k=1}^{\infty} \frac{1}{k} \chi(p)^k p^{-k\sigma} \right)$$

for $\sigma \ge 1$. Hence

$$L(\sigma, \chi) \exp\left(-\sum_p \chi(p) p^{-\sigma} \right) \to \exp\left(\sum_{k=2}^{\infty} \frac{1}{k} \chi(p)^k p^{-k} \right) \ne 0$$

as $\sigma \to 1^+$. If $\chi \ne \chi_0$ the sum

$$\Sigma(\sigma) = \sum_p \chi(p) p^{-\sigma}$$

tends to a finite limit as σ decreases to 1. Certainly $\exp(\Sigma(\sigma))$ tends to a finite limit $w_0 \ne 0$ as $\sigma \to 1^+$ by Proposition 3.19, and the inverse image under exp of a region containing w_0 and determined by inequalities of the form $0 < r \le |w| \le er$ and $|\mathrm{Arg}(w) - \alpha| \le \pi/4$ is a disjoint union of closed rectangles. Furthermore the sum function $\Sigma(\sigma)$ is continuous on $\sigma > 1$, so $\Sigma(J)$ is connected for each interval $J = (1, b)$, thus $\Sigma(J)$ is entirely contained in a single one of the above closed rectangles for b sufficiently small. But exp is continuous and invertible when restricted to such a rectangle, which moreover is compact, so the inverse is continuous, hence the limit exists.

If $\chi = \chi_0$ on the other hand, the sum clearly tends to infinity. Now

$$\lim_{\sigma\to 1^+} \frac{\sum_{p\equiv a \bmod q} p^{-\sigma}}{\sum_p p^{-\sigma}} = \lim_{\sigma\to 1^+} \left(\sum_p p^{-\sigma}\right)^{-1} \sum_p \frac{p^{-\sigma}}{\phi(q)} \sum_{\chi \bmod q} \overline{\chi(a)}\chi(p)$$

$$= \frac{1}{\phi(q)} \sum_{\chi \bmod q} \overline{\chi(a)} \lim_{\sigma\to 1^+} \frac{\sum_p \chi(p) p^{-\sigma}}{\sum_p p^{-\sigma}} = \frac{1}{\phi(q)}$$

since all terms in the last sum over χ are zero, except the term with $\chi = \chi_0$. □

This is an equidistribution result; each of the $\phi(q)$ eligible arithmetic progressions with difference q captures its "fair share" of the primes in the sense of Dirichlet density. Clearly this implies that

$$\sum_{p\equiv a \bmod q} \frac{1}{p} = +\infty$$

if $\gcd(q, a) = 1$ and so Proposition 3.17 is proved.

3.7. Gauss sums and primitive characters

The finite Fourier expansions

$$\tau(\chi, n) \stackrel{\text{def}}{=} \sum_{m=1}^{q} \chi(m) e(mn/q)$$

with χ a Dirichlet character are called *Gauss sums*. The special case $\chi = \chi_0$ modulo q yields the *Ramanujan sums*

$$c_q(n) \stackrel{\text{def}}{=} \tau(\chi_0, n) = \sum_{\substack{1\le m\le q \\ \gcd(q,m)=1}} e(mn/q).$$

We will not regard the latter as Gauss sums proper. The calculation

$$c_q(n) = \sum_{1\le m\le q} \left(\sum_{\substack{j|m \\ j|q}} \mu(j)\right) e(mn/q) = \sum_{j|q} \mu(j) \sum_{1\le jk\le q} e(jkn/q)$$

$$= \sum_{j|q} \mu(j) \sum_{1\le k\le q/j} e(n/(q/j))^k = \sum_{\substack{d|n \\ d|q}} d\, \mu\left(\frac{q}{d}\right)$$

yields the very convenient formula

$$c_q(n) = \sum_{\substack{d|n \\ d|q}} d\, \mu\left(\frac{q}{d}\right).$$

Proposition 3.21. *If* $\gcd(q,n)=1$ *then* $\tau(\chi,n)=\overline{\chi(n)}\tau(\chi,1)$.

Proof. We have

$$\begin{aligned}\tau(\chi,n) &= \sum_{m=1}^{q}\chi(m)e(mn/q) = \sum_{k=1}^{q}\chi(k\overline{n})e(k\overline{n}n/q)\\ &= \chi(\overline{n})\sum_{k=1}^{q}\chi(k)e(k/q) = \overline{\chi(n)}\tau(\chi,1)\end{aligned}$$

by the change of variable $m=k\overline{n}$ where $\overline{n}$ is the multiplicative inverse of n modulo q. □

A Gauss sum modulo q has $\phi(q)$ nonzero terms, each of modulus equal to one. As these terms are fairly evenly spread around the unit circle, we would expect substantial cancellation. The next result shows that under a suitable condition, we have *square root cancellation* in Gauss sums. Square root cancellation in a sum of N terms on the unit circle means roughly that the modulus of the sum is not much larger than $\sqrt{N}$. A sequence of such sums with $N\to+\infty$ terms would be said to exhibit square root cancellation if the moduli of the sums are $O_\varepsilon(N^{1/2+\varepsilon})$ for each $\varepsilon>0$.

Proposition 3.22. *If* $\tau(\chi,n)=\overline{\chi(n)}\tau(\chi,n)$ *for all* n, *then* $|\tau(\chi,1)|=\sqrt{q}$.

Proof. Choosing $\hat{f}=\chi$ in the Fourier inversion theorem for $\mathbb{Z}/q\mathbb{Z}$ we see that

$$f(n)=\frac{\tau(\chi,n)}{\sqrt{q}}=\frac{\tau(\chi,1)}{\sqrt{q}}\overline{\chi(n)}.$$

Now

$$\sum_{m=1}^{q}|\hat{f}(m)|^2=\sum_{m=1}^{q}|\chi(m)|^2=\phi(q)$$

while

$$\sum_{m=1}^{q}|f(m)|^2=\sum_{m=1}^{q}\left|\frac{\tau(\chi,1)}{\sqrt{q}}\overline{\chi(m)}\right|^2=\frac{|\tau(\chi,1)|^2}{q}\phi(q).$$

The desired identity follows from the Plancherel formula. □

The condition of Proposition 3.22 is satisfied for any Gauss sum $\tau(\chi,1)$ with χ a nonprincipal character with prime modulus, by Proposition 3.21 and the fact that the sum of a nonprincipal Dirichlet character over a complete collection of residue classes is zero.

Suppose that q,q' are positive integers and $q'|q$. There is a well-defined epimorphism $\varphi:(\mathbb{Z}/q\mathbb{Z})^\times\to(\mathbb{Z}/q'\mathbb{Z})^\times$ given by $n+q\mathbb{Z}\mapsto n+q'\mathbb{Z}$, which is not mono unless $q'=q$. If χ' is a group character of $(\mathbb{Z}/q'\mathbb{Z})^\times$, then $\chi=\chi'\circ\varphi$ is a group character of $(\mathbb{Z}/q\mathbb{Z})^\times$. We can lift these group characters to

Dirichlet characters on $\mathbb{Z}$, and use the same symbol for the group character and the associated Dirichlet character. Then

$$\chi'(n) = \begin{cases} \chi'(n + q'\mathbb{Z}) & \text{if } \gcd(q', n) = 1, \\ 0 & \text{if } \gcd(q', n) \geq 2 \end{cases}$$

and

$$\chi(n) = \begin{cases} \chi'(n + q'\mathbb{Z}) & \text{if } \gcd(q, n) = 1, \\ 0 & \text{if } \gcd(q, n) \geq 2. \end{cases}$$

Whenever the latter relationship holds, the Dirichlet character χ is said to be *induced* by the Dirichlet character χ'. It is important to have some criterion to recognize when a character χ is induced by another character χ' with smaller modulus. If χ is a Dirichlet character modulo q induced by some Dirichlet character modulo q', then $\chi(n_1) = \chi(n_2)$ for all integers n_1 and n_2 with $n_1 \equiv n_2 \pmod{q'}$ and $\gcd(q, n_1) = \gcd(q, n_2) = 1$. In this case q' is said to be a *quasiperiod* of χ. But it is also true that if a Dirichlet character χ modulo q has a quasiperiod $q'|q$, then χ is induced by some Dirichlet character modulo q'. For if $\gcd(q', n) = 1$, then there exists some integer n_1 with $n_1 \equiv n \pmod{q'}$ and $\gcd(q, n_1) = 1$. This is obvious if every prime dividing q also divides q', for then we can choose $n_1 = n$. In the contrary case, let P be the product of those primes that divide q but not q'. Then any solution n_1 of the congruences

$$n_1 \equiv n \pmod{q'}, \qquad n_1 \equiv 1 \pmod{P}$$

satisfies the conditions on n_1, and such solutions exist by the Chinese remainder theorem. Now define $\chi'(n) = \chi(n_1)$ for $\gcd(q', n) = 1$ and $\chi'(n) = 0$ otherwise. Then χ' is well-defined because q' is a quasiperiod of χ, and χ' is a Dirichlet character that induces χ. This finally yields the desired criterion: A Dirichlet character is induced by a Dirichlet character with smaller modulus if and only if it has a quasiperiod smaller than its modulus.

Evidently among all the characters that induce a given character χ modulo q, there will be one whose modulus q' is minimal. This minimal modulus is called the *conductor* of χ and is denoted by $\mathsf{f} = \mathsf{f}_\chi$. Characters whose conductor equals their modulus are called *primitive,* and the others are called *imprimitive.* A character modulo q is primitive if it has no quasiperiod smaller than q. The principal character modulo $q = 1$ is primitive, while the principal characters modulo $q \geq 2$ are imprimitive, because they are all induced by the principal character modulo $q = 1$. For some purposes it is inconvenient that there is a primitive character, though for $q = 1$ only, that is also principal. Therefore this character is sometimes excluded from the definition of a primitive character.

For some moduli there are no primitive characters at all. The modulus $q = 2$ is an example, for the principal character is the only character to this modulus, and is imprimitive. The primitive characters modulo q are easily counted by Möbius inversion. Each of the $\phi(q)$ characters modulo q is induced by a unique primitive character modulo q' for some $q'|q$, and conversely all these primitive characters modulo q' with $q'|q$ induce distinct characters modulo q. Thus

$$\sum_{q'|q} \phi_2(q') = \phi(q),$$

where $\phi_2(q')$ denotes the number of primitive characters modulo q'. Since ϕ is multiplicative, so is the arithmetic function ϕ_2. Then

$$\phi_2(q) = \prod_{\substack{p|q \\ p^2\nmid q}} (p-2) \prod_{\substack{p^2|q \\ p^\alpha \| q}} (p-1)^2 p^{\alpha-2}$$

by Möbius inversion. In particular, the moduli q for which there are no primitive characters are those of the form $q = 2m$ with m odd.

We introduce the notation

$$\tau(\chi) \stackrel{\text{def}}{=} \tau(\chi, 1)$$

for brevity.

Proposition 3.23. *If χ is a primitive Dirichlet character, then $\tau(\chi, n) = \overline{\chi(n)}\tau(\chi)$ for all n.*

Proof. By Proposition 3.21 it is enough to treat the case $\gcd(q, n) = d \geq 2$. Then $\chi(n) = 0$, so we must establish that $\tau(\chi, n) = 0$ for all these n. Assuming to the contrary that there is an n_0 for which $\tau(\chi, n_0) \neq 0$, we deduce that χ cannot be primitive. Put $n' = n/d$ and $q' = q/d$, then $q' < q$. Let n_1 and n_2 be arbitrary integers for which $n_1 \equiv n_2 \pmod{q'}$ and $\gcd(q, n_1) = \gcd(q, n_2) = 1$, and let $n_1 = n_2 + cq'$ with c an integer. Then

$$\begin{aligned}
\overline{\chi(n_1)}\tau(\chi, n_0) &= \sum_{m=1}^{q} \chi(\overline{n_1}m)e(mn'/q') = \sum_{k=1}^{q} \chi(k)e(n_1kn'/q') \\
&= \sum_{k=1}^{q} \chi(k)e(n_2kn'/q' + ckn') = \sum_{k=1}^{q} \chi(k)e(n_2kn'/q') \\
&= \sum_{m=1}^{q} \chi(\overline{n_2}m)e(mn'/q') = \overline{\chi(n_2)}\tau(\chi, n_0)
\end{aligned}$$

because the sums are over complete collections of residue classes modulo q. Now $\chi(n_1) = \chi(n_2)$ since $\tau(\chi, n_0) \neq 0$, so χ has the quasiperiod $q' < q$ and is thus not primitive. □

By Proposition 3.22 and Proposition 3.23, we see that $|\tau(\chi)| = \sqrt{q}$ for any primitive Dirichlet character χ modulo q. Proposition 3.23 is also important in connection with the problem of expressing the Dirichlet characters modulo q in terms of the additive characters $e(mn/q)$ modulo q. The identity

$$\chi(n) = \frac{1}{q} \sum_{m=1}^{q} \overline{\tau(\overline{\chi}, m)} e(mn/q),$$

is easily established by changing the order of summation in the double sum on the right-hand side. But if χ is primitive, Proposition 3.22 and Proposition 3.23 yield

$$\chi(n) = \frac{1}{\tau(\overline{\chi})} \sum_{m=1}^{q} \overline{\chi(m)} e(mn/q).$$

The last formula often comes in useful when treating sums involving Dirichlet characters. We are going to use it to prove an important bound for character sums.

An analogous formula expresses the additive characters modulo q in terms of the Dirichlet characters modulo q by

$$e(n/q) = \frac{1}{\phi(q)} \sum_{\chi \bmod q} \tau(\chi)\overline{\chi(n)}$$

if q and n are coprime.

Proposition 3.24 (Pólya-Vinogradov). *If χ is a primitive Dirichlet character modulo $q \geq 3$ then*

$$\left| \sum_{A<n\leq B} \chi(n) \right| < \sqrt{q}\log(q)$$

for arbitrary integers A and B. The inequality holds with a factor $\sqrt{8/3}$ on the right-hand side if χ is merely assumed not to be principal.

Proof. We have

$$\begin{aligned}
\sum_{A<n\leq B} \chi(n) &= \frac{1}{\tau(\overline{\chi})} \sum_{A<n\leq B} \sum_{-q/2<m\leq q/2} \overline{\chi(m)} e(mn/q) \\
&= \frac{1}{\tau(\overline{\chi})} \sum_{0<|m|<q/2} \overline{\chi(m)} \sum_{n=A}^{B} e(mn/q)
\end{aligned}$$

since χ is primitive. The terms corresponding to $m = 0$, and $m = q/2$ if q is even, are omitted because $\chi(m) = 0$ then. Now

$$\left|\sum_{A<n\le B} e(mn/q)\right| = \left|\frac{e((B-A)m/q)-1}{e(m/q)-1}\right| \le \frac{2}{|e(m/q)-1|}$$
$$= \frac{1}{|\sin(\pi m/q)|} \le \frac{1}{(2/\pi)|\pi m/q|} = \frac{q}{2|m|}$$

for $0 < |m| < q/2$. Hence

$$\left|\sum_{A<n\le B} \chi(n)\right| \le \frac{2}{\sqrt{q}} \sum_{m<q/2} \frac{q}{2m} = \sqrt{q} \sum_{m<q/2} \frac{1}{m} < \sqrt{q}\log(q)$$

since $|\tau(\overline{\chi})| = \sqrt{q}$ and $|\overline{\chi(m)}| \le 1$.

If χ is an imprimitive character that is not principal, it is induced by a primitive character χ' modulo q' with $q = q'r$ for some positive integer r. Then $\chi(n) = \chi'(n)$ if n and r are coprime, and $\chi(n) = 0$ otherwise. Thus

$$\left|\sum_{A<n\le B} \chi(n)\right| = \left|\sum_{A<n\le B} \chi'(n) \sum_{d|n, d|r} \mu(d)\right|$$
$$= \left|\sum_{d|r} \mu(d)\chi'(d) \sum_{A/d<m\le B/d} \chi'(m)\right|$$
$$< \sum_{d|r} |\mu(d)| \left|\sum_{A/d<m\le B/d} \chi'(m)\right| \le \sqrt{q'}\log(q') \sum_{d|r} |\mu(d)|$$
$$= \sqrt{q'}\log(q')2^{\omega(r)} = \sqrt{q'4^{\omega(r)}}\log(q')$$
$$\le \sqrt{q'8r/3}\log(q') \le \sqrt{8/3}\sqrt{q}\log(q)$$

by the inequality for primitive characters, and the fact that Dirichlet characters are totally multiplicative. □

We give Vinogradov's application of the Pólya-Vinogradov inequality to the smallest quadratic nonresidue. Since the square of a small integer is a small square, there are always some small quadratic residues modulo any odd prime. But for quadratic nonresidues modulo odd primes p, it is by no means obvious that there are always small ones.

Proposition 3.25. *The inequality*

$$n(p) \ll p^{1/(2\sqrt{e})} \log^{2/\sqrt{e}}(p)$$

holds for the least positive quadratic nonresidue $n(p)$ modulo odd primes p.

Proof. Put $P = p^{1/2}\log^2(p)$ and let T be some integer with $p^{1/4} \le T \le P$. If there are no quadratic nonresidues modulo p less than T, then every nonresidue less than P has a prime divisor greater than T, for every nonresidue is divisible by some prime q that is also a nonresidue. Denoting the number of nonresidues modulo p less than P by $\mathcal{N}$, we see that

$$\mathcal{N} \le \sum_{T<q\le P}\left[\frac{P}{q}\right] \le P\sum_{T<q\le P}\frac{1}{q} = P\log\left(\frac{\log(P)}{\log(T)}\right) + O\left(\frac{P}{\log(T)}\right),$$

with q ranging over primes, by Proposition 1.10. Suppose that $0 < A < B < p$ and denote by $\mathcal{N}$ the number of quadratic nonresidues modulo p in the interval $(A, B]$ and by $\mathcal{R}$ the number of quadratic residues modulo p in the same interval. Then $\mathcal{N}+\mathcal{R} = B-A$, while $-\sqrt{p}\log(p) < \mathcal{N}-\mathcal{R} < \sqrt{p}\log(p)$ on applying the Pólya-Vinogradov inequality with the Legendre symbol for χ. Thus

$$\mathcal{N} > \frac{B-A}{2} - \frac{1}{2}\sqrt{p}\log(p),$$

and choosing $A = 0$ and $B = P$ yields

$$P\log\left(\frac{\log(P)}{\log(T)}\right) + O\left(\frac{P}{\log(T)}\right) > \frac{P}{2} - \frac{1}{2}\sqrt{p}\log(p).$$

Dividing through by P and using the value of P in terms of p and the inequality for T we obtain

$$\log\left(\frac{\log(P)}{\log(T)}\right) > \frac{1}{2} + O\left(\frac{1}{\log(p)}\right),$$

and so

$$\frac{\log(P)}{\log(T)} > \sqrt{e} + O\left(\frac{1}{\log(p)}\right),$$

on exponentiating and using $e^u = 1 + O(u)$ for small u. Solving the last inequality for $\log(T)$ yields $\log(T) < \sqrt{e}\log(P) + O(1)$. Then exponentiate one more time to obtain $T < P^{\sqrt{e}}e^{O(1)} \ll p^{1/(2\sqrt{e})}\log^{2/\sqrt{e}}(p)$, which proves the desired inequality after observing that $1/4 < 1/(2\sqrt{e})$. □

3.8. ⋆ The character group

Dirichlet's approach to the arithmetic functions χ did not rely on the theory of group characters. Indeed the concept of a group was scarcely known in the 1830s, and as yet only in the form of groups of "substitutions" (permutations) of solutions of polynomial equations. Dirichlet constructed his characters as arithmetic functions by means of primitive roots. We present his construction as an alternative to the treatment in Sections 3.4 and 3.5.

The Dirichlet characters modulo q can be made into an abelian group, taking the pointwise product as the multiplication, the principal character χ_0 as the neutral element, and the complex conjugate $\overline{\chi}$ as the inverse.

Clearly this group has order $\phi(q)$ since $(\mathbb{Z}/q\mathbb{Z})^\times$ has $\phi(q)$ distinct group characters. The construction of the $\phi(q)$ Dirichlet characters by means of primitive roots displays the character group explicitly.

A *primitive root* g modulo m is a generator of the group of reduced residue classes under multiplication, if such a generator exists, i.e., if this group is cyclic. It is proved in the Summary that this happens for $m = 1, 2, 4, p^k, 2p^k$ where p is any odd prime, and only for these values of m. Now suppose that g is a primitive root modulo m and that n is represented both as the j-th and the k-th power of g modulo m, so that $g^j \equiv n \equiv g^k \pmod{m}$. Then $g^{k-j} \equiv 1 \pmod{m}$ and so $k \equiv j \pmod{\phi(m)}$, and g has to have $\phi(m)$ distinct powers modulo m since it is a primitive root modulo m. Fix an integer ℓ and consider the function χ defined by

$$\chi(g^k) = e(k\ell/\phi(m))$$

on the reduced residue classes modulo m. This function is well defined by the observation above, and it is a homomorphism into $\mathbb{C}^\times$. Extending it to all residue classes modulo m by setting it equal to zero on the remaining residue classes yields a Dirichlet character modulo m. Clearly the choices $\ell = 1, 2, \dots, \phi(m)$ yield distinct characters, and so we have determined all the characters modulo m. Note that the choice $\ell = \phi(m)$ yields the principal character χ_0 modulo m that is one on all integers coprime with m.

Now suppose that $q = p_1^{\alpha_1} \cdots p_r^{\alpha_r}$ with $p_1 < \cdots < p_r$ distinct primes. Unless $p_1 = 2$ with $\alpha_1 \geq 3$, the above considerations imply that there are $\phi(p_j^{\alpha_j})$ characters modulo $p_j^{\alpha_j}$ for $1 \leq j \leq r$. Supposing χ_j to be a character modulo $p_j^{\alpha_j}$ for $1 \leq j \leq r$, we may consider the product $\chi = \chi_1 \chi_2 \cdots \chi_r$. This is a totally multiplicative arithmetic function, periodic with period q, and with $\chi(n) = 0$ precisely when $\gcd(q, n) \geq 2$. Thus χ is a character modulo q. Since there are $\phi(p_j^{\alpha_j})$ distinct choices for χ_j, this will yield us $\phi(p_1^{\alpha_1})\phi(p_2^{\alpha_2}) \cdots \phi(p_r^{\alpha_r}) = \phi(q)$ characters modulo q if all the products are distinct. So suppose that $\chi_1 \chi_2 \cdots \chi_r = \eta_1 \eta_2 \cdots \eta_r$ with χ_j and η_j characters modulo $p_j^{\alpha_j}$. Assuming $\gcd(q, n) = 1$ and rewriting gives $\chi_j \overline{\eta_j}(n) = (\eta_1 \cdots \eta_{j-1} \eta_{j+1} \cdots \eta_r)\overline{(\chi_1 \cdots \chi_{j-1} \chi_{j+1} \cdots \chi_r)}(n)$. Thus $\chi_j \overline{\eta_j}(n)$ restricted to those n for which $\gcd(q, n) = 1$ is a periodic arithmetic function with coprime periods $p_j^{\alpha_j}$ and $q/p_j^{\alpha_j}$. But then it is constant, and equal to $\chi_j \overline{\eta_j}(1) = 1$ for these n, so $\eta = \chi$. We reach the conclusion that given fixed primitive roots g_j modulo $p_j^{\alpha_j}$ for $1 \leq j \leq r$, the Dirichlet characters modulo q are of the form

$$\chi(g_1^{k_1} \cdots g_r^{k_r}) = e\left(\frac{k_1 \ell_1}{\phi(p_1^{\alpha_1})} + \cdots + \frac{k_r \ell_r}{\phi(p_r^{\alpha_r})} \right)$$

for $\ell_1, \cdots, \ell_r$ integers. In particular, the group of Dirichlet characters modulo q is isomorphic to $(\mathbb{Z}/q\mathbb{Z})^\times$.

The above considerations suffice to determine the character group unless $p_1 = 2$ with $\alpha_1 \geq 3$. In the latter case there is no primitive root modulo $p_1^{\alpha_1}$ because the group $(\mathbb{Z}/2^\alpha\mathbb{Z})^\times$ is not cyclic when $\alpha \geq 3$. But then the group is generated by the residue classes of -1 and 5 modulo 2^α. To see this, observe that

$$\binom{2^k}{m} = \frac{(2^k-(m-1))\cdots(2^k-1)2^k}{1\cdot 2\cdots m},$$

where 2^k divides the numerator of the fraction, while the exponent to which 2 occurs in the denominator equals

$$\left[\frac{m}{2}\right] + \left[\frac{m}{4}\right] + \cdots \leq m.$$

Now

$$\begin{aligned} 5^{2^{\alpha-3}} &\equiv (1+4)^{2^{\alpha-3}} \equiv 1 + \cdots + \binom{2^{\alpha-3}}{m}\cdot 4^m + \cdots \\ &\equiv 1 + 2^{\alpha-1} \pmod{2^\alpha} \end{aligned}$$

by the Binomial Theorem. Thus

$$5^{2^{\alpha-3}} \not\equiv 1 \pmod{2^\alpha} \quad \text{while} \quad 5^{2^{\alpha-2}} \equiv 1 \pmod{2^\alpha},$$

so the integers $5, 5^2, \ldots, 5^{2^{\alpha-2}}$ are pairwise incongruent modulo 2^α, and the integers $-5, -5^2, \ldots, -5^{2^{\alpha-2}}$ likewise. All the integers in the first of these two sequences are congruent to 1 modulo 4 while in the second sequence they are congruent to -1 modulo 4. Hence we have $2^{\alpha-1} = \phi(2^\alpha)$ pairwise incongruent integers modulo 2^α of the form $(-1)^h 5^k$. Thus $(\mathbb{Z}/2^\alpha\mathbb{Z})^\times$ is isomorphic to $(\mathbb{Z}/2\mathbb{Z}) \oplus (\mathbb{Z}/2^{\alpha-2}\mathbb{Z})$ when $\alpha \geq 3$. The characters modulo 2^α for $\alpha \geq 3$ are of the form

$$\chi((-1)^h 5^{k_1}) = (-1)^{hj} e\left(\frac{k_1\ell_1}{2^{\alpha-2}}\right)$$

with j and ℓ_1 integers. So the group of Dirichlet characters modulo q is isomorphic to $(\mathbb{Z}/q\mathbb{Z})^\times$ in all cases, and we can in principle write down the Dirichlet characters modulo q for any positive integer q.

We now deduce the properties of Dirichlet characters that are needed to establish infinitude of primes in arithmetic progressions from the description of the Dirichlet characters in terms of primitive roots. For coprime positive integers q and a and an integer n, let $\overline{a}$ be some inverse of a modulo q and put

$$\overline{a}n \equiv g_1^{k_1}\cdots g_r^{k_r} \pmod{q}$$

where g_j for $1 \le j \le r$ are primitive roots as above. Then

$$\begin{aligned}\sum_{\chi \bmod q} \overline{\chi(a)}\chi(n) = \sum_{\chi \bmod q} \chi(\bar{a}n) &= \sum_{\ell_1=1}^{\phi(p_1^{\alpha_1})} e\left(\frac{k_1\ell_1}{\phi(p_1^{\alpha_1})}\right) \cdots \sum_{\ell_r=1}^{\phi(p_r^{\alpha_r})} e\left(\frac{k_r\ell_r}{\phi(p_r^{\alpha_r})}\right) \\ &= \left(\begin{cases}\phi(p_1^{\alpha_1}) & \text{if } k_1 = \phi(p_1^{\alpha_1}) \\ 0 & \text{if } k_1 < \phi(p_1^{\alpha_1})\end{cases}\right) \\ &\quad \cdots \left(\begin{cases}\phi(p_r^{\alpha_r}) & \text{if } k_r = \phi(p_r^{\alpha_r}) \\ 0 & \text{if } k_r < \phi(p_r^{\alpha_r})\end{cases}\right) \\ &= \begin{cases}\phi(q) & \text{if } n \equiv a \pmod q, \\ 0 & \text{if } n \not\equiv a \pmod q,\end{cases}\end{aligned}$$

and so we obtain the harmonic analysis

$$\mathrm{I}_{q,a}(n) = \frac{1}{\phi(q)} \sum_{\chi \bmod q} \overline{\chi(a)}\chi(n)$$

of the indicator function $\mathrm{I}_{q,a}$ of the arithmetic progression $qm + a$, $m \in \mathbb{Z}$ in terms of Dirichlet characters when $\gcd(q, a) = 1$. In the same way

$$\begin{aligned}\sum_{n=1}^{q} \chi(n) = \sum_{k_1=1}^{\phi(p_1^{\alpha_1})} e\left(\frac{k_1\ell_1}{\phi(p_1^{\alpha_1})}\right) \cdots \sum_{k_r=1}^{\phi(p_r^{\alpha_r})} e\left(\frac{k_r\ell_r}{\phi(p_r^{\alpha_r})}\right) &= \left(\begin{cases}\phi(p_1^{\alpha_1}) & \text{if } \ell_1 = \phi(p_1^{\alpha_1}) \\ 0 & \text{if } \ell_1 < \phi(p_1^{\alpha_1})\end{cases}\right) \\ &\quad \cdots \left(\begin{cases}\phi(p_r^{\alpha_r}) & \text{if } \ell_r = \phi(p_r^{\alpha_r}) \\ 0 & \text{if } \ell_r < \phi(p_r^{\alpha_r})\end{cases}\right) \\ &= \begin{cases}\phi(q) & \text{if } \chi = \chi_0, \\ 0 & \text{if } \chi \ne \chi_0,\end{cases}\end{aligned}$$

where χ_0 is the principal character modulo q that equals one on all integers coprime with q.

The above computations are incomplete if $p_1 = 2$ and $\alpha_1 \ge 3$. However

$$\sum_{j=1}^{2} \sum_{\ell_1=1}^{2^{\alpha_1-2}} (-1)^{hj} e\left(\frac{k_1\ell_1}{2^{\alpha_1-2}}\right) = \begin{cases}\phi(2^{\alpha_1}) & \text{if } h = 2 \text{ and } k_1 = 2^{\alpha_1-2}, \\ 0 & \text{otherwise,}\end{cases}$$

so the same calculations go through when appropriately modified.

There is a third way to construct Dirichlet characters, as homomorphisms of a finite abelian group into the circle group. This is presented in many books, for example, in *Introduction to Analytic Number Theory* by K. Chandrasekharan.

3.9. Notes

The Euler product formula for the Riemann zeta function [**Eul37**] dates to 1737. The convergence and equality for $\sigma > 1$ of the series and the product for $\zeta(s)$ was established by Dirichlet [**Dir37a**] in 1837. The proof of Proposition 3.1 is modeled on the one in *Die Analytische Zahlentheorie* by Bachmann [**Bac94**]. In his lectures [**Kro01**] of 1875-76 Kronecker singled out total multiplicativity and proved Proposition 3.2 for this case, assuming the unconditional (i.e. absolute) convergence of the series and products involved. The proof of Proposition 3.2 for multiplicative functions was given by Bachmann, but the result in full generality may well have been known earlier. The work of Cashwell and Everett on the Dirichlet ring is in [**CE59**].

Dirichlet applied the series named after him to primes in arithmetic progressions [**Dir37a, Dir37b, Dir39**], to binary quadratic forms [**Dir38a, Dir39, Dir40**], and to similar problems over the Gaussian integers [**Dir41a, Dir41b, Dir42a**].

The basic convergence theory of Dirichlet series is due to E. Cahen [**Cah94**] in his doctoral dissertation, though he was anticipated in some respects by J. W. R. Dedekind [**Dir42c**], J. L. W. V. Jensen [**Jen84, Jen88**], W. Scheibner [**Sch60**], and T. J. Stieltjes [**Sti85**]. Proposition 3.3 is due to Cahen [**Cah94**], the existence of the abscissa of convergence to Jensen [**Jen84**], Proposition 3.4 and in 3.6 in essentials to Dedekind [**Dir42c**], and Proposition 3.5 to Kronecker in his lectures [**Kro01**] of 1875-76. These lectures were published only in 1901, and so the 1894 treatise [**Bac94**] of Bachmann may be the first occasion on which the result appeared in print.

Dirichlet series, and especially the more general kind

$$\sum_{n=1}^{\infty} a_n e^{-\lambda_n s},$$

have also been studied as a topic in analysis. See *Dirichlet Series* [**Hel05**] by H. Helson and *Dirichlet Series: Principles and Methods* [**Man72**] by S. Mandelbrojt.

The material in the section on harmonics predates the development of Fourier analysis. The values of the basis functions $e(mn/q)$ are roots of unity and expressed in a different notation our reasonings belong to *cyclotomy;* the study of the roots of unity and the branch of polynomial algebra associated with them. In particular, the classical Gauss sum is in part VII of the *Disquisitiones.*

The viewpoint on representation theory based on linear operators and equivalence is due to A. E. Noether [**Noe29**]. Proposition 3.7 is in the paper [**Mas99**] of Maschke, and the averaging device used in the proof is due to E. H. Moore [**Moo98**]. This particular proof was chosen for its utility to harmonic analysis on finite groups. The first part of Proposition 3.8 is due to W. Burnside [**Bur04**], the second part and Proposition 3.9 to I. Schur [**Sch05**].

F. G. Frobenius defined characters and representations of finite groups and proved Proposition 3.10 and Proposition 3.11 in [**Fro96, Fro97**]. The proof of Proposition 3.11 given here is due to I. Schur [**Sch05**]. The fundamental results on the irreducible representations and characters, their relation to the regular representation, and Proposition 3.13 are due to Burnside [**Bur00, Bur03, Bur04**].

Propositions 3.12 and 3.14 are specializations of results for compact Lie groups due to F. Peter and H. K. H. Weyl [**PW27**]. In the even more special case of finite abelian groups these results were known much earlier, but the nonabelian case is quite different in that the dual does not have the structure of a group. Proposition 3.15 is due to Frobenius [**Fro97**] and Proposition 3.16 to H. M. Weber [**Web82**].

Dirichlet [**Dir37a, Dir37b**] proved his theorem for the special case of a prime modulus q in 1837. His proof for the case of a general modulus relied on the class number formula for positive definite binary quadratic forms [**Dir39**] that he published in 1839. The proof given here of the crucial fact that $L(1,\chi) \neq 0$ is due to Mertens [**Mer95a, Mer95b**], with simplifications by H. N. Shapiro [**Sha50**].

Dirichlet's theorem inaugurated the very important theory of the distribution of primes in arithmetic progressions. Results from this theory are generally of great utility in number theory. Probably the earliest indication of this was Legendre's proposed proof in [**Leg85**] of the Law of Quadratic Reciprocity, dating to 1785. He split the task of proving the result into several cases, all of which he could handle, with the exception of one case for which he required the existence of primes in arithmetic progressions. With the work of Dirichlet, Legendre's proof of quadratic reciprocity became viable, but by that time easier approaches had been found, and today it seems unlikely that a proof of Dirichlet's theorem will be discovered that is anywhere near as easy as the easiest proofs of quadratic reciprocity.

The more general topic of primes in particular integer sequences has not as yet developed in any extensive way. Dirichlet [**Dir40**] initiated, and H. M. Weber [**Web82**] completed, the determination of the conditions under which a binary quadratic form

$$Q(x,y) = ax^2 + bxy + cy^2$$

over the integers takes infinitely many prime values. Clearly it is necessary that the coefficients should be coprime; then the quadratic form is called *primitive*. If the form splits into rational linear factors, it may or may not represent infinitely many primes, but this question can be decided by means of Dirichlet's theorem on primes in arithmetic progressions. So we are free to suppose that $Q(x,y)$ is irreducible over the rationals; that is to say, that its discriminant $D = b^2 - 4ac$ is not a square. The Dirichlet-Weber theorem states that if $Q(x,y)$ is primitive and irreducible, then it takes infinitely many prime values. (Actually, Dirichlet and Weber considered only forms whose middle coefficient is even.) A. Meyer [**Mey88**] found the conditions for a primitive, irreducible binary quadratic form to take infinitely many prime values in an arithmetic progression. The case of inhomogenous quadratic polynomials in two variables was considered by Iwaniec [**Iwa78**], who showed that if such a polynomial is irreducible, has coprime coefficients, takes arbitrarily large odd values, and does not reduce to a polynomial in one variable by a change of variable, then it takes infinitely many prime values. The latter two conditions are to be expected, for the polynomial $p(x,y) = x^2 + x + y^2 + y + 2$ is irreducible and has coprime coefficients, yet takes only even values on the integers. While the polynomial $q(x,y) = x^2 + 2xy + y^2 + 1$ may be rewritten as $q(z) = z^2 + 1$ by the change of variable $z = x + y$, and the question whether $m^2 + 1$ takes infinitely many prime values is a well-known unsolved problem.

A result of I. I. Piatetski-Shapiro [**PS53**] from 1953 implies that the rather sparser sequence $[n^c]$ contains infinitely many primes when $1 < c < 12/11$. The range of c was gradually widened by many authors. The current best range is $1 < c < 242/205$, due to J. Rivat and J. Wu [**RW01**].

A substantial advance was achieved by Friedlander and Iwaniec [**FI98**] when they established that the sparse polynomial sequence given by $m^2 + n^4$ contains infinitely many primes. Subsequently Heath-Brown [**HB01**] proved that $m^3 + 2n^3$ takes infinitely many prime values. General binary irreducible cubic forms were handled by Heath-Brown and B. Z. Moroz in [**HBM02**], and in [**HBM04**] they were able to show that inhomogenous cubic polynomials of a suitable kind in two variables also take infinitely many prime values.

The above encompass all those naturally defined sequences currently known to contain infinitely many primes. But in these few cases that have been settled one knows more than the mere existence of infinitely many primes; there are estimates for the counting function of the primes in the sequence. To save space, we shall not quote any of these estimates.

There is extensive and convincing heuristic information available about the prime values of polynomials. Hypothesis H of A. Schinzel [**SS58**] subsumes a good many such questions concerning primes of particular form. It states that if $P_1(x), P_2(x), \ldots, P_r(x)$ are irreducible polynomials in $\mathbb{Z}[x]$ and their product has no fixed prime divisor, then the values $P_1(n), P_2(n), \ldots, P_r(n)$ should be simultaneously prime for infinitely many integers n. Dirichlet's theorem on primes in arithmetic progressions is the simplest nontrivial instance of Hypothesis H, for the conditions are satisfied by choosing $P_1(x) = qx + a$ with q and a coprime. That the polynomial $m^2 + 1$ takes infinitely many prime values is also a particular case of Hypothesis H. And by choosing $P_1(x) = x$ and $P_2(x) = x + 2$ the infinitude of twin primes would follow. Indeed Hypothesis H implies the existence of infinitely many prime tuples of any prescribed form that is not excluded by congruential considerations, for example triples $n, n+2, n+6$. There is also a quantitative version of Hypothesis H, due to P. T. Bateman and R. A. Horn [**BH62**]. This predicts the frequency with which $P_1(n), P_2(n), \ldots, P_r(n)$ simultaneously take prime values. To save space, we shall not reproduce this conjecture here.

The topic of primes in sequences that grow faster than polynomially is naturally even more difficult, and no positive result is known for any such sequence that is naturally defined. The sequence $2^m - 1$ of Mersenne numbers has a certain historical interest; the primes in this sequence are in one-to-one correspondence with the even perfect numbers, a connection established by Euclid and Euler. A polynomial factorization shows that the exponent m must be prime for a Mersenne number to be prime. A simple probabilistic heuristic based on the fact that the density of primes near a large value of x is about $1/\log(x)$ indicates that the sequence $2^p - 1$ contains infinitely many primes, and even predicts an asymptotic estimate for their counting function.

A polynomial factorization shows that $2^m + 1$ can only be prime if the exponent m is a power of 2. The Fermat numbers

$$F_n = 2^{2^n} + 1$$

are prime for $n = 0, 1, 2, 3, 4$, but no further Fermat primes have been found. Gauss discovered that the Fermat primes determine which regular polygons can be constructed with compass and straightedge. He has recorded that it was this success that made him decide to pursue mathematics in preference to philology. A probabilistic heuristic similar to the one for the Mersenne primes indicates that there may well be only finitely many Fermat primes. Sequences that grow faster than polynomially are so difficult to deal with that in the absence of a polynomial factorization, even to show that almost all elements are composite becomes hard. But C. Hooley [**Hoo76**] found by means of a sieve method an unconditional proof that almost all the Cullen numbers $n{\cdot}2^n + 1$ are composite.

The general Gauss sum with characters first appears implicitly in the work of Dirichlet on the analytic theory of binary quadratic forms. The Ramanujan sums were introduced in [**Ram18**]. Proposition 3.21 to 3.23 and the associated classification into primitive and imprimitive characters are in the *Handbuch* of Landau. The latter credits the distinction between primitive and imprimitive characters to R. O. S. Lipschitz [**Lip89**]. Gauss [**Gau11**] determined the argument of the classical Gauss sum τ_p after lengthy efforts. The argument is zero if $p \equiv 1 \pmod 4$ and $\pi/2$ if $p \equiv 3 \pmod 4$. He also noted that the determination of the argument implies the Law of Quadratic Reciprocity. There are many ways to determine the sign of the classical Gauss sum. See the notes to Chapter 1 of [**BEW98**] for a discussion of and references for various approaches.

The arguments of Gauss sums $\tau(\chi)$ for other characters χ modulo primes p do not have the same very regular behavior as for τ_p. D. R. Heath-Brown and S. J. Patterson [**HBP79**] investigated the argument of the Gauss sum $\tau(\chi^{(p)})$ for a cubic character $\chi^{(p)}$ modulo $p \equiv 1 \pmod 3$, and proved that the sequence $\tau(\chi^{(p)})/\sqrt{p}$ is equidistributed on the unit circle as $p \to +\infty$. Equidistribution of the sequence means that each arc on the unit circle captures, in the limit, a fraction of the elements of the sequence that is proportional to its length. This result of Heath-Brown and Patterson settled an old question of E. E. Kummer.

Proposition 3.24 is due to G. Pólya [**Pól18**] and I. M. Vinogradov [**Vin18b**]. Our proof is due to I. Schur [**Sch18**]. The constant in the Pólya-Vinogradov inequality has been improved. Put

$$c_{\pm} = \limsup_{q} \max_{\chi(-1)=\pm 1} \max_{A<B} \frac{\left|\sum_{A<n\le B} \chi(n)\right|}{\sqrt{q}\log(q)},$$

where the maximum over χ is taken over primitive characters of modulus q and with $\chi(-1) = \pm 1$ as the case may be. In this problem there is a distinction between characters with $\chi(-1) = 1$, called even, and with $\chi(-1) = -1$, called odd. Landau [**Lan18a**] established the bounds $c_+ \le 1/(2\pi\sqrt{2})$ and $c_- \le 1/(2\pi)$. In unpublished work P. T. Bateman noted the improvement $c_+ \le 1/\pi^2$. An improvement by a factor of 2/3 in each of Landau's bounds was obtained by A. J. Hildebrand [**Hil88**]. Montgomery and Vaughan [**MV77**] have shown that if $L(s,\chi) \ne 0$ for all $\sigma > 1/2$ and all characters χ, then the factor $\log(q)$ may be replaced by $\log\log(q)$ in the Pólya-Vinogradov inequality. If true, this would be best possible by work of R. E. A. C. Paley [**Pal32**].

Proposition 3.25 is due to Vinogradov [**Vin19**]. It was substantially improved by D. A. Burgess [**Bur57**], who replaced the exponent $1/(2\sqrt{e})$ in Vinogradov's bound by $1/(4\sqrt{e})+\varepsilon$ for any $\varepsilon > 0$. To obtain this improvement, Burgess [**Bur62**] applied a special case of his inequality for short character sums.

Exercises

(1) Show that $6/\pi^2 < \phi(n)\sigma(n)/n^2 \leq 1$ and explain why this inequality is sharp. That is, explain why the constants cannot be improved.

(2) Sum the formal Dirichlet series with coefficients λ and μ^2 in terms of ζ.

(3) Expand the functions $\zeta^2(s)/\zeta(2s)$ and $\zeta^4(s)/\zeta(2s)$ into formal Dirichlet series.

(4) Prove the identity

$$d(n) = \sum_{d^2|n} 2^{\omega(n/d^2)},$$

for example by means of formal Dirichlet series (L. Kronecker).

(5) Let $d_k(n)$ denote the number of ways of writing n as a product of k positive integers, with the order of the factors taken into account. Express the formal Dirichlet series of the arithmetic function $d_k(n)$ in terms of $\zeta(s)$ (A. Piltz).

(6) Expand $(2-\zeta(s))^{-1}$ into a formal Dirichlet series and identify the coefficients in terms of the functions d_k.

(7) a) Critique the computation

$$\begin{aligned}\prod_p \left(1-\frac{1}{p}\right)^{-1} &= \prod_p \left(1+\frac{1}{p}+\frac{1}{p^2}+\cdots\right) \\ &= \sum \frac{1}{p_1^{\alpha_1}\cdots p_r^{\alpha_r}} = \sum_{n=1}^{\infty}\frac{1}{n} = +\infty\end{aligned}$$

as a supposed proof of the infinitude of primes.
b) Rework the argument in part a) into a correct proof by contradiction.
c) Find a proof by considering $\prod_{p\leq x}(1-p^{-1})^{-1}$ as $x \to +\infty$.
d) Find a proof by considering $\prod_p \left(1-p^{-\sigma}\right)^{-1}$ as $\sigma \to 1^+$.

(8) Let f be a multiplicative and g an additive arithmetic function. Show

$$\sum_{n=1}^{\infty} f(n)g(n)n^{-s} = \left(\sum_{n=1}^{\infty} f(n)n^{-s}\right)\sum_p \frac{\sum_{k=1}^{\infty} f(p^k)g(p^k)p^{-ks}}{\sum_{k=0}^{\infty} f(p^k)p^{-ks}}$$

as an identity between formal Dirichlet series.

(9) Expand the infinite product

$$\prod_{k=0}^{\infty}\left(1+x^{2^k}\right)$$

into a formal power series, and sum the series.

(10) † Expand the rational function $f(x) = (1-x)^{-1}(1-x^2)^{-1}(1-x^3)^{-1}$ into a formal power series and interpret the coefficients of this series as the number of solutions of a particular Diophantine problem. Find an exact formula for the number of solutions.

(11) The series

$$\sum_{n=1}^{\infty} \frac{f(n)x^n}{1-x^n}$$

is called the *Lambert series* of the arithmetic function f. Expand it into a formal power series. Sum the Lambert series for $\phi(n)$ in closed form.

(12) a) The *partition function* $p(n)$ gives the number of ways of writing n as a sum of positive integer summands without regard to order. Thus $p(4) = 5$ since $1+1+1+1 = 1+1+2 = 1+3 = 2+2 = 4$. Prove the formal power series identity

$$\sum_{n=0}^{\infty} p(n)x^n = \prod_{m=1}^{\infty}(1-x^m)^{-1}$$

for the generating function of $p(n)$.
b) Show that

$$\sum_{n=0}^{\infty} p(n)x^n = \exp\left(\sum_{k=1}^{\infty} \frac{\sigma(k)}{k}x^k\right)$$

as an identity between formal power series.
c) Show that $\sigma(k)/k \le \log(k)/2 + O(1)$.
d) Show that

$$p(n) \ll e^{cn^{2/3}}$$

for some positive constant c.

(13) Show

$$\limsup_{x\to+\infty} \frac{1}{\log\log(x)} \sum_{p\le x} \frac{1}{p} \le 1$$

without using the results of Chebyshev or Mertens.

(14) Determine the abscissa of convergence of the formal Dirichlet series of $\zeta^2(s)$, $\zeta'(s)$ and $-\zeta'(s)/\zeta(s)$.

(15) Denote by $c(n)$ the number of ways of writing n as a product of integers ≥ 2 without regard to order. The study of $c(n)$ and allied arithmetic functions dates from the 1920s and is called *factorisatio numerorum*. Establish the bound

$$\sum_{n \leq x} c(n) \ll x e^{2\sqrt{\log(x)}}$$

by noting that for each fixed $x \geq 1$ the indicator function of the interval $1 \leq n \leq x$ is dominated by $(x/n)^\sigma$ for any value of the parameter $\sigma \geq 1$, and choosing σ in terms of x. (This is the *Rankin trick*. One cannot expect the Rankin trick to give an optimal bound, but it often yields a useful one.) F. Luca, A. Mukhopadhyay and K. Srinivas [**LMS10**] have much more precise information.

(16) Show that the Dirichlet series

$$\sum_{n=1}^{\infty} \frac{d(n^k)^m}{n^s}$$

has the abscissa of convergence $\sigma_c = 1$ for positive integers k and m.

(17) For every choice of the real parameter α in the Dirichlet series

$$\sum_{n=1}^{\infty} e^{2\pi i \alpha n} n^{-s},$$

determine the abscissas of convergence and absolute convergence.

(18) Suppose that $A(s)$ is the sum of a convergent Dirichlet series and that $A(s)$ is not identically zero. Show that there exists some α so that $A(s) \neq 0$ for $\sigma > \alpha$.

(19) Show that the Dirichlet series

$$\sum_{n=1}^{\infty} a_n n^{-s} \quad \text{and} \quad \sum_{n=1}^{\infty} a_n (n+1)^{-s}$$

converge in the same points.

(20) Show that a Dirichlet series is convergent (somewhere) if and only if the coefficients are of polynomial growth.

(21) Show that a nonconstant rational function does not have a Dirichlet series expansion.

(22) Show that $\tau_p^2 = (-1|p)p$ for each odd prime p. This pins down the value of the Gauss sum up to sign. The determination of the sign was achieved by Gauss after four years during which not a week passed without his attempting the problem, according to a research diary that he kept.

(23) Calculate the trace and the determinant of the linear operator

$$A(f)(n) = \sum_{k=1}^{p} e(kn/p) f(k)$$

on the space of arithmetic functions with period an odd prime p.

(24) Find the sum of $(n|p)(n+1|p)$ from 1 to p for odd primes p. Then show that there are $(p-4-(-1|p))/4$ pairs of consecutive quadratic residues modulo p between 1 and p.

(25) Show that if f is a periodic function on $\mathbb{Z}$ with period q, then

$$\left| \frac{1}{\sqrt{q}} \sum_{A<n\leq B} f(n) - \frac{B-A}{q} \hat{f}(0) \right| \leq \log(q) \max_{1\leq m\leq q-1} |\hat{f}(m)|,$$

where $\hat{f}$ is the Fourier transform of f. Apply this with $f(n) = (n|p)$ to obtain a special case of the Pólya-Vinogradov inequality.

(26) Show that if G is a group and $\rho : G \to \mathrm{GL}(V)$ a finite-dimensional representation, the map $\det(\rho) : G \to \mathrm{GL}(\mathbb{C})$ given by $\det(\rho)(g) = \det(\rho(g))$ is a representation of degree one.

(27) Show that equivalence of representations is an equivalence relation.

(28) † The dihedral group D_8 is the group of eight symmetries of a square. It is generated by a counterclockwise rotation of 90 degrees around the center of the square, and a reflection around a diagonal of the square. Find four inequivalent representations of degree one and one of degree two for this group.

(29) Suppose that G is a finite group and ρ a finite-dimensional complex representation of G. Show that if $g \in G$ is any element, the Jordan canonical form of $\rho(g)$ contains no nontrivial Jordan blocks.

(30) Show that

$$\rho(n) = \begin{bmatrix} 1 & n \\ 0 & 1 \end{bmatrix}$$

is a representation of $\mathbb{Z}$ under addition and that it is not completely reducible.

(31) Show that any finite-dimensional unitary representation of an arbitrary group is an orthogonal direct sum of unitary irreducible subrepresentations.

(32) Show that if $m_1\rho_1 \oplus \cdots \oplus m_r\rho_r$ and $m_1'\rho_1' \oplus \cdots \oplus m_s'\rho_s'$ are decompositions of the same representation, with the ρ_j and the ρ_k' pairwise inequivalent irreducible representations, then the ρ_j are equivalent to the ρ_k' in some order.

(33) Use Schur orthogonality to show that the collection of all matrix entries of a complete collection of inequivalent irreducible finite-dimensional complex representations of a group, relative to a fixed set of ordered bases for the representation spaces, is linearly independent. Conclude that if the group is finite, then it has only finitely many irreducible finite-dimensional complex representations up to equivalence.

(34) Find the conjugacy classes of S_3 and the number of irreps.

(35) Show that any nonabelian group of order eight must have four inequivalent irreducible representations of degree one and one irreducible representation of degree two.

(36) The *character table* of a finite group displays the values that the irreducible characters take on the conjugacy classes of the group, with the characters listed vertically on the left. Find the character table of S_3.

(37) Show that the rows of the character table form an orthonormal set with respect to the Hermitian inner product $\langle \cdots | \cdots \rangle_{L^2(G)}$. Note that if we write out the inner product by conjugacy classes rather than by elements, each term must be weighted by the number of elements in the corresponding conjugacy class.

(38) Show that the columns of the character table form an orthogonal set with respect to the standard Hermitian inner product in complex Euclidean space. This is called the *Second Frobenius orthogonality theorem.*

(39) Compute the trace of the Fourier transform in terms of the irreducible characters, and compute the trace of $\widehat{\chi_\rho}(\rho)$ for any irreducible representation ρ.

(40) Define the convolution of two complex-valued functions ψ_1 and ψ_2 on a finite group G by

$$(\psi_1 \star \psi_2)(g) = \frac{1}{|G|^{1/2}} \sum_{h \in G} \psi_1(h)\psi_2(gh^{-1})$$

for $g \in G$. Prove the convolution theorem $\widehat{\psi_1 \star \psi_2} = \widehat{\psi_1}\widehat{\psi_2}$ for the Fourier transform on G. Specialize the result to the case of a finite abelian group A.

(41) Apply the structure theorem for finitely generated abelian groups to describe the homomorphisms $\chi : A \to \mathbb{T}$ of any finite abelian group A into the circle group. These are the characters of A. Define the *dual group* $\hat{A}$ to be the group of characters under multiplication, with the trivial character as unit element and $\overline{\chi}$ as the inverse of χ. Show that the dual $\hat{A}$ is isomorphic to A. Find a canonical isomorphism between A and its double dual.

(42) Compute the Fourier transform $\hat{1}(\rho_0)$ of the constant function 1 on the trivial representation ρ_0 and show that $\hat{1}(\rho) = 0$ when $\rho_0 \neq \rho \in \hat{G}$.

(43) Show that the complex characters of finite groups are sums of roots of unity.

(44) Show that a finite group must be abelian if all its irreducible representations have degree one.

(45) a) The integers

$$F_n = 2^{2^n} + 1$$

for $n = 0, 1, 2, \ldots$ are called *Fermat numbers*. They are obtained from the sequence $2^m + 1$ by excluding terms that are obviously composite. Show that the Fermat numbers are pairwise coprime. Deduce the infinitude of primes by means of the Fermat numbers (C. Goldbach).
b) Use the small theorem of Fermat to show that if p is a prime divisor of F_n and $p - 1 = 2^j m$ with m odd, then $j > n$. Show that for any fixed $k \in \mathbb{N}$ there are infinitely many primes p congruent to 1 modulo 2^k (Euler, V. A. Lebesgue).

(46) Find the Dirichlet characters modulo $q = 5$.

(47) For each real number $\sigma > 1$, put a probability measure P_σ on the set $\mathbb{N}$ of positive integers by $P_\sigma(\{n\}) = n^{-\sigma}/\zeta(\sigma)$. What is $\lim_{\sigma \to 1^+} P_\sigma(A)$ when A is a subset of $\mathbb{N}$ for which the limit exists? Calculate $P_\sigma(A)$ and its limit when A is the set of squarefree integers.

(48) Calculate the Dirichlet density of the positive integers with an odd number of prime factors (counted with multiplicity) in the natural numbers.

(49) Calculate the Dirichlet density of the integers with an odd number of prime factors (counted with multiplicity) lying in an arithmetic progression $n \equiv a \pmod{q}$, relative to the full set of integers with an odd number of prime factors.

(50) How many primitive Dirichlet characters modulo 20 are there? Find them.

(51) Show that $(n|3)(n|5)$ is a Dirichlet character modulo 45 and find its conductor.

(52) Establish the identities

$$\chi(n) = \frac{1}{q} \sum_{m=1}^{q} \overline{\tau(\overline{\chi}, m)} e(mn/q)$$

and

$$e(n/q) = \frac{1}{\phi(q)} \sum_{\chi \bmod q} \tau(\chi) \overline{\chi(n)},$$

the latter only for q and n coprime.

(53) When $B - A$ is smaller than $q^{1/2}\log(q)$, the Pólya-Vinogradov inequality is worse than estimating the character sum trivially by the triangle inequality. Bounding short character sums is a difficult problem that has drawn much attention. Show that

$$\frac{1}{\phi(q)} \sum_{\chi \bmod q} \left| \sum_{A<n\leq B} \chi(n) \right|^2 \leq B - A$$

so short character sums are seen to be small at least on average. Also show that

$$\max_{\chi \neq \chi_0} \left| \sum_{A<n\leq B} \chi(n) \right| \geq \frac{1}{2}\sqrt{\phi(q) - 1}$$

for suitable A and B. Conclude that there is only limited room to improve the Pólya-Vinogradov inequality for general A and B.

(54) Find a closed-form expression for the Ramanujan sum in terms of known arithmetic functions (O. Hölder).

(55) Show that

$$\sum_{n=1}^{q} \chi(n)\omega(n) \ll \sqrt{q}\log(q)$$

for nonprincipal Dirichlet characters χ modulo q.

(56) Show that

$$\sum_{\substack{A<n\leq B \\ n\equiv 1 \bmod q}} \left(\frac{n}{p}\right) \ll (pq)^{1/2}\log(pq)$$

if p and q are odd primes with $p > q$. What can you say about the least quadratic nonresidue modulo p in the arithmetic progression $n \equiv 1 \pmod{q}$? (A more general situation was considered by I. M. Vinogradov in 1918. The bound may be improved by work of D. A. Burgess.)

(57) Find the Dirichlet characters modulo $q = 8$.

(58) A Dirichlet character χ is called *even* if $\chi(-1) = 1$. Determine for each q the order of the subgroup of even Dirichlet characters modulo q.

(59) Determine for each q the order of the subgroup of real Dirichlet characters modulo q.

Chapter 4

The Circle Method

4.1. Diophantine equations

Though Diophantine analysis has a much wider scope today, in the beginning it covered polynomial equations to be solved in integers, or in rational numbers. This subject was developed extensively in the third century by Diophantus of Alexandria, and equations to be solved in integers or rational numbers are called *Diophantine* after him. He found solutions to various geometrically inspired equations, in most cases requiring certain lengths in a geometric figure to be rational. Isolated Diophantine problems had occurred much earlier, for example to find right-angled triangles all of whose sides are integers. Examples were known in Babylonia and in Egypt, and the Babylonians may have possessed a general method of finding such triangles two millennia before Diophantus. Euclid of Alexandria gave the complete solution in integers of the equation $x^2+y^2 = z^2$ associated with this problem.

Linear Diophantine equations are familiar from elementary number theory, being closely connected with linear congruences. Algorithms to solve such equations occur already in ancient treatises from China and India. But finding the solutions of Diophantine equations of higher degree often turns out to be very difficult, and sometimes impossible in practice. Indeed in full generality, even the simpler problem of deciding whether a Diophantine equation has a solution was shown to be algorithmically undecidable, by Y. V. Matiyasevich. This is the solution to Hilbert's tenth problem in the negative sense.

Early treatments of Diophantine equations relied on what amounts to polynomial algebra, together with divisibility arguments. As an example, to solve $x^2 + y^2 = z^2$ in positive integers, it is enough to assume x, y, z pairwise coprime, since any common factor of two of the three unknowns must also divide the third, and may be divided away. Now x and y cannot both be odd, for then the left-hand side of the equation is congruent to 2 modulo 4, and thus not equal to any square. Without loss of generality, x may be taken to be odd and y to be even. Rewriting the equation as $(z+x)(z-x) = y^2$ and putting $y = 2t$ we see that

$$\frac{z+x}{2} \cdot \frac{z-x}{2} = t^2.$$

Here the two fractions on the left-hand side are coprime integers, for their sum equals z and their difference equals x. Then both of them are squares, since their product is a square. But $(z+x)/2 = u^2$ and $(z-x)/2 = v^2$ yield $x = u^2 - v^2, y = 2uv$ and $z = u^2 + v^2$. Clearly all positive solutions with x, y, z coprime in pairs and y even are obtained by choosing $u > v > 0$ to be coprime positive integers, one even and one odd.

A variety of more substantial techniques of Diophantine analysis have been developed, beginning with the method of infinite descent, due to P. de Fermat. Analytic number theory also has something to contribute to this endeavor. A technique related to Fourier analysis called the Circle Method may be used to estimate the number of solutions in integers of various polynomial equations over $\mathbb{Z}$ when the number of unknowns is sufficiently large. An example is the well-known Waring's Problem, dating to 1770, to show that for each integer $k \geq 2$ there exists some positive integer s such that every positive integer n has a representation $n = x_1^k + \cdots + x_s^k$ in nonnegative integers. Denoting the least such s by $g(k)$ the problem is to prove that $g(k)$ is finite for all $k \geq 2$. This was achieved by D. Hilbert in 1909. Because the integer $3^k - 1$ can only be expressed as a sum of k-th powers if the summands are 1 and 2^k, it is easy to see that $g(k)$ must grow exponentially. But the least number $G(k)$ such that every sufficiently large integer n is the sum of at most $G(k)$ k-th powers grows much more slowly than $g(k)$. It seems reasonable to regard $G(k)$ as more fundamental than $g(k)$, since the latter is determined by the special difficulty of representing comparatively small integers as sums of k-th powers. The Circle Method has been the main tool to study $G(k)$ since the technique was invented around 1920. The current best upper bound for $G(k)$ implies that $\limsup_{k \to +\infty} G(k)/(k \log(k)) \leq 1$. The bound is due to T. D. Wooley.

For Diophantine systems involving only homogenous polynomials, the problem of finding all rational solutions and the problem of finding all integral solutions are essentially equivalent. Any rational solution generates a proportional integral solution, by multiplying the rational solution by the least common multiple of the denominators of its components. But the situation for inhomogenous Diophantine equations is entirely different.

The equation $X^2 + Y^2 = 1$ has only the integral solutions $(X, Y) = (\pm 1, 0), (0, \pm 1)$, but it has rational solutions $(X, Y) = (x/z, y/z)$ with integers x, y, z satisfying $z > 0$ and $x^2 + y^2 = z^2$; of these there are infinitely many essentially different ones, as we have already seen.

A fundamental observation about systems $P_1 = \cdots = P_m = 0$ of polynomial equations over the integers is that solvability over $\mathbb{R}$ and solvability of the congruences $P_1 \equiv \cdots \equiv P_m \equiv 0 \pmod q$ for all positive integers q are necessary conditions for existence of a solution. But solvability modulo all prime powers p^α implies solvability modulo all q, by the Chinese remainder theorem.

Polynomials $P(x_1, \ldots, x_s) = c_1 x_1^k + \cdots + c_s x_s^k$ are called *diagonal forms.* We shall spend the rest of the chapter developing the Circle Method to the point where we can count representations

$$c_1 x_1^k + \cdots + c_s x_s^k = n$$

of integers n by diagonal forms when the number s of variables is large in terms of the degree k. This problem reduces to a linear Diophantine equation when $k = 1$, so it is natural to assume that $k \geq 2$. But the case $k = 2$ is also rather special. The strongest results on the number of representations by quadratic forms proved by means of the Circle Method require an elaboration of the method, due to H. D. Kloosterman, which we cannot cover here. To avoid some changes in details that are necessary when $k = 2$, we eschew the opportunity to obtain weaker results on quadratic forms, and make the standing assumption that $k \geq 3$. Moreover, we restrict both the variables $x_1, \ldots, x_s$ and coefficients $c_1, \ldots, c_s$ to be positive integers, to ensure that there are only finitely many solutions. The alternative would be to count the number of solutions in bounded boxes, which is always finite. But this would slightly complicate the exposition by introducing the size of the box as an extra free parameter.

Denote by $M_n(q)$ the number of mutually incongruent solutions of the congruence

$$c_1 x_1^k + \cdots + c_s x_s^k \equiv n \pmod q.$$

By an earlier remark, $M_n(q) > 0$ for all q is a necessary condition for solvability of the associated Diophantine equation, and it is equivalent to ask that $M_n(p^\alpha) > 0$ for all prime powers p^α. The left-hand side of the above congruence takes q^s values modulo q, but there are only q different possibilities for these values. So $M_n(q)$ is expected on average to equal q^{s-1}. The actual number of solutions may be larger or smaller than this. The congruence $x_1^4 + 13x_2^4 \equiv 6 \pmod{17}$ has no solutions at all, for both terms on the left-hand side are congruent to one of $0, \pm 1, \pm 13$ modulo 17.

The exponential sums

$$S(q,a)=\sum_{x=1}^{q}e\left(\frac{ax^k}{q}\right)$$

play an important role in the study of both the above Diophantine equation and the associated congruences.

Proposition 4.1. *If p is a prime and $a \not\equiv 0 \pmod p$, then $|S(p,a)| \le (k-1)\sqrt{p}$.*

Proof. Denote by $\nu(m)$ the number of mutually incongruent solutions of the congruence $x^k \equiv m \pmod p$. Clearly ν is a periodic arithmetic function of period p. Since

$$S(p,a)=\sum_{m=1}^{p}\nu(m)e\left(\frac{am}{p}\right)$$

it will be enough to study ν. Now $\nu(p)=1$ and so there is no loss in considering ν as a function on the multiplicative group of reduced residue classes modulo p. Thus we can Fourier analyze ν in terms of Dirichlet characters. Then

$$\nu(m)=\frac{1}{p-1}\sum_{\chi}\hat{\nu}(\chi)\chi(m)$$

where

$$\begin{aligned}\hat{\nu}(\chi)&=\sum_{m=1}^{p-1}\nu(m)\overline{\chi(m)}=\sum_{m=1}^{p-1}\Bigg(\sum_{x^k\equiv m \bmod p}1\Bigg)\overline{\chi(m)}\\&=\sum_{x=1}^{p-1}\overline{\chi(x^k)}=\overline{\sum_{x=1}^{p-1}\chi(x)^k}=\begin{cases}p-1 & \text{if } \chi^k=\chi_0,\\ 0 & \text{if } \chi^k\neq\chi_0.\end{cases}\end{aligned}$$

Next

$$\nu(m)=\frac{1}{p-1}\sum_{\chi}\hat{\nu}(\chi)\chi(m)=\sum_{\chi^k=\chi_0}\chi(m)$$

for $1\le m\le p-1$ and so

$$\begin{aligned}S(p,a)&=1+\sum_{m=1}^{p-1}\Bigg(\sum_{\chi^k=\chi_0}\chi(m)\Bigg)e\left(\frac{am}{p}\right)\\&=\sum_{m=1}^{p}e\left(\frac{am}{p}\right)+\sum_{\chi^k=\chi_0\neq\chi}\sum_{m=1}^{p-1}\chi(m)e\left(\frac{am}{p}\right)\\&=\sum_{\chi^k=\chi_0\neq\chi}\tau(\chi,a).\end{aligned}$$

Thus

$$|S(p,a)| \le \sum_{\chi^k=\chi_0\neq\chi} |\tau(\chi,a)| = \sqrt{p} \sum_{\chi^k=\chi_0\neq\chi} 1$$

by Propositions 3.22 and 3.23. We need to bound the number of Dirichlet characters χ modulo p that satisfy $\chi^k = \chi_0$. For this purpose we use the fact that $(\mathbb{Z}/p\mathbb{Z})^\times$ is a cyclic group for p prime. That is to say, there exists a primitive root g modulo p, so that the powers of g represent all the reduced residue classes modulo p. In particular, any Dirichlet character χ modulo p is uniquely determined by its value $\chi(g)$ on g. Now the condition $\chi^k = \chi_0$ implies that $\chi(g)^k = 1$ and so $\chi(g)$ is a k-th root of unity, of which there are only k. □

As a lead-in to the treatment of the Diophantine problem $c_1x_1^k + \cdots + c_sx_s^k = n$, we use the bound in Proposition 4.1 to estimate the number of solutions of the associated congruences $c_1x_1^k + \cdots + c_sx_s^k \equiv n \pmod p$ for prime moduli. The technique used is structurally similar to the Circle Method, though the details are far simpler.

Proposition 4.2. *Suppose that p is a prime, $c_1, \ldots, c_s$ are integers not divisible by p, and n an arbitrary integer. Then the number $M_n(p)$ of mutually incongruent solutions of the congruence*

$$c_1x_1^k + \cdots + c_sx_s^k \equiv n \pmod p$$

satisfies the estimate $|M_n(p) - p^{s-1}| \le (k-1)^s p^{s/2}$.

Proof. The identity

$$\sum_{x_1=1}^{q} \cdots \sum_{x_s=1}^{q} \sum_{a=1}^{q} e\left(\frac{a(c_1x_1^k + \cdots + c_sx_s^k - n)}{q}\right) = qM_n(q)$$

holds for any positive integer q because

$$\sum_{a=1}^{q} e\left(\frac{am}{q}\right) = \begin{cases} q & \text{if } q|m, \\ 0 & \text{otherwise.} \end{cases}$$

Now

$$\begin{aligned} qM_n(q) &= \sum_{a=1}^{q} \left(\sum_{x_1=1}^{q} e\left(\frac{ac_1x_1^k}{q}\right)\right) \cdots \left(\sum_{x_s=1}^{q} e\left(\frac{ac_sx_s^k}{q}\right)\right) e\left(-\frac{an}{q}\right) \\ &= \sum_{a=1}^{q} S(q,ac_1)\cdots S(q,ac_s) e\left(-\frac{an}{q}\right) \end{aligned}$$

and thus, after choosing $q = p$ prime,

$$pM_n(p) = p^s + \sum_{a=1}^{p-1} S(p,ac_1)\cdots S(p,ac_s) e\left(-\frac{an}{p}\right)$$

since all terms in the inner sums equal 1 for $a = p$. Now

$$\begin{aligned}|pM_n(p) - p^s| &\leq \sum_{a=1}^{p-1} |S(p, ac_1)| \cdots |S(p, ac_s)| \left| e\left(-\frac{an}{p}\right) \right| \\ &\leq (p-1)(k-1)p^{1/2} \cdots (k-1)p^{1/2} = (p-1)(k-1)^s p^{s/2}\end{aligned}$$

by Proposition 4.1. □

The Higher Arithmetic by H. Davenport is a concise introduction to number theory with good coverage of classical Diophantine problems. Another introductory treatment with a Diophantine orientation is *Introduction to Number Theory* by T. Nagell. A more advanced treatment of number theory with a Diophantine emphasis is *Number Theory* by Z. I. Borevich and I. R. Shafarevich.

The second edition of *Equations over Finite Fields, An Elementary Approach* by W. M. Schmidt is a standard reference on counting the number of solutions of congruences. There is also a book-length expository paper *Équations et Variétés Algébriques sur un Corps Fini* by J.-R. Joly.

4.2. The major arcs

We now turn to the task of estimating the number of solutions of the Diophantine equation

$$c_1 x_1^k + \cdots + c_s x_s^k = n$$

in positive integers $x_1, \ldots, x_s$, where the coefficients $c_1, \ldots, c_s$ are positive integers, and $k \geq 3$ an integer.

It is clear that any solution has to lie in the box defined by the inequalities

$$1 \leq x_1 \leq P_1, \ldots, 1 \leq x_s \leq P_s$$

with $P_j = (n/c_j)^{1/k}$ for $j = 1, \ldots, s$. The relation

$$\int_{\mathfrak{U}} e(\ell\alpha)\, d\alpha = \begin{cases} 1 & \text{if } \ell = 0, \\ 0 & \text{if } \ell \neq 0, \end{cases}$$

holds for $\ell \in \mathbb{Z}$ and any interval $\mathfrak{U} \subseteq \mathbb{R}$ of length equal to 1. Denote by $R(n)$ the number of integer solutions of

$$c_1 x_1^k + \cdots + c_s x_s^k = n, \quad 1 \leq x_j \leq P_j, \quad 1 \leq j \leq s$$

for fixed $c_1, \ldots, c_s$ and k as above. Then

$$R(n) = \sum_{1 \leq x_1 \leq P_1} \cdots \sum_{1 \leq x_s \leq P_s} \int_{\mathfrak{U}} e((c_1 x_1^k + \cdots + c_s x_s^k - n)\alpha)\, d\alpha$$

in precisely the same way as for the congruence treated in Proposition 4.2 Interchanging sums and integral yields

$$R(n) = \int_{\mathfrak{U}} \left(\sum_{x_1 \le P_1} e(c_1 x_1^k \alpha) \right) \cdots \left(\sum_{x_s \le P_s} e(c_s x_s^k \alpha) \right) e(-n\alpha)\, d\alpha.$$

In the similar formula

$$pM_n(p) = \sum_{m=0}^{p-1} \left(\sum_{x_1=0}^{p-1} e(mc_1 x_1^{k_1}/p) \right) \cdots \left(\sum_{x_s=0}^{p-1} e(mc_s x_s^{k_s}/p) \right) e(-mn/p)$$

for the number of solutions of the related congruence, splitting off the term with $m = 0$ in the outer sum yielded the main term p^{s-1} of the estimate for $M_n(p)$ while the other terms contributed the error term. The analogue of this step for the integral over $\mathfrak{U}$ is the key feature of the Circle Method. Putting

$$f_j(\alpha) = \sum_{m \le P_j} e(\alpha c_j m^k)$$

for $1 \le j \le s$, it turns out that for α close to a rational number with small denominator, $f_j(\alpha)$ can be well approximated by a simpler expression. We shall partition $\mathfrak{U}$ into two sets $\mathfrak{M}$ and $\mathfrak{m}$, each of them a union of intervals. The set $\mathfrak{M}$ is called the set of *major arcs* and is a set of points in $\mathfrak{U}$ that are close to rationals a/q with small denominators. The set $\mathfrak{m}$ of *minor arcs* is the complement of $\mathfrak{M}$ in $\mathfrak{U}$. Consider the exponential sum

$$f(\alpha) = \sum_{m \le P} e(\alpha c m^k)$$

where $c \ge 1$ and $k \ge 3$ are integers. Fix a real parameter $\rho > 0$ and an integer $Q \ge 1$ and choose $\mathfrak{U} = (\rho, 1 + \rho]$. Then define

$$\mathfrak{M}(q,a) = \{ \alpha \in \mathfrak{U} \,|\, |\alpha - a/q| \le \rho \}$$

for a and q coprime integers satisfying $1 \le a \le q \le Q$, and let $\mathfrak{M}$ be the union of all these intervals $\mathfrak{M}(q,a)$. If $\alpha \in \mathfrak{M}$, then α is within ρ of a/q for some such integers a and q as above. The exponential sum $f(\alpha)$ has an approximation in terms of

$$S(q, ac) = \sum_{x=1}^{q} e\left(\frac{acx^k}{q} \right)$$

and

$$v(\beta) = \int_0^P e(\beta c u^k)\, du$$

for $\alpha \in \mathfrak{M}(q,a)$.

Proposition 4.3. *If* $1 \le a \le q \le Q$ *with* $\gcd(q,a) = 1$*, then*

$$f(\alpha) = q^{-1}S(q, ac)v(\alpha - a/q) + O(Q + \rho Q n)$$

for all $\alpha \in \mathfrak{M}(q,a) = \{\, \alpha \in \mathfrak{U} \,|\, |\alpha - a/q| \le \rho \,\}$.

Proof. The sum over $1 \le m \le P$ defining $f(\alpha)$ transforms to a double sum by

$$\begin{aligned} f(\alpha) &= \sum_{q\ell + r \le P} e(\alpha c(q\ell + r)^k) \\ &= \sum_{r=1}^{q} \sum_{q\ell + r \le P} e(ac(q\ell + r)^k/q) e((\alpha - a/q)c(q\ell + r)^k) \\ &= \sum_{r=1}^{q} e(acr^k/q) \sum_{0 \le \ell \le (P-r)/q} e((\alpha - a/q)c(q\ell + r)^k), \end{aligned}$$

after making the change of summation index $m = q\ell + r$, using division with remainder.

We are going to approximate the innermost sum by an integral, applying the Euler summation formula. Thus

$$\begin{aligned} \sum_{0 \le \ell \le (P-r)/q} e((\alpha - a/q)c(q\ell + r)^k) &= \int_0^{[(P-r)/q]} e((\alpha - a/q)c(qt + r)^k)\, dt \\ &+ \frac{e(\alpha - a/q)r^k}{2} + \frac{e(\alpha - a/q)(q[(P-r)/q] + r)^k}{2} \\ &+ \int_0^{[(P-r)/q]} S(t)(\alpha - a/q)ck(qt + r)^{k-1} q e((\alpha - a/q)c(qt + r)^k)\, dt, \end{aligned}$$

where

$$\left| \frac{e(\alpha - a/q)r^k}{2} + \frac{e(\alpha - a/q)(q[(P-r)/q] + r)^k}{2} \right| \le 1$$

and

$$\begin{aligned} &\left| \int_0^{[(P-r)/q]} S(t)(\alpha - a/q)ck(qt + r)^{k-1} q e((\alpha - a/q)c(qt + r)^k)\, dt \right| \\ &\qquad \le \int_0^{[(P-r)/q]} \frac{1}{2} |\alpha - a/q| ck(qt + r)^{k-1} q\, dt = O(\rho n). \end{aligned}$$

Next

$$\begin{aligned}\int_0^{[(P-r)/q]} & e((\alpha-a/q)c(qt+r)^k)\,dt \\ &= \int_r^{q[(P-r)/q]+r} e((\alpha-a/q)cu^k)q^{-1}\,du \\ &= q^{-1}\int_0^P e((\alpha-a/q)cu^k)\,du + O(1),\end{aligned}$$

using the change of variable $u = qt + r$. Now

$$\sum_{0\le \ell\le (P-r)/q} e((\alpha-a/q)c(q\ell+r)^k) = q^{-1}v(\alpha-a/q) + O(1+\rho n)$$

and finally

$$f(\alpha) = q^{-1}S(q, ac)v(\alpha - a/q) + O(Q + \rho Q n)$$

since $q \le Q$. □

At this stage we impose the condition $2\rho Q^2 < 1$ to ensure that distinct intervals $\mathfrak{M}(q, a)$ and $\mathfrak{M}(q', a')$ be disjoint. The inequality

$$\left|\frac{a}{q} - \frac{a'}{q'}\right| \ge \frac{1}{qq'} \ge \frac{1}{Q^2} \ge 2\rho$$

implies that the distance between the centers of the two intervals exceeds the sum of their radii. Since $Q \ge 1$ and $2\rho QQ < 1$ implies $2\rho Q < 1$ and thus $\rho < Q^{-1} - \rho$, it is easy to see that $\mathfrak{M} \subseteq \mathfrak{U}$ and so

$$\begin{aligned}R(n) &= \int_{\mathfrak{U}} f_1(\alpha)\cdots f_s(\alpha)e(-n\alpha)\,d\alpha \\ &= \int_{\mathfrak{M}} f_1(\alpha)\cdots f_s(\alpha)e(-n\alpha)\,d\alpha + \int_{\mathfrak{m}} f_1(\alpha)\cdots f_s(\alpha)e(-n\alpha)\,d\alpha\end{aligned}$$

with

$$\int_{\mathfrak{M}} f_1(\alpha)\cdots f_s(\alpha)e(-n\alpha)\,d\alpha = \sum_{\substack{1\le a\le q\le Q \\ \gcd(q,a)=1}} \int_{\mathfrak{M}(q,a)} f_1(\alpha)\cdots f_s(\alpha)e(-n\alpha)\,d\alpha$$

where

$$f_j(\alpha) = \sum_{m\le P_j} e(\alpha c_j m^k)$$

has the approximation

$$f_j(\alpha) = q^{-1}S(q, ac_j)v_j(\alpha - a/q) + O(Q + \rho Q n)$$

on $\mathfrak{M}(q, a)$. Here

$$v_j(\beta) = \int_0^{P_j} e(\beta c_j u^k)\,du$$

for $1 \le j \le s$. Define

$$J_\rho(n) = \int_{|\beta| \le \rho} v_1(\beta) \cdots v_s(\beta) e(-n\beta)\, d\beta$$

and

$$A_n(q) = \sum_{\substack{a=1 \\ \gcd(q,a)=1}}^{q} q^{-s} S(q, ac_1) \cdots S(q, ac_s) e\left(-\frac{an}{q}\right).$$

The next step is to approximate to the integral over the major arcs $\mathfrak{M}$ in terms of $J_\rho(n)$ and $A_n(q)$.

Proposition 4.4. *The estimate*

$$\int_{\mathfrak{M}} f_1(\alpha) \cdots f_s(\alpha) e(-n\alpha)\, d\alpha = J_\rho(n) \sum_{q=1}^{Q} A_n(q) \\ + O(n^{(s-1)/k} \rho^2 Q^3 n + \rho^{s+1} Q^{s+2} n^s)$$

holds for $2\rho Q^2 < 1$ *and* $\rho n \ge 1$.

Proof. We observe that $f_j(\alpha) = O(n^{1/k})$ and

$$f_j(\alpha) = q^{-1} S(q, ac_j) v_j(\alpha - a/q) + O(\rho Q n)$$

by $\rho n \ge 1$, so

$$f_1(\alpha) \cdots f_s(\alpha) = q^{-s} S(q, ac_1) \cdots S(q, ac_s) v_1(\alpha - a/q) \cdots v_s(\alpha - a/q) \\ + O(n^{(s-1)/k} \rho Q n + \rho^s Q^s n^s)$$

on $\mathfrak{M}(q, a)$. Now

$$\begin{aligned} \int_{\mathfrak{M}(q,a)} f_1(\alpha) \cdots f_s(\alpha) e(-n\alpha)\, d\alpha &= q^{-s} S(q, ac_1) \cdots S(q, ac_s) \\ &\quad \times \int_{\mathfrak{M}(q,a)} v_1(\alpha - a/q) \cdots v_s(\alpha - a/q) e(-n\alpha)\, d\alpha \\ &\quad + O(n^{(s-1)/k} \rho^2 Q n + \rho^{s+1} Q^s n^s) \\ &= q^{-s} S(q, ac_1) \cdots S(q, ac_s) e(-an/q) \\ &\qquad \times \int_{|\beta| \le \rho} v_1(\beta) \cdots v_s(\beta) e(-n\beta)\, d\beta \\ &\qquad + O(n^{(s-1)/k} \rho^2 Q n + \rho^{s+1} Q^s n^s) \\ &= q^{-s} S(q, ac_1) \cdots S(q, ac_s) e(-an/q) J_\rho(n) \\ &\qquad + O(n^{(s-1)/k} \rho^2 Q n + \rho^{s+1} Q^s n^s), \end{aligned}$$

since the length of $\mathfrak{M}(q, a)$ is 2ρ. Note the change of variable $\alpha = a/q + \beta$. Summing over the $O(Q^2)$ pairs of integers a and q with $1 \le a \le q \le Q$

and $\gcd(q,a)=1$ yields the desired estimate, since the condition $2\rho Q^2<1$ implies that the intervals $\mathfrak{M}(q,a)$ are pairwise disjoint. □

Proposition 4.5. *The estimate*

$$J_\rho(n)=\frac{C_{k,s}}{(c_1\cdots c_s)^{1/k}}n^{s/k-1}+O(\rho^{1-s/k})$$

holds with some constant $C_{k,s}>0$ if $s>k$.

Proof. Note that

$$v_j(\beta)=\int_0^{(n/c_j)^{1/k}} e(\beta c_j u^k)\,du=\frac{n^{1/k}}{c_j^{1/k}}\int_0^1 e(\beta n w^k)\,dw$$

by the change of variable $u=(n/c_j)^{1/k}w$, and thus

$$\begin{aligned}J_\rho(n)&=\frac{n^{s/k}}{(c_1\cdots c_s)^{1/k}}\int_{|\beta|\le\rho}\left(\int_0^1 e(n\beta w^k)\,dw\right)^s e(-n\beta)\,d\beta\\&=\frac{n^{s/k-1}}{(c_1\cdots c_s)^{1/k}}\int_{|\gamma|\le n\rho}\left(\int_0^1 e(\gamma w^k)\,dw\right)^s e(-\gamma)\,d\gamma\end{aligned}$$

by the change of variable $\gamma=n\beta$. The estimate

$$\left|\int_0^1 e(\gamma w^k)\,dw\right|\ll\min(1,|\gamma|^{-1/k})$$

holds, for we may assume without loss of generality that $\gamma>0$, and then

$$\begin{aligned}\int_0^1 e(\gamma w^k)\,dw&=k^{-1}\gamma^{-1/k}\int_0^\gamma e(t)t^{1/k-1}\,dt=k^{-1}\gamma^{-1/k}\\&\quad\times\left(\frac{e(t)-1}{2\pi i}t^{1/k-1}\Big|_{t=0}^{t=\gamma}-\int_0^\gamma\frac{e(t)-1}{2\pi i}\left(\frac{1}{k}-1\right)t^{1/k-2}\,dt\right)\\&=k^{-1}\gamma^{-1/k}\frac{e(\gamma)-1}{2\pi i}+\frac{\gamma^{-1/k}}{2\pi i k}\left(1-\frac{1}{k}\right)\int_0^\gamma(e(t)-1)t^{1/k-2}\,dt\end{aligned}$$

by the change of variable $t=\gamma w^k$ and an integration by parts, and

$$\left|\int_1^\gamma(e(t)-1)t^{1/k-2}\,dt\right|\le\int_1^\infty 2t^{1/k-2}\,dt<\infty$$

since $1/k<1$, while

$$\left|\int_0^1(e(t)-1)t^{1/k-2}\,dt\right|\le\int_0^1|e(t)-1|t^{1/k-2}\,dt<\infty$$

because $e(t)-1=O(t)$ as $t\to 0^+$, and $1/k>0$. Then the improper integral

$$C_{k,s}=\int_{-\infty}^\infty\left(\int_0^1 e(\gamma w^k)\,dw\right)^s e(-\gamma)\,d\gamma$$

converges if $s > k$, and

$$\left| J_\rho(n) - \frac{n^{s/k-1}}{(c_1\cdots c_s)^{1/k}} C_{k,s} \right| \ll \frac{n^{s/k-1}}{(c_1\cdots c_s)^{1/k}} \int_{\gamma \geq n\rho} \gamma^{-s/k}\, d\gamma \ll \rho^{1-s/k},$$

also for $s > k$. Thus the asymptotic formula is established, but it remains to show that $C_{k,s} > 0$.

The limit

$$C_{k,s} = \lim_{t\to 0^+} \int_{-\infty}^{\infty} e^{-t^2\gamma^2} \left(\int_0^1 e(\gamma w^k)\, dw \right)^s e(-\gamma)\, d\gamma$$

holds because

$$\begin{aligned}
&\left| C_{k,s} - \int_{-\infty}^{\infty} e^{-t^2\gamma^2} \left(\int_0^1 e(\gamma w^k)\, dw \right)^s e(-\gamma)\, d\gamma \right| \\
&\quad \leq 2 \int_0^{\infty} (1 - e^{-t^2\gamma^2}) \left| \int_0^1 e(\gamma w^k)\, dw \right|^s d\gamma \leq 2 \int_0^1 (1 - e^{-t^2\gamma^2})\, d\gamma \\
&\qquad + 2 \int_1^{t^{-1/2}} (1 - e^{-t^2\gamma^2}) \gamma^{-s/k}\, d\gamma + 2 \int_{t^{-1/2}}^{\infty} (1 - e^{-t^2\gamma^2}) \gamma^{-s/k}\, d\gamma \\
&\quad \leq 2(1 - e^{-t^2}) + 2(1 - e^{-t}) \int_1^{\infty} \gamma^{-s/k}\, d\gamma + 2 \int_{t^{-1/2}}^{\infty} \gamma^{-s/k}\, d\gamma \to 0
\end{aligned}$$

as $t \to 0^+$. Next

$$\begin{aligned}
&\int_{-\infty}^{\infty} e^{-t^2\gamma^2} \left(\int_0^1 e(\gamma w^k)\, dw \right)^s e(-\gamma)\, d\gamma \\
&\quad = \int_{-\infty}^{\infty} e^{-t^2\gamma^2} \int_0^1 \cdots \int_0^1 e(\gamma(w_1^k + \cdots + w_s^k))\, dw_1 \cdots dw_s\, e(-\gamma)\, d\gamma \\
&\quad = \int_0^1 \cdots \int_0^1 \int_{-\infty}^{\infty} e^{-t^2\gamma^2} e(\gamma(w_1^k + \cdots + w_s^k - 1))\, d\gamma\, dw_1 \cdots dw_s \\
&\quad = \int_0^1 \cdots \int_0^1 \int_{-\infty}^{\infty} e^{-t^2\gamma^2} \cos(2\pi\gamma(w_1^k + \cdots + w_s^k - 1))\, d\gamma\, dw_1 \cdots dw_s \\
&\quad = \int_0^1 \cdots \int_0^1 \frac{\sqrt{\pi}}{t} e^{-\pi^2(w_1^k + \cdots + w_s^k - 1)^2/t^2}\, dw_1 \cdots dw_s
\end{aligned}$$

by the known definite integral

$$\int_{-\infty}^{\infty} e^{-a^2x^2} \cos(bx)\, dx = \frac{\sqrt{\pi}}{a} e^{-b^2/(4a^2)},$$

is valid for $a > 0$. Defining for each $t > 0$ a set E_t by $0 \leq w_1, \ldots, w_s \leq 1$ and $1 \leq w_1^k + \cdots + w_s^k \leq 1 + t$, one sees that

$$\int \cdots \int_{E_t} dw_1 \cdots dw_s \geq \frac{tV_{k,s}}{2}$$

for all $t > 0$ sufficiently small, where $V_{k,s}$ is the $(s-1)$-dimensional volume of the piece of hypersurface in $\mathbb{R}^s$ given by $0 \le w_1, \ldots, w_s \le 1$ and $w_1^k + \ldots + w_s^k = 1$. Now

$$\int_0^1 \cdots \int_0^1 \frac{\sqrt{\pi}}{t} e^{-\pi^2(w_1^k+\cdots+w_s^k-1)^2/t^2} dw_1 \cdots dw_s$$
$$\ge \int \cdots \int_{E_t} \frac{\sqrt{\pi}}{t} e^{-\pi^2(w_1^k+\cdots+w_s^k-1)^2/t^2} dw_1 \cdots dw_s$$
$$\ge \int \cdots \int_{E_t} \frac{\sqrt{\pi}}{t} e^{-\pi^2 t^2/t^2} dw_1 \cdots dw_s \ge \frac{\sqrt{\pi}}{2} e^{-\pi^2} V_{k,s}$$

and so $C_{k,s} \ge \sqrt{\pi} e^{-\pi^2} V_{k,s}/2 > 0$. □

The constant $C_{k,s}$ may be calculated exactly in terms of the volume $V_{k,s}$, and indeed explicitly in terms of a special function from classical analysis.

To make further progress with the integral over the major arcs it is necessary to discuss the convergence and other properties of the *singular series*

$$\mathfrak{S}(n) = \sum_{q=1}^{\infty} A_n(q)$$

belonging to the Diophantine problem $c_1 x_1^k + \cdots + c_s x_s^k = n$.

4.3. The singular series

Proposition 4.6. *Suppose that $k \ge 3$ and $\alpha \ge 2$ and $a \not\equiv 0 \pmod{p}$. If $\alpha > k$, then*

$$S(p^\alpha, a) = p^{k-1} S(p^{\alpha-k}, a)$$

while

$$S(p^\alpha, a) = p^{\alpha-1}$$

if $2 \le \alpha \le k$ and $k \not\equiv 0 \pmod{p}$.

Proof. Let τ be the exponent to which p occurs in the prime factorization of k. We are going to make a change of summation index

$$x = p^{\alpha-\tau-1} y + z$$

in the exponential sum $S(p^\alpha, a)$. For this to be useful, it is necessary to establish that $\alpha - \tau - 1 \ge 1$. The latter inequality is clear if $k \not\equiv 0 \pmod{p}$, for then $\tau = 0$. While if $\alpha > k$, then $\alpha > k \ge p^\tau \ge \tau + 1$. Now

$$x^k = \sum_{j=0}^{k} \binom{k}{j} z^{k-j} p^{j(\alpha-\tau-1)} y^j$$

by the Binomial Theorem, and $3(\alpha-\tau-1)\geq\alpha$ because $\alpha\geq 2^\tau+1$, thus

$$x^k \equiv z^k + kz^{k-1}p^{\alpha-\tau-1}y + \frac{k(k-1)}{2}p^{2(\alpha-\tau-1)}y^2 \pmod{p^\alpha}.$$

If $\tau=0$, then $2(\alpha-\tau-1)\geq\alpha$ because $\alpha\geq 2$. If $p\neq 2$, then $2(\alpha-\tau-1)\geq\alpha$ because $\alpha\geq 3^\tau+1$. If $p=2$ and $\tau=1$, then $2(\alpha-\tau-1)\geq\alpha$ because $k\geq 3$ implies that $\alpha>k\geq 6$ in that case. While if $p=2$ and $\tau\geq 2$, then $2^{\tau-1}|k(k-1)/2$ and $\tau-1+2(\alpha-\tau-1)\geq\alpha$ because $\alpha\geq 2^\tau+1$. Thus the congruence

$$x^k \equiv z^k + kz^{k-1}p^{\alpha-\tau-1}y \pmod{p^\alpha}$$

holds in all cases. Then

$$S(p^\alpha,a) = \sum_{z=1}^{p^{\alpha-\tau-1}} e\left(\frac{az^k}{p^\alpha}\right)\sum_{y=1}^{p^{\tau+1}} e\left(\frac{akz^{k-1}y}{p^{\tau+1}}\right),$$

where the innermost sum is zero unless z is divisible by p, in which case it equals $p^{\tau+1}$. Hence

$$\begin{aligned} S(p^\alpha,a) &= p^{\tau+1}\sum_{z=1}^{p^{\alpha-\tau-1}} e\left(\frac{az^k}{p^\alpha}\right) = p^{\tau+1}\sum_{w=1}^{p^{\alpha-\tau-2}} e\left(\frac{ap^kw^k}{p^\alpha}\right) \\ &= p^{\tau+1}\frac{p^{\alpha-\tau-2}}{p^{\alpha-k}}\sum_{w=1}^{p^{\alpha-k}} e\left(\frac{aw^k}{p^{\alpha-k}}\right) \end{aligned}$$

on making the change of the summation index $z=pw$. But the last sum equals $p^{\alpha-k}$ if $\alpha\leq k$, for then each term equals 1. While it equals $S(p^{\alpha-k},a)$ if $\alpha>k$. □

The exponential sums $S(q,a)$ have a kind of multiplicative property that will enable us to find a bound that is valid for general q.

Proposition 4.7. *The identity*

$$S(q_1q_2,a) = S(q_1,aq_2^{k-1})S(q_2,aq_1^{k-1})$$

holds if q_1 and q_2 are coprime.

Proof. By the theory of linear Diophantine equations, there exists for each integer x modulo q_1q_2 precisely one integer y modulo q_2 and precisely one integer z modulo q_1 for which $q_1y+q_2z=x$, and

$$x^k = (q_1y+q_2z)^k \equiv q_1^ky^k + q_2^kz^k \pmod{q_1q_2}.$$

Thus

$$
\begin{aligned}
S(q_1q_2, a) &= \sum_{x=1}^{q_1q_2} e\left(\frac{ax^k}{q_1q_2}\right) = \sum_{z=1}^{q_1}\sum_{y=1}^{q_2} e\left(\frac{aq_1^k y^k + q_2^k z^k}{q_1q_2}\right) \\
&= \sum_{z=1}^{q_1}\sum_{y=1}^{q_2} e\left(\frac{aq_1^{k-1}y^k}{q_2}\right) e\left(\frac{aq_2^{k-1}z^k}{q_1}\right) \\
&= \sum_{z=1}^{q_1} e\left(\frac{aq_2^{k-1}z^k}{q_1}\right) \sum_{y=1}^{q_2} e\left(\frac{aq_1^{k-1}y^k}{q_2}\right) \\
&= S(q_1, aq_2^{k-1})S(q_2, aq_1^{k-1})
\end{aligned}
$$

as y and z vary independently over the residue classes modulo q_2 and q_1 respectively. □

Proposition 4.8. *For each integer $k \geq 3$ there is some constant $B_k > 0$ so that the bound $|S(q,a)| \leq B_k q^{1-1/k}$ holds for any coprime q and a.*

Proof. Let p^α be an arbitrary prime power and b any integer coprime with p. Then there is a nonnegative integer m so that $1 \leq \alpha - km \leq k$ and

$$S(p^\alpha, b) = p^{(k-1)m} S(p^{\alpha-km}, b),$$

by Proposition 4.6. Because k has only finitely many prime divisors, the inequality

$$
\begin{aligned}
|S(p^{\alpha-km}, b)| &= p^{\alpha-km-1} = p^{\alpha-km-(\alpha-km)(\alpha-km)^{-1}} \\
&= \left(p^{\alpha-km}\right)^{1-(\alpha-km)^{-1}} \leq \left(p^{\alpha-km}\right)^{1-1/k}
\end{aligned}
$$

holds for $2 \leq \alpha \leq k$ with at most finitely many exceptions p, by Proposition 4.6. But this inequality also holds for $\alpha = 1$ with at most finitely many exceptions p, for $k \geq 3$ implies that $|S(p,b)| \leq (k-1)p^{1/2} \leq p^{1-1/k}$ holds for all primes p sufficiently large, by Proposition 4.1. Denote by M the maximum of $|S(p^\alpha, b)|$ as p ranges over the finitely many, say N, exceptional primes p identified above, the integer α ranges over the interval $1 \leq \alpha \leq k$ and b ranges over the integers coprime with p in the interval $1 \leq b \leq p^\alpha$. Then

$$
\begin{aligned}
|S(p^\alpha, b)| &= p^{(k-1)m} |S(p^{\alpha-km}, b)| \\
&\leq Mp^{(k-1)m} \leq Mp^{(k-1)\alpha/k} = M\left(p^\alpha\right)^{1-1/k}
\end{aligned}
$$

since $km \leq \alpha$.

If $q = p_1^{\alpha_1} \cdots p_r^{\alpha_r}$ is the prime factorization of an arbitrary positive integer q coprime with a, then $S(q,a) = S(p_1^{\alpha_1}, a_1) \cdots S(p_r^{\alpha_r}, a_r)$ for some integers

$a_1, \ldots, a_r$ coprime with $p_1^{\alpha_1}, \ldots, p_r^{\alpha_r}$ respectively, by Proposition 4.7. Then

$$\begin{aligned}|S(q,a)| &= |S(p_1^{\alpha_1}, a_1)|\cdots|S(p_r^{\alpha_r}, a_r)| \\ &\leq M^N \,(p_1^{\alpha_1})^{1-1/k} \ldots (p_r^{\alpha_r})^{1-1/k} = M^N q^{1-1/k}\end{aligned}$$

by the inequalities found above. □

Proposition 4.9. *The estimate* $|A_n(q)| \ll q^{1-s/k}$ *holds uniformly in* n.

Proof. We need to estimate the sums $S(q, ac_j)$, but though a and q are coprime, c_j and q may have a common factor, so that Proposition 4.8 does not immediately apply. But if $d_j = \gcd(q, c_j)$, then

$$S(q, ac_j) = \sum_{x=1}^{q} e\left(\frac{ac_j d_j^{-1} x^k}{q d_j^{-1}}\right) = d_j \sum_{x=1}^{q d_j^{-1}} e\left(\frac{ac_j d_j^{-1} x^k}{q d_j^{-1}}\right) = d_j S\left(\frac{q}{d_j}, \frac{ac_j}{d_j}\right)$$

and so $|S(q, ac_j)| \leq d_j B_k (q/d_j)^{1-1/k}$ for $j = 1, 2, \ldots, s$. Thus

$$\begin{aligned}|A_n(q)| &= \left| \sum_{\substack{a=1 \\ \gcd(q,a)=1}}^{q} q^{-s} S(q, ac_1)\cdots S(q, ac_s)\, e\left(-\frac{an}{q}\right)\right| \\ &\leq q^{-s} d_1 B_k (q/d_1)^{1-1/k} \cdots d_s B_k (q/d_s)^{1-1/k} \sum_{\substack{a=1 \\ \gcd(q,a)=1}}^{q} 1 \\ &\leq q^{-s} d_1 \cdots d_s B_k^s q^{s-s/k} (d_1 \cdots d_s)^{-1+1/k} q \\ &= B_k^s (d_1 \cdots d_s)^{1/k} q^{1-s/k} \leq B_k^s (c_1 \cdots c_s)^{1/k} q^{1-s/k}\end{aligned}$$

establishes the desired bound uniformly in n. □

Recalling the singular series

$$\mathfrak{S}(n) = \sum_{q=1}^{\infty} A_n(q)$$

belonging to our Diophantine problem, we see by Proposition 4.9 that this series converges absolutely if $s \geq 2k+1$, and moreover $\mathfrak{S}(n)$ is a bounded function of n by the uniformity. From now on we require that $s \geq 2k+1$, and in fact we shall later have to take s much larger than this.

Proposition 4.10. $A_n(q)$ *is a multiplicative function of* q.

Proof. By the theory of linear Diophantine equations, there exists for each integer a modulo $q_1 q_2$ with q_1 and q_2 coprime precisely one integer a_1 modulo q_2 and precisely one integer a_2 modulo q_1 for which $q_1 a_2 + q_2 a_1 = a$. If a

and q_1q_2 are coprime, then a_1 is prime to q_1 and a_2 is prime to q_2, and vice versa. Thus

$$\begin{aligned}
A_n(q_1q_2) &= \sum_{\substack{a=1\\ \gcd(q_1q_2,a)=1}}^{q_1q_2} (q_1q_2)^{-s} S(q_1q_2, ac_1)\cdots S(q_1q_2, ac_s)\, e\left(-\frac{an}{q_1q_2}\right)\\
&= \sum_{\substack{a_1=1\\ \gcd(q_1,a_1)=1}}^{q_1} \sum_{\substack{a_2=1\\ \gcd(q_2,a_2)=1}}^{q_2} q_1^{-s}q_2^{-s} S(q_1, (q_1a_2+q_2a_1)c_1q_2^{k-1})\\
&\quad \times S(q_2, (q_1a_2+q_2a_1)c_1q_1^{k-1})\cdots S(q_1, (q_1a_2+q_2a_1)c_sq_2^{k-1})\\
&\quad \times S(q_2, (q_1a_2+q_2a_1)c_sq_1^{k-1})\, e\left(-\frac{(q_1a_2+q_2a_1)n}{q_1q_2}\right)\\
&= \sum_{\substack{a_1=1\\ \gcd(q_1,a_1)=1}}^{q_1} \sum_{\substack{a_2=1\\ \gcd(q_2,a_2)=1}}^{q_2} q_1^{-s}q_2^{-s} S(q_1, a_1c_1q_2^k) S(q_2, a_2c_1q_1^k)\\
&\quad \cdots S(q_1, a_1c_sq_2^k) S(q_2, a_2c_sq_1^k)\, e\left(-\frac{a_1n}{q_1}\right) e\left(-\frac{a_2n}{q_2}\right)\\
&= \sum_{\substack{a_1=1\\ \gcd(q_1,a_1)=1}}^{q_1} \sum_{\substack{a_2=1\\ \gcd(q_2,a_2)=1}}^{q_2} q_1^{-s}q_2^{-s} S(q_1, a_1c_1) S(q_2, a_2c_1)\cdots\\
&\quad \times S(q_1, a_1c_s) S(q_2, a_2c_s)\, e\left(-\frac{a_1n}{q_1}\right) e\left(-\frac{a_2n}{q_2}\right)\\
&= \left(\sum_{\substack{a_1=1\\ \gcd(q_1,a_1)=1}}^{q_1} q_1^{-s} S(q_1, a_1c_1)\cdots S(q_1, a_1c_s)\, e\left(-\frac{a_1n}{q_1}\right)\right)\\
&\quad \times \left(\sum_{\substack{a_2=1\\ \gcd(q_2,a_2)=1}}^{q_2} q_2^{-s} S(q_2, a_2c_1)\cdots S(q_2, a_2c_s)\, e\left(-\frac{a_2n}{q_2}\right)\right)\\
&= A_n(q_1)A_n(q_2)
\end{aligned}$$

if q_1 and q_2 are coprime. Note that

$$S(q, ab^k) = \sum_{x=1}^{q} e\left(\frac{ab^kx^k}{q}\right) = \sum_{y=1}^{q} e\left(\frac{ay^k}{q}\right) = S(q, a)$$

if b and q are coprime, by the change $y = bx$ of summation index. □

It is decisive to have $\mathfrak{S}(n)$ bounded below as a function of n when bounding from below the number of solutions of $c_1x_1^k + \cdots + c_sx_s^k = n$ by the Circle Method. To investigate this question we note that the singular series has the Euler product

$$\mathfrak{S}(n) = \prod_p \sum_{\alpha=0}^{\infty} A_n(p^\alpha)$$

by Propositions 3.2, 4.9 and 4.1. The inequality

$$|A_n(q)| \le B_k^s (c_1 \cdots c_s)^{1/k} q^{1-1/k}$$

from the proof of Proposition 4.9 implies that there exists some prime $p_0 = p_0(c,k,s)$ where $c = c_1 \cdots c_s$, so that

$$\frac{1}{2} \le \left| \prod_{p>p_0} \sum_{\alpha=0}^{\infty} A_n(p^\alpha) \right| \le \frac{3}{2}.$$

Thus it will be enough to consider the factors of the Euler product when $p \le p_0$. We establish a link between these factors and the number $M_n(q)$ of mutually incongruent solutions of the congruences $c_1 x_1^k + \cdots + c_s x_s^k \equiv n \pmod{q}$.

Proposition 4.11. $\displaystyle\sum_{\alpha=0}^{\infty} A_n(p^\alpha) = \lim_{\beta\to+\infty} M_n(p^\beta) p^{-\beta(s-1)}.$

Proof. The formula

$$qM_n(q) = \sum_{a=1}^{q} S(q, ac_1) \cdots S(q, ac_s)\, e\left(-\frac{an}{q}\right)$$

was obtained in the course of the proof of Proposition 4.2. Now

$$\begin{aligned}
M_n(q) &= \frac{1}{q} \sum_{d|q} \sum_{\substack{a=1 \\ \gcd(q,a)=d}}^{q} S(q, ac_1) \cdots S(q, ac_s)\, e\left(-\frac{an}{q}\right) \\
&= \frac{1}{q} \sum_{d|q} \sum_{\substack{a'=1 \\ \gcd(q',a')=1}}^{q'=q/d} S(dq', da'c_1) \cdots S(dq', da'c_s)\, e\left(-\frac{da'n}{dq'}\right) \\
&= \frac{1}{q} \sum_{d|q} \sum_{\substack{a'=1 \\ \gcd(q',a')=1}}^{q'} d\, S(q', a'c_1) \cdots dS(q', a'c_s)\, e\left(-\frac{a'n}{q'}\right) \\
&= \frac{1}{q} \sum_{d|q} d^s (q/d)^s A_n(q/d) = q^{s-1} \sum_{d|q} A_n(q/d)
\end{aligned}$$

by putting $q = dq'$ and $a = da'$. Choosing $q = p^\beta$ and letting $\beta \to +\infty$ completes the proof. □

From Proposition 4.11 it is clear that the factors in the Euler product can be bounded from below by suitably bounding $M_n(p^\beta)$ from below. We shall actually need to show that the number of solutions of the congruence $c_1x_1^k+\cdots+c_sx_s^k \equiv n \pmod{p^\beta}$ is, in the limit and uniformly in n, a positive fraction of the number $(p^\beta)^{s-1}$ of solutions expected on average.

Proposition 4.12. *Let p be a prime and τ the exponent to which p occurs in the prime factorization of k. Put $\gamma = \tau + 1$ if $p > 2$ and $\gamma = \tau + 2$ if $p = 2$. If $m \not\equiv 0 \pmod p$ and $y^k \equiv m \pmod{p^\gamma}$ is solvable, then $x^k \equiv m \pmod{p^\beta}$ is solvable for every $\beta > \gamma$.*

Proof. For the case when $p > 2$ we may choose a primitive root g modulo p^β and express $m \equiv g^\ell \pmod{p^\beta}$ and $x \equiv g^w \pmod{p^\beta}$ with an unknown w. Then

$$g^{kw} \equiv g^\ell \pmod{p^\beta}$$

and so $kw \equiv \ell \pmod{(p-1)p^{\beta-1}}$ by Euler's theorem. By assumption there is some solution y to the congruence $y^k \equiv m \pmod{p^\gamma}$, and since g is also a primitive root modulo p^γ, we may express $y \equiv g^z \pmod{p^\gamma}$. Then $kz \equiv \ell \pmod{(p-1)p^{\gamma-1}}$ in the same way as above, so

$$\gcd(k,(p-1)p^{\gamma-1}) = \gcd(k,(p-1)p^{\tau}) = \gcd(k,(p-1)p^{\beta-1})$$

divides ℓ. Consequently $kw \equiv \ell \pmod{(p-1)p^{\beta-1}}$ has a solution w.

For the case when $p = 2$ we may express $m \equiv (-1)^{\ell_1}5^{\ell_2} \pmod{p^\beta}$ and $x \equiv (-1)^{w_1}5^{w_2} \pmod{p^\beta}$ with unknown w_1, w_2. Then

$$(-1)^{kw_1}5^{kw_2} \equiv (-1)^{\ell_1}5^{\ell_2} \pmod{2^\beta}$$

and so it is sufficient to solve $kw_1 \equiv \ell_1 \pmod 2$ and $kw_2 \equiv \ell_2 \pmod{2^{\beta-2}}$. By assumption there is some solution y to the congruence $y^k \equiv m \pmod{2^\gamma}$, and we may express $y \equiv (-1)^{z_1}5^{z_2} \pmod{2^\gamma}$. The representation of any odd integer as $(-1)^{\ell_1}5^{\ell_2}$ modulo 2^γ is unique in ℓ_1 modulo 2 and in ℓ_2 modulo $2^{\gamma-2}$. Then $kz_1 \equiv \ell_1 \pmod 2$ and $kz_2 \equiv \ell_2 \pmod{2^{\gamma-2}}$ by the uniqueness, and so

$$\gcd(k,2^{\gamma-2}) = \gcd(k,2^{\tau}) = \gcd(k,2^{\beta-2})$$

divides ℓ_2. Consequently $kw_2 \equiv \ell_2 \pmod{2^{\beta-2}}$ has a solution w_2. □

Proposition 4.13. *Let $c \geq 1$ an integer, and $s \geq 2k+1$. Then there is some constant $C(c,k,s) > 0$ so that if the congruence*

$$c_1x_1^k + \cdots + c_sx_s^k \equiv n \pmod{p^\gamma}$$

with $c_1\cdots c_s = c$ has for any prime p and γ as in Proposition 4.12 a solution with not all of $c_1x_1^k, \ldots, c_sx_s^k$ divisible by p, then $\mathfrak{S}(n) \geq C(c,k,s)$.

Proof. Suppose without loss of generality that $x_1 = y_1, \ldots, x_s = y_s$ is a solution with $c_1 y_1^k \not\equiv 0 \pmod{p}$. From this one solution we may obtain at least $(p^{\beta-\gamma})^{s-1}$ solutions of

$$c_1 x_1^k + \cdots + c_s x_s^k \equiv n \pmod{p^\beta}$$

by choosing $x_2, \ldots, x_s$ with $0 < x_j \le p^\beta$ and $x_j \equiv y_j \pmod{p^\beta}$ for $2 \le j \le s$, and solve

$$c_1 x_1^k \equiv n - c_2 x_2^k - \cdots - c_s x_s^k \pmod{p^\beta}$$

for x_1. This is possible by Proposition 4.12 since $c_1 a_1^k \not\equiv 0 \pmod{p}$. Thus $M_n(p^\beta) \ge p^{(\beta-\gamma)(s-1)}$ and so

$$\sum_{\alpha=0}^{\infty} A_n(p^\alpha) = \lim_{\beta \to +\infty} M_n(p^\beta) p^{-\beta(s-1)} \ge p^{-\gamma(s-1)}.$$

Then

$$\begin{aligned} |\mathfrak{S}(n)| &= \left| \prod_{p \le p_0} \sum_{\alpha=0}^{\infty} A_n(p^\alpha) \right| \left| \prod_{p > p_0} \sum_{\alpha=0}^{\infty} A_n(p^\alpha) \right| \\ &\ge \frac{1}{2} \prod_{p \le p_0} p^{-\gamma_{k,p}(s-1)} = C(c,k,s) > 0 \end{aligned}$$

where $p_0 = p_0(c,k,s)$ and $\gamma_{k,p}$ equals the γ defined in Proposition 4.12. □

4.4. Weyl sums

A *Weyl sum* is an exponential sum of the form

$$S = \sum_{m=1}^{N} e(P(m))$$

where P is a polynomial with real coefficients. The sums

$$f(\alpha) = \sum_{m \le N} e(\alpha c m^k)$$

that we need for our application of the circle method are Weyl sums. The terms of a Weyl sum lie on the unit circle, and if they spread out fairly evenly, it is reasonable to hope for a substantial amount of cancellation, so that the modulus of the sum is considerably smaller than the number of terms.

Our estimates for Weyl sums are going to yield a nontrivial bound for the integral

$$\int_{\mathfrak{m}} f_1(\alpha) \cdots f_s(\alpha) e(-n\alpha)\, d\alpha$$

over the minor arcs. This integral contributes to the error term in the asymptotic estimate for $R(n)$, and is difficult to treat compared to the other contributions to the error term.

We start by considering a Weyl sum

$$S = \sum_{m=1}^{N} e(\alpha m + \beta)$$

for a linear polynomial. Clearly

$$\begin{aligned} |S| &= \left| \sum_{m=1}^{N} e(\alpha m) \right| = \left| \frac{e(\alpha) - e((N+1)\alpha)}{1 - e(\alpha)} \right| \\ &\leq \frac{2}{|1 - e(\alpha)|} = \frac{1}{|\sin(\pi\alpha)|} \leq \frac{1}{2\|\alpha\|} \end{aligned}$$

where $\|\alpha\|$ is the distance from α to the nearest integer. Here we use the inequality $|\sin(u)| \geq 2|u|/\pi$, valid for $|u| \leq \pi/2$. Now $|S| \leq \min(N, 1/(2\|\alpha\|))$ follows by the triangle inequality. The calculations

$$\begin{aligned} |S|^2 &= \left| \sum_{m=1}^{N} e(P(m)) \right|^2 = \sum_{n=1}^{N} e(P(n)) \sum_{m=1}^{N} e(-P(m)) \\ &= \sum_{n=1}^{N} \sum_{m=1}^{N} e(P(n) - P(m)) \\ &= \sum_{|h| \leq N-1} \sum_{m=\max(1,1-h)}^{\min(N,N-h)} e(\Delta_h P(m)), \end{aligned}$$

where $\Delta_h P(m) = P(m+h) - P(m)$ is a forward difference, and

$$\begin{aligned} &\sum_{|h| \leq N-1} \sum_{m=\max(1,1-h)}^{\min(N,N-h)} e(\Delta_h P(m)) \\ &\quad = N + \sum_{h=1}^{N-1} \sum_{m=1}^{N-h} e(\Delta_h P(m)) + \sum_{h=1}^{N-1} \sum_{m=1+h}^{N} e(\Delta_h P(m)) \\ &\quad = N + \sum_{h=1}^{N-1} \sum_{m=1}^{N-h} c(\Delta_h P(m)) + \sum_{h=1}^{N-1} \sum_{n=1}^{N-h} e(\Delta_h P(m)) \\ &\quad = N + \sum_{h=1}^{N-1} \sum_{m=1}^{N-h} e(\Delta_h P(m)) + \sum_{h=1}^{N-1} \sum_{m=1}^{N-h} \overline{e(\Delta_h P(m))} \end{aligned}$$

yield

$$|S|^2 = N + 2\,\mathrm{Re}\left(\sum_{h=1}^{N-1}\sum_{m=1}^{N-h} e(\Delta_h P(m))\right).$$

The forward difference $\Delta_h P(m) = P(m+h) - P(m)$ is a polynomial of degree one lower than $P(m)$. The process of bounding the square of a Weyl sum in this way is called *Weyl differencing*. By repeated Weyl differencing any Weyl sum can ultimately be bounded in terms of Weyl sums of linear polynomials, for which we already have a bound.

Proposition 4.14 (Weyl's inequality). *Suppose that P is a real polynomial of degree k whose leading coefficient α has a rational approximation a/q with*

$$\left|\alpha - \frac{a}{q}\right| \leq \frac{c}{q^2},$$

where $q \geq 1$ and $\gcd(q,a) = 1$ and $c > 0$ is a fixed constant. Then

$$\left|\sum_{m=1}^{N} e(P(m))\right| \ll_{c,\varepsilon} N^{1+\varepsilon}\left(N^{-1} + q^{-1} + qN^{-k}\right)^{1/K}$$

for any $\varepsilon > 0$, where $K = 2^{k-1}$.

Proof. Weyl differencing gives

$$|S|^2 \leq N + 2\sum_{h=1}^{N-1}\left|\sum_{m=1}^{N-h} e(\Delta_h P(m))\right|.$$

The sum under the absolute value sign is also a Weyl sum, over an interval $[1, N-h]$ that is contained in the interval $[1, N]$. Extending the notation so that $S_k(P)$ denotes a Weyl sum for a polynomial P of degree k, over some interval of summation contained in $[1, N]$, the last inequality takes the form

$$|S_k(P)|^2 \ll N + \sum_{h=1}^{N-1} |S_{k-1}(\Delta_h P)|.$$

To show that

$$|S_k(P)|^{2^j} \ll N^{2^j-1} + N^{2^j-j-1}\sum_{h_1=1}^{N-1}\cdots\sum_{h_j=1}^{N-1} |S_{k-j}(\Delta_{h_j}\cdots\Delta_{h_1}P)|$$

for each $j \geq 1$, we use an inductive calculation. The last inequality is already established for $j = 1$. Assuming it is valid for some particular value of j, we shall see that it is also valid when this is incremented by 1. For

$$\begin{aligned}|S_k(P)|^{2^{j+1}} &\ll N^{2^{j+1}-2} + N^{2^{j+1}-2j-2}\left(\sum_{h_1=1}^{N-1}\cdots\sum_{h_j=1}^{N-1}|S_{k-j}(\Delta_{h_j}\cdots\Delta_{h_1}P)|\right)^2 \\ &\ll N^{2^{j+1}-2} + N^{2^{j+1}-2j-2}N^j\sum_{h_1=1}^{N-1}\cdots\sum_{h_j=1}^{N-1}|S_{k-j}(\Delta_{h_j}\cdots\Delta_{h_1}P)|^2\end{aligned}$$

by the inequality $(A+B)^2 \leq 2A^2+2B^2$ and the Cauchy-Schwarz inequality. Then

$$\begin{aligned}|S_k(P)|^{2^{j+1}} &\ll N^{2^{j+1}-2} + N^{2^{j+1}-2j-2}N^j \\ &\quad\times\sum_{h_1=1}^{N-1}\cdots\sum_{h_j=1}^{N-1}\left(N+\sum_{h_{j+1}=1}^{N-1}|S_{k-j-1}(\Delta_{h_{j+1}}\cdots\Delta_{h_1}P)|\right) \\ &\ll N^{2^{j+1}-2j-2}N^jN^{j+1} + N^{2^{j+1}-2j-2}N^j \\ &\quad\times\sum_{h_1=1}^{N-1}\cdots\sum_{h_{j+1}=1}^{N-1}|S_{k-j-1}(\Delta_{h_{j+1}}\cdots\Delta_{h_1}P)| \\ &= N^{2^{j+1}-1} + N^{2^{j+1}-(j+1)-1} \\ &\quad\times\sum_{h_1=1}^{N-1}\cdots\sum_{h_{j+1}=1}^{N-1}|S_{k-j-1}(\Delta_{h_{j+1}}\cdots\Delta_{h_1}P)|.\end{aligned}$$

Weyl differencing $j = k-1$ times thus yields

$$|S_k(P)|^K \ll N^{K-1} + N^{K-k}\sum_{h_1=1}^{N-1}\cdots\sum_{h_{k-1}=1}^{N-1}|S_1(\Delta_{h_{k-1}}\cdots\Delta_{h_1}P)|,$$

where $\Delta_{h_{k-1}}\cdots\Delta_{h_1}P$ is now linear. Since $\Delta_h ax^d = dahx^{d-1}+\cdots$, it is clear that $(\Delta_{h_{k-1}}\cdots\Delta_{h_1}P)(m) = k!\alpha h_1\cdots h_{k-1}m+\beta$ where β is some constant. Thus

$$|S_1(\Delta_{h_{k-1}}\cdots\Delta_{h_1}P)| \ll \min\left(N, \frac{1}{\|k!\alpha h_1\cdots h_{k-1}\|}\right)$$

and so

$$|S_k(P)|^K \ll N^{K-1} + N^{K-k}\sum_{h_1=1}^{N-1}\cdots\sum_{h_{k-1}=1}^{N-1}\min\left(N, \frac{1}{\|k!\alpha h_1\cdots h_{k-1}\|}\right).$$

Now

$$\sum_{h_1=1}^{N-1}\cdots\sum_{h_{k-1}=1}^{N-1}\min\left(N,\frac{1}{\|k!\alpha h_1\cdots h_{k-1}\|}\right)$$
$$=\sum_{\ell=1}^{k!(N-1)^{k-1}}\min\left(N,\frac{1}{\|\ell\alpha\|}\right)\sum_{k!h_1\cdots h_{k-1}=\ell}1$$

by collecting terms. But

$$\sum_{k!h_1\cdots h_{k-1}=\ell}1\leq d(\ell)^{k-1}\ll\ell^{K\varepsilon/2}$$

for any $\varepsilon>0$, by Proposition 1.15, since $h_i|\ell$ for $1\leq i\leq k-1$. Thus

$$|S_k(P)|^K\ll N^{K-1}+N^{K-k+K\varepsilon/2}\sum_{\ell=1}^{k!N^{k-1}}\min\left(N,\frac{1}{\|\ell\alpha\|}\right),$$

while

$$\sum_{\ell=1}^{k!N^{k-1}}\min\left(N,\frac{1}{\|\ell\alpha\|}\right)=\sum_{\substack{1\leq\ell\leq k!N^{k-1}\\ \|\alpha\ell\|\leq 1/N}}N+\sum_{j=1}^{[N/2]-1}\sum_{\substack{1\leq\ell\leq k!N^{k-1}\\ j/N<\|\alpha\ell\|\leq(j+1)/N}}\frac{1}{\|\alpha\ell\|}$$
$$\leq N\left(\sum_{\substack{1\leq\ell\leq k!N^{k-1}\\ \|\alpha\ell\|\leq 1/N}}1+\sum_{j=1}^{[N/2]-1}\frac{1}{j}\sum_{\substack{1\leq\ell\leq k!N^{k-1}\\ j/N<\|\alpha\ell\|\leq(j+1)/N}}1\right).$$

We shall bound the number of integers ℓ that satisfy $1\leq\ell\leq k!N^{k-1}$ and $j/N\leq\|\alpha\ell\|\leq(j+1)/N$, for each j with $0\leq j\leq[N/2]-1$. By assumption there is some η with $|\eta|\leq c$ so that $\alpha=a/q+\eta/q^2$ and thus

$$\|\alpha\ell\|=\left\|\left(\frac{a}{q}+\frac{\eta}{q^2}\right)(bq+r)\right\|=\left\|ab+\frac{ar}{q}+\frac{\eta b}{q}+\frac{\eta r}{q^2}\right\|=\left\|\frac{ar}{q}+\frac{\eta b}{q}+\frac{\eta r}{q^2}\right\|$$

by division with remainder $\ell=bq+r$, where $0\leq b\leq k!N^{k-1}/q$ and $0\leq r<q$. It will be enough to find for each b an upper bound for the number of remainders r for which

$$\frac{j}{N}-\frac{c}{q}\leq\left\|\frac{ar}{q}+\frac{\eta b}{q}\right\|\leq\frac{j+1}{N}+\frac{c}{q}.$$

The points $ar/q+\eta b/q$ for $r=0,1,2,\ldots,q-1$ may be translated into the interval $[-1/2,1/2]$ by subtracting suitable integers, and then they form an arithmetic progression of q terms contained in this interval, with difference

$1/q$, since the integers ar constitute a complete collection of residues modulo q. The number of these points contained in the set

$$\left[-\frac{j+1}{N}-\frac{c}{q}, -\frac{j}{N}+\frac{c}{q}\right] \bigcup \left[\frac{j}{N}-\frac{c}{q}, \frac{j+1}{N}+\frac{c}{q}\right]$$

is at most

$$2\left(1+\left(\frac{j+1}{N}+\frac{c}{q}-\left(\frac{j}{N}-\frac{c}{q}\right)\right)\Big/ q^{-1}\right) = 2+4c+\frac{2q}{N}.$$

Thus

$$\begin{aligned}\sum_{\substack{1\le \ell\le k!N^{k-1}\\ j/N<\|\alpha\ell\|\le (j+1)/N}} 1 &\le \left(1+\frac{k!N^{k-1}}{q}\right)\left(2+4c+\frac{2q}{N}\right)\\ &\ll \frac{q}{N}+\frac{N^{k-1}}{q}+N^{k-2}\end{aligned}$$

and so

$$\begin{aligned}\sum_{\ell=1}^{k!N^{k-1}} \min\left(N, \frac{1}{\|\ell\alpha\|}\right) &\le N\left(\frac{q}{N}+\frac{N^{k-1}}{q}+N^{k-2}\right)\left(1+\sum_{j=1}^{[N/2]-1}\frac{1}{j}\right)\\ &\ll \left(q+\frac{N^k}{q}+N^{k-1}\right)N^{K\varepsilon/2}.\end{aligned}$$

Then

$$\begin{aligned}|S_k(P)|^K &\ll N^{K-1}+N^{K-k+K\varepsilon/2}\left(q+\frac{N^k}{q}+N^{k-1}\right)N^{K\varepsilon/2}\\ &\ll N^{K+K\varepsilon}\left(\frac{1}{N}+\frac{1}{q}+\frac{q}{N^k}\right)\end{aligned}$$

completes the proof. □

A more specialized result applying to the sums

$$f(\alpha)=\sum_{m\le N} e(\alpha c m^k)$$

permits us to treat the minor arcs more successfully than we could achieve by Weyl's inequality alone. This improvement is due to L. K. Hua.

Proposition 4.15 (Hua's lemma). *The inequality*

$$\int_0^1 |f(\alpha)|^{2^j}\, d\alpha \ll N^{2^j-j+\varepsilon}$$

holds for any $\varepsilon>0$ *and* $j=1,2,\ldots,k$.

Proof. We first reduce to the case $c = 1$ by the calculation

$$\int_0^1 |f(\alpha)|^{2^j}\, d\alpha = \int_0^1 \left|\sum_{m\le N} e(\alpha c m^k)\right|^{2^j} d\alpha$$
$$= \int_0^c \left|\sum_{m\le N} e(\beta m^k)\right|^{2^j} c^{-1}\, d\beta = \int_0^1 \left|\sum_{m\le N} e(\beta m^k)\right|^{2^j} d\beta,$$

using the change of variable $\beta = \alpha c$ and the periodicity of $e(x)$.

The proof now goes by induction on j. The calculation

$$\int_0^1 |f(\alpha)|^2\, d\alpha = \int_0^1 \left(\sum_{m\le N} e(\alpha m^k)\right)\overline{\left(\sum_{n\le N} e(\alpha n^k)\right)}\, d\alpha$$
$$= \sum_{m\le N}\sum_{n\le N}\int_0^1 e(\alpha(m^k - n^k))\, d\alpha = N$$

establishes the inequality for $j = 1$. The inequality

$$|S_k(P)|^{2^j} \ll N^{2^j-1} + N^{2^j-j-1}\sum_{h_1=1}^{N-1}\cdots\sum_{h_j=1}^{N-1} |S_{k-j}(\Delta_{h_j}\cdots\Delta_{h_1}P)|$$

from the previous proof implies

$$|S_k(P)|^{2^j} \ll N^{2^j-2} + N^{2^j-2j}N^{j-1}\sum_{h_1=1}^{N-1}\cdots\sum_{h_{j-1}=1}^{N-1} |S_{k-(j-1)}(\Delta_{h_{j-1}}\cdots\Delta_{h_1}P)|^2$$

by decrementing j by one, squaring and applying $(A+B)^2 \le 2A^2 + 2B^2$ and the Cauchy-Schwarz inequality. Now

$$|S_{k-(j-1)}(\Delta_{h_{j-1}}\cdots\Delta_{h_1}P)|^2 \ll N + \sum_{h_j=1}^{N-1} \mathrm{Re}\left(S_{k-j}(\Delta_{h_j}\cdots\Delta_{h_1}P)\right)$$

by Weyl differencing, and so

$$|S_k(P)|^{2^j} \ll N^{2^j-1} + N^{2^j-j-1}\sum_{h_1=1}^{N-1}\cdots\sum_{h_j=1}^{N-1} \mathrm{Re}\left(S_{k-j}(\Delta_{h_j}\cdots\Delta_{h_1}P)\right).$$

In the present case $P(x) = \alpha x^k$ and $S_k(P) = f(\alpha)$. Multiplying by $|f(\alpha)|^{2^j}$ and integrating on both sides of the inequality yields

$$\int_0^1 |f(\alpha)|^{2^{j+1}}\, d\alpha \ll N^{2^j-1}\int_0^1 |f(\alpha)|^{2^j}\, d\alpha + N^{2^j-j-1}$$
$$\times \sum_{h_1=1}^{N-1}\cdots\sum_{h_j=1}^{N-1}\operatorname{Re}\left(\int_0^1 S_{k-j}(\Delta_{h_j}\cdots\Delta_{h_1}P)|f(\alpha)|^{2^j}\, d\alpha\right).$$

The Binomial Theorem implies that

$$\Delta_h x^\ell = hq(x;h)$$

with a polynomial q of degree $\ell - 1$ in x with coefficients that are positive integers when h is a positive integer. Thus

$$\Delta_{h_j}\cdots\Delta_{h_1}\alpha x^k = \alpha h_1\cdots h_j p_j(x; h_1,\ldots,h_j)$$

where p_j is a polynomial of degree $k-j$ with coefficients that are positive integers when $h_1,\ldots,h_j$ are positive integers. Now

$$S_{k-j}(\Delta_{h_j}\cdots\Delta_{h_1}\alpha x^k) = \sum_m e(\alpha h_1\cdots h_j p_j(m;h_1,\ldots,h_j))$$

where the summation is over an integer subinterval of $[1,N]$, and $|f(\alpha)|^{2^j}$ equals

$$\sum_{m_1=1}^{N}\cdots\sum_{m_{2^j-1}=1}^{N}\sum_{n_1=1}^{N}\cdots\sum_{n_{2^j-1}=1}^{N} e(\alpha(m_1^k+\cdots+m_{2^{j-1}}^k - n_1^k - \cdots - n_{2^{j-1}}^k)),$$

so for fixed integers $1 \le h_1,\ldots,h_j \le N$ the integral

$$\int_0^1 S_{k-j}(\Delta_{h_j}\cdots\Delta_{h_1}P)|f(\alpha)|^{2^j}\, d\alpha$$

is an integer bounded above by the number of solutions of

$$h_1\cdots h_j p_j(m,h_1,\ldots,h_j) = n_1^k+\cdots+n_{2^{j-1}}^k - m_1^k - \cdots - m_{2^{j-1}}^k$$

in integers $1 \le m, m_1,\ldots,m_{2^{j-1}}, n_1,\ldots,n_{2^{j-1}} \le N$. To estimate the number of solutions, we note that p_j is a positive and increasing function of m in the interval $[1,N]$ since all the coefficients are positive. In particular both sides of the equation are positive, and if all the other unknowns in the equation are fixed, then there is at most one value for m that fits. There are no more than $O(N^{2^j})$ admissible values for the unknowns $n_1,\ldots,m_{2^{j-1}}$. The integers $h_1,\ldots,h_j$ are divisors of the right-hand side

$$r = n_1^k+\cdots+n_{2^{j-1}}^k - m_1^k - \cdots - m_{2^{j-1}}^k$$

of the equation. Thus there are no more than $d(r)^j$ possibilities for these divisors, where $r \leq 2^{j-1}N^k$. Then

$$\sum_{h_1=1}^{N-1} \cdots \sum_{h_j=1}^{N-1} \mathrm{Re}\left(\int_0^1 S_{k-j}(\Delta_{h_j}\cdots\Delta_{h_1}P)|f(\alpha)|^{2^j}\,d\alpha\right)$$
$$\ll N^{2^j}((2^{j-1}N^k)^{\varepsilon/jk})^j \ll N^{2^j+\varepsilon}$$

by Proposition 1.15. Applying

$$\int_0^1 |f(\alpha)|^{2^j}\,d\alpha \ll N^{2^j-j+\varepsilon}$$

in the calculation

$$\begin{aligned}\int_0^1 |f(\alpha)|^{2^{j+1}}\,d\alpha &\ll N^{2^j-1}\int_0^1 |f(\alpha)|^{2^j}\,d\alpha \\ &\quad + N^{2^j-j-1}N^{2^j+\varepsilon}\int_0^1 |f(\alpha)|^{2^{j+1}}\,d\alpha \\ &\ll N^{2^j-1}\int_0^1 |f(\alpha)|^{2^j}\,d\alpha + N^{2^j-j-1}N^{2^j+\varepsilon} \\ &\ll N^{2^j-1}N^{2^j-j+\varepsilon} + N^{2^{j+1}-(j+1)+\varepsilon} \ll N^{2^{j+1}-(j+1)+\varepsilon}\end{aligned}$$

establishes the desired inequality inductively. □

4.5. An asymptotic estimate

We require information about the approximation of reals by rationals. It will be enough to prove the following theorem on Diophantine approximation.

Proposition 4.16 (Dirichlet approximation theorem). *For any real numbers $\alpha_1, \ldots, \alpha_l$ and any integer $X \geq 2$ there are integers $a_1, \ldots, a_l$ and q with $1 \leq q < X$ and $|q\alpha_j - a_j| \leq X^{-1/l}$ for $1 \leq j \leq l$.*

Proof. Consider the torus $T = \mathbb{R}^l/\mathbb{Z}^l$ with the distance

$$d(\mathbf{x}+\mathbb{Z}^l, \mathbf{y}+\mathbb{Z}^l) = \max_{1\leq j\leq l}\min_{a\in\mathbb{Z}} |x_j - y_j - a|.$$

For arbitrary $\delta > 0$ let B_r be the open d-ball in T with center $(r\alpha_1, \ldots, r\alpha_l) + \mathbb{Z}^l$ and radius $1/(2\delta)$. The torus T inherits a concept of volume from $\mathbb{R}^l$, and if the B_r are pairwise disjoint then

$$X\delta^l = X\mathrm{vol}(B_0) = \sum_{r=0}^{X-1} \mathrm{vol}(B_r) \leq \mathrm{vol}(T) = 1.$$

Thus the B_r cannot be disjoint if $\delta > X^{-1/l}$, and so there are integers r, s with $0 \leq r < s \leq X-1$ and a_j for $1 \leq j \leq l$ such that $|s\alpha_j - r\alpha_j - a_j| \leq X^{-1/l}$ for $1 \leq j \leq l$. Choose $q = s - r$. □

Proposition 4.17. *For any integers $k \geq 3$ and $s \geq 2^k + 1$ there exist some $\delta' > 0$ so that*

$$R(n) = \frac{C_{k,s}}{(c_1 \cdots c_s)^{1/k}} n^{s/k-1} \mathfrak{S}(n) + O(n^{s/k-1-\delta'})$$

as $n \to +\infty$.

Proof. The estimate

$$\sum_{q=Q+1}^{\infty} |A_n(q)| \ll \sum_{q=Q+1}^{\infty} q^{1-s/k} \leq \int_Q^{\infty} y^{1-s/k}\, dy \ll Q^{2-s/k}$$

follows from Proposition 4.9 when $s \geq 2k+1$, and is uniform in n. Now

$$\begin{aligned}
\int_{\mathfrak{M}} f_1(\alpha) \cdots f_s(\alpha) e(-n\alpha)\, d\alpha &= (\mathfrak{S}(n) + O(Q^{2-s/k})) \\
&\quad \times \left(\frac{C_{k,s}}{(c_1 \cdots c_s)^{1/k}} n^{s/k-1} + O(\rho^{1-s/k}) \right) \\
&\quad + O(n^{(s-1)/k} \rho^2 Q^3 n + \rho^{s+1} Q^{s+2} n^s) \\
&= \frac{C_{k,s}}{(c_1 \cdots c_s)^{1/k}} n^{s/k-1} \mathfrak{S}(n) + O(n^{(s-1)/k} \rho^2 Q^3 n \\
&\quad + \rho^{s+1} Q^{s+2} n^s + \rho^{1-s/k} + n^{s/k-1} Q^{2-s/k})
\end{aligned}$$

because $Q \geq 1$.

For $0 < \delta < 1/3$ choose $\rho = n^{\delta-1}$ and $Q = [n^\delta]$. Then the two conditions $n\rho \geq 1$ and $2\rho Q^2 < 1$ of Proposition 4.4 are satisfied for all n sufficiently large. Moreover

$$\begin{aligned}
\int_{\mathfrak{M}} f_1(\alpha) \cdots f_s(\alpha) e(-\alpha n)\, d\alpha &= \frac{C_{k,s}}{(c_1 \cdots c_s)^{1/k}} n^{s/k-1} \mathfrak{S}(n) \\
&\quad + O(n^{s/k-1+5\delta-1/k} + n^{2s\delta+3\delta-1} + n^{s/k-1-(s/k-2)\delta}) \\
&= \frac{C_{k,s}}{(c_1 \cdots c_s)^{1/k}} n^{s/k-1} \mathfrak{S}(n) + O(n^{s/k-1-\delta'})
\end{aligned}$$

for some $\delta' > 0$, by choosing δ small enough depending on k and s.

It remains to establish a bound of the form

$$\left| \int_{\mathfrak{m}} f_1(\alpha) \cdots f_s(\alpha) e(-\alpha n)\, d\alpha \right| \ll n^{s/k-1-\delta'}$$

for the integral over the minor arcs. Now

$$\begin{aligned}
&\int_{\mathfrak{m}} |f_1(\alpha) \cdots f_s(\alpha)|\, d\alpha \\
&\qquad \leq \max_{\alpha \in \mathfrak{m}} |f_{2^k+1}(\alpha)| \cdots \max_{\alpha \in \mathfrak{m}} |f_s(\alpha)| \int_0^1 |f_1(\alpha)| \cdots |f_{2^k}(\alpha)|\, d\alpha.
\end{aligned}$$

Nothing significant is lost in this bound by replacing the domain $\mathfrak{m}$ of integration by $[0,1]$. The major arcs make the main contribution to $R(n)$, but when n is large their total length $O(\rho Q^2) = O(n^{-1+3\delta})$ is minuscule.

The inequality

$$\begin{aligned}\int_0^1 |f_1(\alpha)|\cdots|f_{2^k}(\alpha)|\,d\alpha &\\ &\leq \left(\int_0^1 |f_1(\alpha)|^{2^k}\,d\alpha\right)^{1/2^k} \cdots \left(\int_0^1 |f_{2^k}(\alpha)|^{2^k}\,d\alpha\right)^{1/2^k}\\ &\ll (P_1^{2^k-k+\varepsilon})^{1/2^k}\cdots(P_{2^k}^{2^k-k+\varepsilon})^{1/2^k} \ll n^{2^k/k-1+\varepsilon/k}\end{aligned}$$

is a consequence of Hölder's inequality and Hua's lemma.

Let α be an arbitrary point on the minor arcs $\mathfrak{m}$. Choosing $X = \rho^{-1}$ in Dirichlet's theorem on Diophantine approximation, we obtain

$$\left|\alpha - \frac{a}{q}\right| \leq \frac{1}{q\rho^{-1}}$$

with a and q integers and $1 \leq q \leq \rho^{-1}$. We may without loss of generality assume that $\gcd(q,a) = 1$. For if d is the greatest common divisor of a and q, then a may be replaced by a/d and q by q/d in the inequality, since q/d is no larger than q.

Recalling that $\mathfrak{U} = (\rho, 1+\rho]$ and noting that $\mathfrak{M}(1,1) = [1-\rho, 1+\rho]$ one sees that $\mathfrak{m} \subseteq (\rho, 1-\rho)$ and so $1 \leq a \leq q$. Now

$$\left|\alpha - \frac{a}{q}\right| \leq \frac{\rho}{q}$$

with $1 \leq a \leq q$ implies that $q > Q$ since otherwise $\alpha \in \mathfrak{M}$ by $q \geq 1$.

We apply Weyl's inequality to bound f_j on the minor arcs. The inequality

$$\left|\alpha - \frac{a}{q}\right| \leq \frac{1}{q^2}$$

is a consequence of $q \leq \rho^{-1}$. Multiplying through by c_j yields

$$\left|\alpha c_j - \frac{ac_j}{q}\right| \leq \frac{c_j}{q^2}.$$

Here a and q are coprime, but the greatest common divisor d_j of c_j and q may be larger than one. Then

$$\left|\alpha c_j - \frac{a'}{q'}\right| \leq \frac{c_j d_j^{-2}}{q'^2} \leq \frac{c_j}{q'^2}.$$

with $\gcd(q', a') = 1$, after putting $a' = ac_j/d_j$ and $q' = q/d_j$. Weyl's inequality yields

$$\begin{aligned}|f_j(\alpha)| &\ll ((n/c_j)^{1/k})^{1+\varepsilon}(((n/c_j)^{1/k})^{-1} \\ &\qquad + (q/d_j)^{-1} + (q/d_j)((n/c_j)^{1/k})^{-k})^{1/K} \\ &\ll n^{1/k+\varepsilon/k}(n^{-1/k} + q^{-1} + qn^{-1})^{1/K} \ll n^{1/k+\varepsilon/k-\delta/K},\end{aligned}$$

for $[n^\delta] = Q < q \le \rho^{-1} = n^{1-\delta}$. Since

$$\begin{aligned}&\left|\int_{\mathfrak{m}} f_1(\alpha)\cdots f_s(\alpha)\,d\alpha\right| \\ &\quad \le \max_{\alpha\in\mathfrak{m}} |f_{2^k+1}(\alpha)|\cdots \max_{\alpha\in\mathfrak{m}} |f_s(\alpha)| \int_0^1 |f_1(\alpha)|\cdots|f_{2^k}(\alpha)|\,d\alpha \\ &\quad \ll (n^{1/k+\varepsilon/k-\delta/K})^{s-2^k} n^{2^k/k-1+\varepsilon/k} \\ &\quad = n^{s/k-1+\varepsilon(s+1-2^k)/k-\delta(s-2^k)/K},\end{aligned}$$

and ε may be chosen as small as we wish, while $s > 2^k$, the inequality holds with some $\delta' > 0$. □

To complete our treatment of the singular series, we need to show that under suitable conditions the congruence $c_1x_1^k + \cdots + c_sx_s^k \equiv n \pmod{p^\gamma}$ has a solution with not all of $c_1x_1^k, \ldots, c_sx_s^k$ divisible by p. For this purpose we prove a result on the addition of residue classes. Define

$$A + B = \{\,\underline{a} + \underline{b} \in \mathbb{Z}/q\mathbb{Z} \mid \underline{a} \in A\,,\ \underline{b} \in B\,\}$$

for any subsets $A, B \subseteq \mathbb{Z}/q\mathbb{Z}$.

Proposition 4.18 (Cauchy-Davenport-Chowla theorem). *Let $A, B \subseteq \mathbb{Z}/q\mathbb{Z}$ with $\underline{0} \in B$ and $\gcd(q, \underline{b}) = 1$ for $\underline{0} \neq \underline{b} \in B$. If $A + B \neq \mathbb{Z}/q\mathbb{Z}$, then $|A + B| \ge |A| + |B| - 1$.*

Proof. By induction on $|B|$. The inequality is trivial for $|B| = 1$. If $|B| \ge 2$ then there exists some $c \in A$ such that $\{c\} + B \not\subseteq A$. Otherwise $A + B = A$, and then

$$\sum_{\underline{a}\in A} (\underline{a} + \underline{b}) = \sum_{\underline{a}\in A} \underline{a},$$

for some $\underline{b} \in B$, so $b|A| \equiv 0 \pmod{q}$. Thus $|A| = q$, which contradicts $A + B \neq \mathbb{Z}/q\mathbb{Z}$.

Define $C = \{\underline{b} \in B \mid \underline{c} + \underline{b} \notin A\}$, $A_1 = A \cup (\{c\} + C)$ and $B_1 = B \setminus C$. Note that $\underline{0} \notin C$ since $\underline{c} \in A$. Moreover,

$$\begin{aligned}A_1 + B_1 &= (A + B_1) \cup ((\{c\} + C) + B_1) \\ &= (A + B_1) \cup ((\{c\} + B_1) + C) \subseteq A + B \neq \mathbb{Z}/q\mathbb{Z}\end{aligned}$$

since $\{\underline{c}\} + B_1 \subseteq A$ by the definition of C. Thus A_1 and B_1 satisfy the same conditions as A and B.

Now $|B_1| = |B| - |C| < |B|$ and so $|A_1| + |B_1| \geq |A_1| + |B_1| - 1$ by the induction hypothesis. Then

$$\begin{aligned}|A+B| &\geq |A_1 + B_1| \geq |A_1| + |B_1| - 1 \\ &= |A_1| - |C| + |B_1| + |C| - 1 = |A| + |B| - 1,\end{aligned}$$

as $|A_1| = |A| + |C|$, since no element of $\{c\} + B$ is contained in A. □

Proposition 4.19. *Suppose that p and γ are as in Proposition 4.12. If $s \geq k^2 + 1$ and the integers $c_1, \ldots, c_s$ are pairwise coprime, then the congruence*

$$c_1 x_1^k + \cdots + c_s x_s^k \equiv n \pmod{p^\gamma}$$

has a solution with not all the terms $c_j x_j^k$ divisible by p.

Proof. Assume without loss of generality that $c_j \not\equiv 0 \pmod p$ for $2 \leq j \leq s$. Choosing $x_s = 1$, it will be enough to show that $\Sigma = c_2 x_2^k + \cdots + c_{s-1} x_{s-1}^k$ takes all values modulo p^γ.

First suppose that p is an odd prime and that $\gamma = 1$. Then Σ takes all values modulo p if

$$(k-1)^{s-2} p^{(s-2)/2} < p^{s-3}$$

by Proposition 4.2. Thus we are free to suppose that

$$s - 1 \geq k^2 \geq (k-1)^{2\frac{k^2-1}{k^2-3}} \geq (k-1)^{2\frac{s-2}{s-4}} \geq p.$$

Applying the Cauchy-Davenport-Chowla theorem repeatedly with the sets $B = \{0, c_2\}, \ldots, \{0, c_{s-1}\}$ modulo p, one sees that Σ takes all values modulo p.

If $\gamma \geq 2$ and p is odd, then $s - 1 \geq k^2 \geq p^{2\tau} \geq p^\gamma$, so Σ takes all values modulo p^γ as above.

If $p = 2$, then $s-1 \geq k^2 \geq 9$ and the Cauchy-Davenport-Chowla theorem implies that Σ takes all values modulo 2^γ for $\gamma \leq 3$. While if $\gamma \geq 4$, then $\tau \geq 2$ and thus $s - 1 \geq k^2 \geq 2^{2\tau} \geq 2^\gamma$, so again Σ takes all values modulo 2^γ. □

The sufficient condition $s \geq k^2 + 1$ is best possible for $k = 4$, but for some other values of k it may be improved upon.

Proposition 4.20. *For any integers $k \geq 3$ and $s \geq 2^k + 1$ and pairwise coprime positive integers $c_1, \ldots, c_s$ the number $R(n)$ of solutions of*

$$c_1 x_1^k + \cdots + c_s x_s^k = n$$

in positive integers $x_1, \ldots, x_s$ satisfies $R(n) \to +\infty$ as $n \to +\infty$.

Proof. If $k = 3$, then $\Sigma = c_2 x_2^3 + \cdots + c_{s-1} x_{s-1}^3$ takes all values modulo 9 with $s = 9$, by the Cauchy-Davenport-Chowla theorem applied repeatedly with the sets $B = \{0, c_2, -c_2\}, \{0, c_3\} \ldots, \{0, c_8\}$, as in the proof of Proposition 4.19. The inequality $2^k + 1 \geq k^2 + 1$ holds for all $k \geq 4$, and so $\mathfrak{S}(n) \geq C(c, k, s) > 0$ for every choice of $k \geq 3$ and pairwise coprime positive integers $c_1, \ldots, c_s$ if $s \geq 2^k + 1$, by Proposition 4.13 and Proposition 4.19. Then $R(n) \to +\infty$ as $n \to +\infty$ by Proposition 4.17. □

Specializing the previous result, we obtain the solution to Waring's Problem.

Proposition 4.21 (Hilbert-Waring theorem). *We have $g(k) < \infty$ for all $k \geq 2$.*

Proof. Choosing $c_1 = \cdots = c_s = 1$ we see from Proposition 4.20 that $G(k) \leq 2^k + 1$ for $k \geq 3$. Then $g(k)$ is finite for all $k \geq 3$ by its definition. Since any biquadrate is a square, $g(2) \leq g(4)$ so $g(2)$ is also finite. □

The present exposition purposely ignored the case $k = 2$. There is an introduction to the Circle Method exemplified with diagonal quadratic forms in the notes *Analytic methods for the distribution of rational points on algebraic varieties* by D. R. Heath-Brown in the proceedings of the 2005 NATO Advanced Study Institute on Equidistribution in Number Theory.

The notes of lectures that H. Davenport gave at the University of Michigan on applications of the Circle Method to Diophantine problems have been reissued: *Analytic Methods for Diophantine Equations and Diophantine Inequalities*, edited by T. D. Browning and with a Foreword by D. E. Freeman, D. R. Heath-Brown and R. C. Vaughan. The present exposition is based mainly on Davenport's treatment in those notes, though in some particulars it follows the monograph *The Hardy-Littlewood Method* by R. C. Vaughan. The second edition of the latter work is the standard reference on the Circle Method.

The most up to date of Vinogradov's expositions of the Circle Method is available in English translation in his *Selected Works*.

There are further applications of the Circle Method to Diophantine problems in *Diophantine Inequalities* by R. C. Baker, and in *Analytische Methoden für diophantische Gleichungen* by W. M. Schmidt.

There are other techniques for counting solutions of Diophantine equations beside the Circle Method; see *Quantitative Arithmetic of Projective Varieties* by T. D. Browning, and the notes *Counting rational points on algebraic varieties* by D. R. Heath-Brown in the proceedings of the 2002 C.I.M.E. Summer School on Analytic Number Theory.

4.6. Notes

A connection between an exponential sum and the number of solutions of a congruence is seen already in article 358 of the *Disquisitiones* of Gauss. For the case of a general polynomial in an arbitrary number of variables the identity was noted in 1832 by G. Libri, Count of Sommaia [**Lib32**]. His career later took a very unfortunate turn; see his biography *The Life and Times of Guglielmo Libri* [**RM95**] by P. A. Maccioni Ruju and M. Mostert for details of this.

Hardy and Littlewood [**HL22b**] used Gauss sums to count the number of solutions of diagonal congruences. Weakening the more general version of their result given by A. Weil [**Wei49**] yields Propositions 4.1 and 4.2. Weil has historical remarks on counts of solutions to congruences, and his paper is notable for the statement of the celebrated *Weil conjectures* on varieties over finite fields.

The Circle Method originated in papers [**HR17b, HR18**] of Hardy and Ramanujan on the partition function $p(n)$ that counts the number of ways of writing n as a sum of positive integers without regard to order. The generating function

$$f(z) = \sum_{n=0}^{\infty} p(n) z^n$$

has the infinite product representation

$$f(z) = \prod_{m=1}^{\infty} (1 - z^m)^{-1}.$$

The latter shows that $f(z)$ is holomorphic in $|z| < 1$, and in particular the generating power series of $p(n)$ converges in the unit disk. It is easy to see that every root of unity is a singularity of $f(z)$, so the unit circle is a natural boundary. But it is not an undifferentiated mass of singularities; they have a certain individuality. In particular $z = 1$ is the heaviest singularity in the sense that on a circle $|z| = r$ with $r < 1$ close to 1, the modulus $|f(z)|$ attains its largest values near $z = 1$. By an application of the Cauchy integral formula

$$p(n) = \frac{1}{2\pi i} \oint_{|z|=r} f(z) z^{-n-1}\, dz$$

Hardy and Ramanujan found the asymptotic estimate

$$p(n) \sim \frac{e^{\pi\sqrt{2n/3}}}{4n\sqrt{3}}.$$

They achieved this by expressing $f(z)$ in terms of the Dedekind eta function from classical analysis. The behavior of the latter is precisely known through its functional equation, and based on this information they modeled $f(z)$ by a much simpler function $F(z)$ with asymptotically the same singularity at $z = 1$. The Cauchy integral formula with $F(z)$ instead of $f(z)$ yields the asymptotic approximation to $p(n)$, while the contribution from the Cauchy integral formula with $f(z) - F(z)$ turns out to be of lower order because $z = 1$ is the heaviest singularity.

Hardy and Ramanujan went on to take into account the contributions from the singularities at the other roots of unity, and obtained an asymptotic expansion for $p(n)$ with the above asymptotic estimate as its first term. For this asymptotic expansion they were able to bound the truncation error and discovered that for all n sufficiently large, the error can be made less than 0.5 by truncating the expansion suitably. Since $p(n)$ is necessarily an integer, this is a curious instance of exact numerics. Later H. Rademacher [**Rad38**] obtained a rapidly convergent series for $p(n)$ by modifying the argument of Hardy and Ramanujan. There is an exposition of this result, including the requisite theory of the Dedekind eta function, in Rademacher's treatise *Topics in Analytic Number Theory* [**Rad73**].

Hardy and Ramanujan also applied their method to count representations by a fixed number of squares. The generating function

$$g(z) = \sum_{n=-\infty}^{\infty} z^{n^2}$$

of the squares of the integers is related to an elliptic theta function, which has a functional equation, and whose behavior is precisely known. By applying the Cauchy integral formula to $g(z)^s$ for positive integers $s \geq 5$, Hardy and Ramanujan obtained estimates for the number of representations of integers as sums of five or more squares. The Circle Method in their form fails to establish Lagrange's theorem that every positive integer is a sum of four squares. Kloosterman developed a refinement of the Circle Method, adapted it to such problems, and proved results on the number of representations of positive integers by positive definite diagonal quaternary quadratic forms [**Klo26**]. In particular he obtained Lagrange's theorem where the original version of the Circle Method had failed.

Hardy and Littlewood [**HL20a, HL20b**] applied the Circle Method to Waring's Problem. However, the generating function

$$h_k(z) = \sum_{n=0}^{\infty} z^{n^k}$$

of the nonnegative k-th powers is not related to any well-studied function from classical analysis when $k \geq 3$, and no functional equation is known for it. Less powerful tools, such as the Euler-Maclaurin summation formula, only give good information on a collection of short arcs on $|z| = r$ determined by roots of unity of low order. For problems where strong analytic information on the integrand is not available Hardy and Littlewood improved on the original version of the circle method by splitting the circle $|z| = r$ into two sets; the *major arcs* corresponding to roots of unity of low order, and the complement, called the *minor arcs*. Integration over the major arcs gave a main term together with an error term that they could handle well by choosing the major arcs suitably. This left the harder problem of bounding the contribution to the final error term from the integral over the minor arcs. In the case of Waring's Problem Hardy and Littlewood adapted the method of Weyl for bounding exponential sums to this purpose.

Hardy and Littlewood [**HL22a**] also applied the Circle Method to the binary and ternary Goldbach problems; to show that every even integer $n \geq 4$ is the sum of two primes and that every odd integer $n \geq 7$ is the sum of three primes, respectively. The ternary Goldbach problem is easier than the binary one; this is plausible on the face of it by $n = 3 + (n-3)$, though there is more to it than that. When attempting the ternary Goldbach problem, Hardy and Littlewood were faced with a serious obstacle; they possessed no technique for nontrivially bounding exponential sums over primes such as

$$f(\alpha) = \sum_{p \le n} e(\alpha p).$$

A nontrivial bound on the minor arcs for some such sum is needed to benefit from a decomposition into major and minor arcs in the ternary Goldbach problem, and so they were forced to work with major arcs alone. Approximating α by a/q on the major arcs, Hardy and Littlewood required a very strong estimate on primes in arithmetic progressions modulo q, and obtained this estimate conditionally on the conjecture that $L(s, \chi) \neq 0$ for all L-functions, in some half plane $\sigma > 3/4 - \delta$ with $\delta > 0$. This yielded the conditional asymptotic relation

$$R(n) \sim \frac{Cn^2}{\log^3(n)} \prod_{p|n} \frac{(p-1)(p-2)}{1+(p-1)(p-2)}, \qquad C = \prod_{p \ge 3} \left(1 + \frac{1}{(p-1)^3}\right)$$

for the number $R(n)$ of representations of a large odd integer n as a sum of three primes.

Hardy and Littlewood were unable to establish the binary Goldbach conjecture by the Circle Method even on a conditional basis. But applying the method formally, without an estimate for the error, they obtained a hypothetical main term and thus a conjectured asymptotic relation

$$R(n) \sim \frac{Cn}{\log^2(n)} \prod_{2 \neq p|n} \frac{p-1}{p-2}, \qquad C = 2\prod_{p \ge 3} \left(1 - \frac{1}{(p-1)^2}\right)$$

for the number $R(n)$ of representations of a large even integer n as a sum of two primes.

In 1937 I. M. Vinogradov [**Vin37**] solved the ternary Goldbach problem without relying on any unproved conjecture. In 1928 he had given an easier treatment [**Vin28**] of Waring's Problem by the Circle Method, in which infinite power series were replaced by finite exponential sums as generating functions. This significantly simplifies the arguments, and in the absence of good analytic information about the power series close to the circle of convergence, nothing is lost thereby. Now the power series

$$\sum_p z^p \quad \text{or} \quad \sum_p \log(p) z^p \quad \text{or} \quad \sum_n \Lambda(n) z^n$$

that are appropriate for additive problems with primes all have the unit circle as a natural boundary, but no information about their asymptotic behavior there is available beyond what can be extracted from the theory of the distribution of primes itself. Vinogradov attacked the ternary Goldbach problem using exponential sums rather than power series as generating functions and expressed the number $R(n)$ of representations $n = p_1 + p_2 + p_3$ of a large odd integer n as a sum of three primes by the integral

$$R(n) = \int_0^1 f(\alpha)^3 e(-\alpha n)\, d\alpha = \int_0^1 \left(\sum_{p \le n} e(\alpha p)\right)^3 e(-\alpha n)\, d\alpha.$$

Defining a collection of suitably short "major arcs" intervals centered around rationals a/q with small denominators q, and using information on the distribution of primes in arithmetic progressions due to A. Page [**Pag35**], C. L. Siegel [**Sie35**], and A. Z. Walfisz [**Wal36**], Vinogradov obtained the main term of $R(n)$ and an associated error term by integration over the major arcs. He bounded the integral over the minor arcs (the complement in $[0,1]$ of the major arcs) $\mathfrak{m}$ by

$$\begin{aligned}\left|\int_{\mathfrak{m}} f(\alpha)^3 e(-\alpha n)\, d\alpha\right| &\leq \int_{\mathfrak{m}} |f(\alpha)|^3\, d\alpha \\ &\leq \int_0^1 |f(\alpha)|^2\, d\alpha \max_{\alpha\in\mathfrak{m}} |f(\alpha)| = \pi(n) \max_{\alpha\in\mathfrak{m}} |f(\alpha)|\end{aligned}$$

and proved that the main term dominated the sum of the error term from the integral over the major arcs and the integral over the minor arcs. The key advance in his approach, more important than the use of exponential sums as generating functions, was a good bound for $|f(\alpha)|$ on the minor arcs $\mathfrak{m}$. Vinogradov applied the underlying idea of the sieve of Eratosthenes-Legendre to the exponential sum $f(\alpha)$, breaking it up in a complicated way into sums that were easier to bound on $\mathfrak{m}$. The final result was an asymptotic estimate for $R(n)$ with the conditional asymptotic relation of Hardy and Littlewood as the main term, plus an error term that grows more slowly. Thus the ternary Goldbach conjecture holds for all sufficiently large odd integers.

Proposition 4.3 occurs as Lemma 2 in the paper [**Vin28**] of Vinogradov. Proposition 4.4 is a version of the evaluation of the integral over the major arcs from Section 9 of the paper [**HL20b**] of Hardy and Littlewood. Naturally there are differences in detail, since they worked with power series rather than with exponential sums. Stronger versions of Proposition 4.5 are implicit in [**HL20b**] and [**Vin28**], but our version is obtained by weakening the evaluation of the integral from the paper [**Lan30**] of Landau. In order to avoid both the gamma function and Fourier transforms, while keeping the advantages of choosing $v(\beta)$ to be an integral, the constant $C_{k,s}$ is not calculated, but only shown to be positive. See the discussion on page 23 of [**Dav05**] and on page 15 of [**Vau97**]. The classical definite integral

$$\int_{-\infty}^{\infty} e^{-a^2x^2} \cos(bx)\, dx = \frac{\sqrt{\pi}}{a} e^{-b^2/(4a^2)}$$

may be evaluated by calculus, as in an exercise following this chapter. But it is quicker to apply Cauchy's theorem to the integral of $\exp(-a^2z^2)$ around the rectangular contour from $-R$ to R to $R + ib/2a^2$ to $-R + ib/2a^2$ and back to R, and let $R \to +\infty$.

The theory of the singular series is due to Hardy and Littlewood [**HL20a, HL20b, HL21, HL22b**]. Vinogradov [**Vin80**] simplified their treatment. Our exposition follows that of Davenport [**Dav05**].

The technique of bounding exponential sums by squaring and differencing can be traced back to the *Disquisitiones*, where Gauss determined the modulus of the classical Gauss sum. The iterative use of this process to bound exponential sums is due to Weyl [**Wey16**]. Proposition 4.14 in explicit form is due to Hardy and Littlewood [**HL20b**]. Proposition 4.15 is due to L. K. Hua [**Hua38**]. In [**HL20b**] Hardy and Littlewood had estimated the integral over the minor arcs by the inequality

$$\left|\int_{\mathfrak{m}} f(\alpha)^s e(-n\alpha)\, d\alpha\right| \leq \int_0^1 |f(\alpha)|^2\, d\alpha\, (\max_{\alpha\in\mathfrak{m}} |f(\alpha)|)^{s-2}$$

where the integral on the right-hand side is calculated trivially by multiplying out the integrand $f\overline{f}$ and integrating termwise. In [**HL21**] they applied the inequality

$$\left|\int_{\mathfrak{m}} f(\alpha)^s e(-n\alpha)\, d\alpha\right| \leq \int_0^1 |f(\alpha)|^4\, d\alpha\, (\max_{\alpha\in\mathfrak{m}} |f(\alpha)|)^{s-4}$$

where the integral on the right-hand side is bounded slightly less trivially, though still easily, by multiplying out the integrand and integrating termwise. Hua's argument is considerably more complicated, but his lemma forms a natural continuation of Hardy and Littlewood's progression from the exponent 2 to the exponent 4.

The most important objective of Chapter 4 was, of course, a solution to Waring's Problem. Since Waring made his celebrated statement in the 1770 edition of *Meditationes Algebraicae* [**War70**], an enormous amount of work has been expended on this and closely related problems, and it is out of the question to attempt a survey here. Only some highlights of the history of the problem will be touched upon. In modern notation, Waring made four definite assertions: That $g(2) = 4$, that $g(3) = 9$, that $g(4) = 19$ and that $g(k) < \infty$ for all k. A proof that $g(2) = 4$ was published by Lagrange [**Lag70**] in the same year that the *Meditationes Algebraicae* came out. That $g(3) = 9$ is true, and was established by 1912 through the combined efforts of A. Wieferich [**Wie09**] and A. Kempner [**Kem12**]. That $g(4) = 19$ is also true; this was proved as late as 1986 by R. Balasubramanian, J.-M. Deshouillers and F. Dress [**BDD86a, BDD86b**]. That $g(k) < \infty$ is Proposition 4.5.6, established by Hilbert [**Hil09**] in 1909.

Today the function $g(k)$ is known definitely for the first few hundred million k, and presumably for all k. See the Notes to Chapter XXI of [**HW08**] for information and references on this topic.

The function $G(k)$ is as yet known definitely only for two values of k. The classical four-squares theorem of Lagrange implies that $G(2) = 4$, and Davenport [**Dav39**] showed in 1939 that $G(4) = 16$. Landau [**Lan09**] proved in 1909 that $G(3) \leq 8$ and Linnik [**Lin42, Lin43**] in 1942 that $G(3) \leq 7$. They did not employ the Circle Method and did not obtain the asymptotic formula for the number of representations. Vaughan [**Vau86, Vau85**] showed in 1986 that the asymptotic formula holds for eight cubes and sixteen biquadrates. It is still not known whether the asymptotic formula holds for seven cubes, nor whether every sufficiently large integer is a sum of six cubes.

The first upper bound $G(k) \leq (k-2)2^{k-1} + 5$, valid for all k, was obtained by Hardy and Littlewood [**HL22b**] in 1922. In 1934 Vinogradov developed a new technique for estimating Weyl sums, which can replace both Weyl's inequality and Hua's lemma to yield a much better bound for the integral over the minor arcs when k is large. That same year he had already established the bound $G(k) \leq 6k\log(k) + 10k$ by the new method [**Vin34**]. He and others obtained further improvements so that $\limsup_{k\to+\infty} G(k)/(k\log(k)) \leq 2$ was known by 1959. Progress on the growth of $G(k)$ then all but ceased until Vaughan [**Vau89**] introduced a new technique in 1989, which in combination with another advance made by Wooley [**Woo92**] enabled the latter to establish $\limsup_{k\to+\infty} G(k)/(k\log(k)) \le 1$ in 1992.

We used the Circle Method to count solutions of Diophantine equations in many unknowns, but these equations were strictly tied to diagonal forms. It is natural to ask what the method can achieve for more general equations.

The algebraic process of completing squares allows us to bring a quadratic form to diagonal form. So for homogenous polynomials of degree two, a preliminary change of variables makes the Circle Method apply. One may suspect that the method can also be made to work for forms of higher degree with the aid of algebra, and this turns out to be true. In particular the Circle Method yields information about the existence of nontrivial solutions of systems of homogenous equations over the integers $\mathbb{Z}$, or equivalently, over the rational numbers $\mathbb{Q}$. For definiteness, we assume that the equations are defined over the integers.

Building on earlier work by R. D. Brauer [**Bra45**], in 1957 D. J. Lewis [**Lew57**] showed that any cubic form has a nontrivial zero if the number of variables is large enough. Soon after, B. Birch [**Bir57**] showed that any system of homogenous equations has a nontrivial solution if all the equations have odd degree and the number of variables is large enough. More precisely, if the degrees of the equations are $k_1, k_2, \dots, k_r$, all positive odd integers, then there is a number $s_0 = s_0(k_1, \dots, k_r)$ so that, if the number of variables $s \ge s_0$, then there is a nontrivial solution in integers. Since a form of even degree may have only the trivial zero, the theorem of Birch is best possible of its kind.

The most interesting particular case of the theorem of Birch is that of a single homogenous cubic equation. Such an equation must necessarily have a nontrivial solution over $\mathbb{R}$ if there are at least two variables, but need not be solvable nontrivially as a congruence modulo all prime powers. Indeed L. J. Mordell [**Mor37**] found examples of homogenous cubic equations in as many as nine variables that are not solvable nontrivially as congruences modulo some primes. This shows that the least number of variables must be ten or larger in Birch's theorem applied to a single cubic form. After proving weaker results, Davenport [**Dav63**] succeeded in 1963 in showing that any cubic form in at least 16 variables has a nontrivial zero. In 2007 Heath-Brown [**HB07**] achieved the same conclusion if the number of variables is 14 or larger. The gap between this result and that of Mordell is partially filled by an earlier theorem [**HB83**] of Heath-Brown; a nonsingular cubic form has a nontrivial zero if the number of variables is 10 or larger. That the form is nonsingular means that the associated hypersurface in complex projective space is nonsingular in the sense of algebraic geometry, and this is the generic case. A stronger result [**Hoo88**] was established by C. Hooley; a nonsingular homogenous cubic equation that is solvable nontrivially modulo all prime powers has a nontrivial solution if the number of variables is 9 or larger. By work of D. J. Lewis [**Lew52**] any homogenous cubic equation in at least 10 variables is solvable nontrivially as a congruence modulo all prime powers, so the result of Heath-Brown follows from that of Hooley. In [**Hoo91**] Hooley was able to allow singularities of a particular kind.

Dirichlet stated in the paper [**Dir42b**] where he proved Proposition 4.16 that in the one-dimensional case the result was already long known as a consequence of the theory of continued fractions. The validity of the asymptotic formula in Proposition 4.17 for $s \geq 2^k + 1$ is due to Hua [**Hua38**], though the formula itself goes back to the paper [**HL20b**] of Hardy and Littlewood (with trivial differences due to the coefficients c_j). Proposition 4.18 is due to I. Chowla [**Cho35, Cho37**], who improved an earlier result of Cauchy [**Cau13**] and Davenport [**Dav35**] for application to the singular series. Proposition 4.19 is a weaker version of a result of H. Davenport and D. J. Lewis [**DL63**]. In particular Davenport and Lewis determined for which k the sufficient condition $s \geq k^2 + 1$ is best possible. Proposition 4.20 follows directly from the cited work of Davenport and Lewis and of Hua, and Proposition 4.21 is due to Hilbert [**Hil09**].

Exercises

(1) Improve Proposition 4.1 to $|S(p,a)| \leq d\sqrt{p}$ where $d = \gcd(k, p-1)$, and improve and generalize Proposition 4.2 to congruences of the form

$$c_1 x_1^{k_1} + \cdots + c_s x_s^{k_s} \equiv n \pmod{p}$$

with exponents $k_1, \ldots, k_s$ that are not necessarily equal.

(2) It is known by work of J. L. Lagrange that every positive integer is the sum of at most four positive squares. Leverage this information by means of the polynomial identity

$$6(a^2+b^2+c^2+d^2)^2 = \frac{1}{2} \sum_{x \in \{a,b,c,d\} \ni y \neq x} \left((x+y)^4 + (x-y)^4\right)$$

to show that every positive integer is the sum of at most 53 biquadrates (J. Liouville). Complicated polynomial identities were used to solve Waring's Problem for a few more exponents during the nineteenth century, and in 1909 Hilbert settled the problem by *proving the existence* of suitable polynomial identities.

(3) Find all solutions of $x^3 + 7y^3 = 3z^3$ in integers.

(4) Show that the rational solutions of $X^2 + Y^2 = 1$ are dense in the unit circle.

(5) † Show that $x^3 = y^2 + 17$ has no solution in integers (G. C. Gerono).

(6) a) For an arbitrary positive integer $n \geq 4$, write the integers $2, 3, \ldots, n-2$ first in increasing order in a row and then immediately below in decreasing order in another row, so that the sum of each vertical pair (one immediately below the other) is n. For fixed z find an exact expression for the number of pairs remaining after all vertical pairs with an element lying in the arithmetic progressions $p\mathbb{Z}$ with $p \leq z$ have been removed. This sieve is due to J. Merlin.

b) For fixed z find an asymptotic estimate for the number of ways of writing n as a sum $n = n_1 + n_2$ of two integers $n_1, n_2 \geq 2$, neither of which has any prime divisor $p \leq z$.
c) Show that every sufficiently large positive integer n may be written as a sum $n = n_1 + n_2$ of two integers $n_1, n_2 \geq 2$ with $p|n_1 n_2 \Rightarrow p \gg \log(n)$. This is a very weak analogue of the binary Goldbach problem, replacing primes by integers with no very small prime factors. (The reasoning is due to V. Brun; by much more elaborate sieving he replaced the logarithm by a power. This enabled him to show that every sufficiently large even integer is a sum of two integers neither of which has more than nine prime factors.)

(7) Show that there is an integer solution of $ax + by = 1$ with

$$x^2 + y^2 \leq \frac{a^2 + b^2}{4} + \frac{1}{a^2 + b^2}$$

if a and b are integers with $\gcd(a, b) = 1$. (See the paper [**RR09**] by M. S. Risager and Z. Rudnick for a proof that $\sqrt{(x^2 + y^2)/(a^2 + b^2)}$ is equidistributed on $[0, 1/2]$ if (x, y) is a minimal solution and $a^2 + b^2 \leq R^2$ with $R \to +\infty$.)

(8) Suppose that $\mathcal{P}$ is a finite set of polynomials over $\mathbb{Z}$ and denote the number of mutually incongruent solutions of the simultaneous congruences

$$P(\mathbf{x}) \equiv 0 \pmod{q}, \qquad P \in \mathcal{P}$$

by $M_{\mathcal{P}}(q)$. Show that the arithmetic function $M_{\mathcal{P}}$ is multiplicative.

(9) The following proof of the *Chevalley-Warning theorem* is due to J. Ax. The first and slightly weaker version of the result was proved by C. Chevalley, and the full version by E. Warning. The latter should not be confused with the eighteenth-century English algebraist E. Waring.
a) Prove that if p is a prime and u a nonnegative integer, then

$$\sum_{m=1}^{p} m^u \equiv -1 \pmod{p}$$

if $u \geq 1$ and $(p-1)|u$, and that the sum is congruent to zero otherwise.
b) Suppose that $\mathcal{P}$ is a finite set of polynomials over $\mathbb{Z}/p\mathbb{Z}$ in n variables and such that $\deg(P) < n$ for each $P \in \mathcal{P}$. Denote the set of solutions in $(\mathbb{Z}/p\mathbb{Z})^n$ of the system of simultanous congruences $P(\mathbf{x}) \equiv 0 \pmod{p}$ for $P \in \mathcal{P}$ by $V_{\mathcal{P}}$. Show that the polynomial

$$I_{V_{\mathcal{P}}}(\mathbf{x}) = \prod_{P \in \mathcal{P}} (1 - P(\mathbf{x})^{p-1})$$

over $\mathbb{Z}/p\mathbb{Z}$ is $\underline{1}$ on $V_{\mathcal{P}}$ and $\underline{0}$ off $V_{\mathcal{P}}$.

c) Show that p divides $|V_{\mathcal{P}}|$ under the conditions assumed in part b). This is the Chevalley-Warning theorem. Conclude that a quadratic form over $\mathbb{Z}/p\mathbb{Z}$ in three or more variables has a nontrivial zero.

(10) For p a prime, let Q be a polynomial in n variables over $\mathbb{Z}/p\mathbb{Z}$ of total degree $\deg(Q) \leq n$ with $Q(\underline{\mathbf{0}}) = \underline{0}$. Suppose that for a particular one of the variables of Q, the degree of each term of Q is less than or equal to n minus the degree of that particular variable in the term. Show that $Q(\mathbf{x}) = 0$ has a nontrivial zero in $(\mathbb{Z}/p\mathbb{Z})^n$.

(11) a) Show that if k and s are positive integers and

$$f(\alpha) = \sum_{|m| \leq P} e(\alpha m^k),$$

then the integral

$$\int_0^1 |f(\alpha)|^{2s}\, d\alpha$$

equals the number of integer solutions of the equation

$$x_1^k + \cdots + x_s^k = y_1^k + \cdots + y_s^k$$

in the box $|x_1|, \ldots, |x_s|, |y_1|, \ldots, |y_s| \leq P$.
b) Show that

$$P^s \ll \int_0^1 |f(\alpha)|^{2s}\, d\alpha \ll P^{2s-1}$$

for all k and s. When $s \geq 2$ there is room between the upper and lower bounds for improvements. See what you can achieve in this direction.

(12) a) Show that

$$\left(\int_{-\infty}^{\infty} e^{-a^2x^2} \cos(bx)\, dx\right)^2 = \int_{-\infty}^{\infty}\int_{-\infty}^{\infty} e^{-a^2(x^2+y^2)} \cos(b(x+y))\, dx\, dy.$$

b) Show that

$$\int_{-\infty}^{\infty}\int_{-\infty}^{\infty} e^{-a^2(x^2+y^2)} \cos(b(x+y))\, dx\, dy$$
$$= \int_0^{\infty} e^{-a^2r^2} r \left(\int_{\pi/4}^{9\pi/4} \cos\left(br\sqrt{2}\sin(\theta)\right)\, d\theta\right) dr.$$

c) Show that

$$\int_{\pi/4}^{9\pi/4} \cos\left(br\sqrt{2}\sin(\theta)\right) d\theta = \sum_{k=0}^{\infty} (-1)^k \frac{b^{2k}r^{2k}2^k}{(2k)!} \frac{2\pi}{(2i)^{2k}} (-1)^k \binom{2k}{k}.$$

d) Show that

$$\int_{-\infty}^{\infty} e^{-a^2x^2} \cos(bx)\, dx = \frac{\sqrt{\pi}}{a} e^{-b^2/(4a^2)}.$$

This definite integral may also be evaluated, and less laboriously, using complex analysis or the theory of partial differential equations.

(13) Calculate $C_{k,s}$ exactly in terms of $V_{k.s}$.

(14) We imposed the condition $s \geq 2k+1$ so as to have the singular series converge absolutely, which also implied the validity of its Euler product. But show that the factors

$$\sum_{\alpha=0}^{\infty} A_n(p^\alpha)$$

of the Euler product of the singular series converge absolutely already for $s \geq k+1$.

(15) Show that if $s \geq 2k+1$ and $\mathfrak{S}(n) > 0$ then all the congruences

$$c_1 x_1^k + \cdots + c_s x_s^k \equiv n \pmod{q}$$

must have solutions.

(16) Find the singular series of the Diophantine problem

$$c_1 x_1^{k_1} + \cdots + c_s x_s^{k_s} = n,$$

where $c_1, \ldots, c_s, k_1, \ldots, k_s$ are positive integers. Find a condition on $k_1, \ldots, k_s$ for absolute convergence of the singular series.

(17) Suppose that p is an odd prime. Apply the Weyl inequality to bound the classical Gauss sum τ_p. Do you see great potential to improve the Weyl inequality when $\deg(P) = k = 2$?

(18) a) Show that

$$x_1^3 + x_2^3 + x_3^3 + x_4^3 = y_1^3 + y_2^3 + y_3^3 + y_4^3$$

has at most $N^{5+\varepsilon}$ solutions in the box $|x_1|, \ldots, |x_4|, |y_1|, \ldots, |y_4| \leq N$.

b) Show that

$$x_1^3 + x_2^3 + x_3^3 = y_1^3 + y_2^3 + y_3^3$$

has at most $N^{7/2+\varepsilon}$ integer solutions in $|x_1|, |x_2|, |x_3|, |y_1|, |y_2|, |y_3| \leq N$.

(19) Show that

$$\sum_{n=0}^{\infty} R(n,N)^2 \ll N^{5+\varepsilon}$$

where $R(n,N)$ denotes the number of ways of writing n as a sum of four nonnegative cubes each no larger than N^3.

(20) Show that the number $R(n)$ of representations of n by nine cubes satisfies

$$R(n) = C_{3,9} n^2 \mathfrak{S}(n) + O(n^{23/24+\varepsilon'})$$

for every $\varepsilon' > 0$.

(21) Show that

$$\sum_m \left(\sum_{\substack{a_1+a_2=m \\ a_1,a_2\in A}} 1 \right)^2 = \sum_n \left(\sum_{\substack{b_1-b_2=n \\ b_1,b_2\in A}} 1 \right)^2$$

for any finite set A of integers.

(22) † Study the still open *binary Goldbach conjecture* that every even integer $n \geq 4$ is the sum of two primes, by means of the Circle Method. The method has been very successful on the easier *ternary Goldbach conjecture* that every odd integer $n \geq 7$ is a sum of three primes. In 1922 Hardy and Littlewood gave a conditional proof of this by the Circle Method for all large odd n using a strong, and still unproved, conjecture on the zeros of Dirichlet L-functions. In 1937 Vinogradov found an unconditional proof for all large odd n, reducing the ternary Goldbach problem to a finite, though enormous, amount of calculation. The full ternary Goldbach conjecture was established conditionally on the Riemann Hypothesis for Dirichlet L-functions by J.-M. Deshouilliers, G. Effinger, H. te Riele and D. Zinoviev in 1997. Finally H. A. Helfgott established the ternary Goldbach conjecture unconditionally in 2013.
a) Show that

$$R(n) = \sum_{p_1+p_2=n} \log(p_1)\log(p_2) = \int_0^1 f(\alpha)^2 e(-\alpha n)\, d\alpha$$

where

$$f(\alpha) = \sum_{p\leq n} \log(p)e(\alpha p)$$

is an exponential sum over primes.
b) Express $f(a/q)$ in terms of $\vartheta(n;q,r)$, which is the sum of $\log(p)$ with $p \leq n$ and $p \equiv r \pmod{q}$.
c) On the one hand the binary Goldbach problem is unsolved, while on the other hand we have not yet proved any result on the distribution of primes in arithmetic progressions going beyond Dirichlet's theorem. To make your work easier, you may consequently assume the strongest available conjecture

$$\vartheta(n;q,a) - \frac{n}{\phi(q)} \ll_\varepsilon \frac{n^{1/2+\varepsilon}}{q^{1/2}}$$

on the distribution of primes in arithmetic progressions, due to Montgomery. Here the bound is assumed uniform in $q < n$. Determine $f(a/q)^2 e(-an/q)$ up to an error term on this assumption.

d) Find a bound for

$$|f(\alpha)^2 e(-\alpha n) - f(a/q)^2 e(-an/q)|$$

in terms of $|\alpha - a/q|$.

e) Vinogradov's bound

$$f(\alpha) \ll \left(nq^{-1/2} + n^{4/5} + n^{1/2}q^{1/2}\right)\log^4(n)$$

is valid for $|\alpha - a/q| \leq q^{-2}$ with $\gcd(q,a) = 1$ and $q \leq n$. Assess your options for bounding the integral over the minor arcs if you decide to have minor arcs. Hardy and Littlewood had no minor arcs in their conditional treatment of the ternary Goldbach problem.

f) Calculate the singular series of the binary Goldbach problem.

g) Pull together the information that you have gathered to see if you can achieve a solution conditional on the Montgomery conjecture. If this proves unattainable, try to identify heuristically a reasonable main term in an asymptotic estimate for the sum $R(n)$.

(23) † Show that every sufficiently large integer is the sum of a square and seven cubes. (This exercise is taken from *The Hardy-Littlewood Method* by Vaughan. Actually every sufficiently large integer is the sum of seven cubes; this was proved by Linnik in 1942. Vaughan found a proof by the Circle Method in 1986.)

(24) † For each positive integer k denote by $b(k)$ the least positive integer such that every sufficiently large integer n has a representation

$$n = x_1^{k+1} + x_2^{k+2} + \cdots + x_{b(k)}^{k+b(k)}$$

in nonnegative integers $x_1, \ldots, x_{b(k)}$. Show that $b(k)$ exists for each k.

Chapter 5

The Method of Contour Integrals

5.1. The Perron formula

The sum

$$A(s) = \sum_{n=1}^{\infty} a_n n^{-s}$$

of a convergent Dirichlet series is a holomorphic (single-valued analytic) function in the half plane $\sigma > \sigma_c(A)$, and the terms of the Dirichlet series are holomorphic in the whole complex plane, and the series converges uniformly on every compact subset of $\sigma > \sigma_c(A)$ by Proposition 3.3. Generally speaking, the relationship between the convergence properties of a Dirichlet series and the analytic properties of its sum is less clear than what we are used to from the theory of power series. For the latter, there is the convenient principle that the convergence disk extends from the center out to the nearest singularity of the sum. But the sum $A(s)$ of a Dirichlet series may have no singularity on the line of convergence $\sigma = \sigma_c(A)$. For Dirichlet series with nonnegative coefficients there is nonetheless a useful result due to Landau.

Proposition 5.1. *If the Dirichlet series*

$$A(s) = \sum_{n=1}^{\infty} a_n n^{-s}$$

has finite abscissa of convergence $\sigma_c(A)$ and $a_n \geq 0$ for all n, then the point $s = \sigma_c(A)$ is a singularity of $A(s)$.

Proof. By translation and without loss of generality, it is enough to prove that if the series converges at $s = 0$, and $A(s)$ is holomorphic at $s = 0$, then the series converges somewhere to the left of $s = 0$.

By assumption, if we develop $A(s)$ into a power series around $s = 1$, the convergence disk will extend to the left of $s = 0$, so the point $s = -\varepsilon$ lies in the convergence disk for some $\varepsilon > 0$. Now

$$\begin{aligned} A(-\varepsilon) &= \sum_{k=0}^{\infty} \frac{A^{(k)}(1)}{k!}(-\varepsilon - 1)^k = \sum_{k=0}^{\infty} \frac{(-\varepsilon-1)^k}{k!} \sum_{n=1}^{\infty} \frac{a_n}{n}(-\log(n))^k \\ &= \sum_{k=0}^{\infty}\sum_{n=1}^{\infty} \frac{(\varepsilon+1)^k}{k!}\frac{a_n}{n}(\log(n))^k = \sum_{n=1}^{\infty}\sum_{k=0}^{\infty} \frac{(\varepsilon+1)^k}{k!}\frac{a_n}{n}(\log(n))^k \\ &= \sum_{n=1}^{\infty} \frac{a_n}{n} \sum_{k=0}^{\infty} \frac{(-\varepsilon-1)^k}{k!}(-\log(n))^k = \sum_{n=1}^{\infty} \frac{a_n}{n} n^{\varepsilon+1} = \sum_{n=1}^{\infty} a_n n^{-(-\varepsilon)} \end{aligned}$$

since the change of the order of summation is valid by $a_n \geq 0$. Thus the Dirichlet series converges at $s = -\varepsilon$. □

The method of contour integrals is a technique to obtain asymptotic estimates for summatory functions

$$F(x) = \sum_{n \leq x} a_n$$

from the associated Dirichlet series

$$A(s) = \sum_{n=1}^{\infty} a_n n^{-s}$$

by means of residue calculus. Thus it is an alternative to elementary techniques such as the neighborhood method, and is frequently more powerful than these.

We have already observed that

$$-\frac{\zeta'(s)}{\zeta(s)} = \sum_{n=1}^{\infty} \Lambda(n) n^{-s}$$

as a formal identity. But it actually holds as an identity between holomorphic functions in the half plane $\sigma > 1$. This may be deduced by taking the logarithm of the Euler product formula for $\zeta(s)$, differentiating through and using uniform convergence of the resulting Dirichlet series on compact subsets of $\sigma > 1$, or differentiate the Dirichlet series defining $\zeta(s)$, use uniform convergence again, and apply the identity

$$-\zeta'(s) = \left(-\frac{\zeta'(s)}{\zeta(s)}\right)\zeta(s).$$

The summatory function associated with the Dirichlet series of $-\zeta'(s)/\zeta(s)$ is $\psi(x)$. The Prime Number Theorem in the form $\psi(x) \sim x$ can be established by applying the method of contour integrals to $-\zeta'(s)/\zeta(s)$. That is how Hadamard and de la Vallée Poussin first proved the PNT (independently) in 1896.

Proposition 5.2 (Truncated Perron formula). *Suppose that*

$$A(s) = \sum_{n=1}^{\infty} a_n n^{-s}$$

is a convergent Dirichlet series and that $c > \max(0, \sigma_c(A))$. *Then*

$$\left| \sum_{n<x} a_n - \frac{1}{2\pi i} \int_{c-iT}^{c+iT} A(s) \frac{x^s}{s}\, ds \right| \leq 2x^c \sum_{n=1}^{\infty} \frac{|a_n| n^{-c}}{\max(1, T|\log(x/n)|)}$$

for any $T > 0$ *and any* $x > 0$ *that is not an integer.*

Proof. We have

$$\begin{aligned} \frac{1}{2\pi i} \int_{c-iT}^{c+iT} A(s) \frac{x^s}{s}\, ds &= \frac{1}{2\pi i} \int_{c-iT}^{c+iT} \sum_{n=1}^{\infty} a_n n^{-s} \frac{x^s}{s}\, ds \\ &= \sum_{n=1}^{\infty} \frac{a_n}{2\pi i} \int_{c-iT}^{c+iT} \left(\frac{x}{n}\right)^s \frac{ds}{s} \end{aligned}$$

because the Dirichlet series is uniformly convergent on the line segment from $c - iT$ to $c + iT$ by Proposition 3.3.

Suppose that $n > x$, and let $b > c$. Then

$$\left(\int_{c-iT}^{c+iT} + \int_{c+iT}^{b+iT} + \int_{b+iT}^{b-iT} + \int_{b-iT}^{c-iT} \right) \left(\frac{x}{n}\right)^s \frac{ds}{s} = 0$$

by Cauchy's theorem. Now

$$\left| \int_{b+iT}^{b-iT} \left(\frac{x}{n}\right)^s \frac{ds}{s} \right| \leq \left(\frac{x}{n}\right)^b \frac{2T}{b}$$

so the integral from $b + iT$ to $b - iT$ goes to zero as $b \to +\infty$. Moreover

$$\left| \int_{c+iT}^{b+iT} \left(\frac{x}{n}\right)^s \frac{ds}{s} \right| \leq \frac{1}{T} \int_c^{\infty} \left(\frac{x}{n}\right)^u du = \frac{1}{T|\log(x/n)|} \left(\frac{x}{n}\right)^c$$

by the change of variable $s = u + iT$. Then

$$\left| \int_{c-iT}^{c+iT} \left(\frac{x}{n}\right)^s \frac{ds}{s} \right| \leq \frac{2}{T|\log(x/n)|} \left(\frac{x}{n}\right)^c.$$

But if $T|\log(x/n)| \le 1$, then

$$\begin{aligned}
\left|\int_{c-iT}^{c+iT}\left(\frac{x}{n}\right)^s \frac{ds}{s}\right| &= \left(\frac{x}{n}\right)^c \left|\int_{-T}^{T}\left(\frac{x}{n}\right)^{it} \frac{i\,dt}{c+it}\right| \\
&= \left(\frac{x}{n}\right)^c \left|\int_{-T}^{T}\sum_{k=0}^{\infty}\frac{1}{k!}\log^k(x/n)(it)^k \frac{i\,dt}{c+it}\right| \\
&= \left(\frac{x}{n}\right)^c \left|\log\left(\frac{c+iT}{c-iT}\right) + \sum_{k=1}^{\infty}\frac{1}{k!}\log^k(x/n)\int_{-T}^{T}(it)^k \frac{i\,dt}{c+it}\right| \\
&\le \left(\frac{x}{n}\right)^c \left(\pi + \sum_{k=1}^{\infty}\frac{2}{k!}\log^k(x/n)\frac{T^k}{k}\right) \le 2\pi\left(\frac{x}{n}\right)^c.
\end{aligned}$$

If $n < x$ and $a < 0$ then

$$\left(\int_{c-iT}^{c+iT} + \int_{c+iT}^{a+iT} + \int_{a+iT}^{a-iT} + \int_{a-iT}^{c-iT}\right)\left(\frac{x}{n}\right)^s \frac{ds}{s} = 2\pi i r = 2\pi i$$

by the Residue theorem, where $r = 1$ is the residue of $(x/n)^s/s$ in the simple pole $s = 0$. Thus

$$\left|\int_{c-iT}^{c+iT}\left(\frac{x}{n}\right)^s \frac{ds}{s} - 2\pi i\right| \le \frac{2}{T|\log(x/n)|}\left(\frac{x}{n}\right)^c,$$

as above. But if $T|\log(x/n)| \le 1$, then

$$\left|\int_{c-iT}^{c+iT}\left(\frac{x}{n}\right)^s \frac{ds}{s} - 2\pi i\right| \le 2\pi\left(\frac{x}{n}\right)^c + 2\pi \le 4\pi\left(\frac{x}{n}\right)^c.$$

Now

$$\left|\frac{1}{2\pi i}\int_{c-iT}^{c+iT} a_n n^{-s}\frac{x^s}{s}\,ds\right| \le x^c \frac{|a_n|n^{-c}}{\max(1, T|\log(x/n)|)}$$

for $n > x$ and

$$\left|\frac{1}{2\pi i}\int_{c-iT}^{c+iT} a_n n^{-s}\frac{x^s}{s}\,ds - a_n\right| \le 2x^c \frac{|a_n|n^{-c}}{\max(1, T|\log(x/n)|)}$$

for $n < x$. The desired inequality follows. □

Letting $T \to +\infty$ in this inequality yields the Perron formula

$$\sum_{n<x} a_n = \frac{1}{2\pi i}\int_{c-i\infty}^{c+i\infty} A(s)\frac{x^s}{s}\,ds,$$

valid for any $x > 0$ that is not an integer. The path of integration in the formula is the vertical line given by $s = c+it$ with t increasing on the interval $-\infty < t < \infty$. It may seem that the integral here is a Cauchy principal value integral. However, the proof of Proposition 5.2 can be carried through with the path of integration from $c - iT_1$ to $c + iT_2$ and T_1 and T_2 independent, so the integral is an ordinary improper integral.

Though undeniably elegant, the Perron formula is difficult to use as it stands. For the improper integral in the formula is only conditionally convergent, and thus hard to estimate. One way to avoid this issue is to use some truncated formula like the one in Proposition 5.2. Another way is to weight the terms in the coefficient sum. The formula

$$\sum_{n<x}(x-n)a_n = \frac{1}{2\pi i}\int_{c-i\infty}^{c+i\infty} A(s)\frac{x^{s+1}}{s(s+1)}\,ds$$

is easily deduced from the Perron formula, but has an important advantage over it, in that the denominator $s(s+1)$ grows rapidly enough for the improper integral to be absolutely convergent. The calculation

$$\begin{aligned}\sum_{n<x}(x-n)a_n &= \frac{1}{2\pi i}\int_{c-i\infty}^{c+i\infty} A(s)\frac{x^{s+1}}{s}\,ds - \frac{1}{2\pi i}\int_{c-i\infty}^{c+i\infty} A(s-1)\frac{x^{s}}{s}\,ds\\ &= \frac{1}{2\pi i}\int_{c-i\infty}^{c+i\infty} A(s)\frac{x^{s}}{s}\,ds - \frac{1}{2\pi i}\int_{c-1-i\infty}^{c-1+i\infty} A(s)\frac{x^{s+1}}{s+1}\,ds\\ &= \frac{1}{2\pi i}\int_{c-i\infty}^{c+i\infty} A(s)\frac{x^{s}}{s}\,ds - \frac{1}{2\pi i}\int_{c-i\infty}^{c+i\infty} A(s)\frac{x^{s+1}}{s+1}\,ds\\ &= \frac{1}{2\pi i}\int_{c-i\infty}^{c+i\infty} A(s)\frac{x^{s+1}}{s(s+1)}\,ds\end{aligned}$$

is valid for $c > \max(1, 1+\sigma_a(A))$ by Cauchy's theorem. The application of Cauchy's theorem is justified by

$$\left|\int_{c-1+iT}^{c+iT} A(s)\frac{x^{s+1}}{s+1}\,ds\right| \le \frac{\max_{-1\le s-c-iT\le 0}|A(s)|}{T}x^c \to 0$$

as $T \to +\infty$.

Integration by parts is often helpful in estimating awkward integrals. The calculation

$$\begin{aligned}\sum_{n<x} a_n\log(x/n) &= \frac{1}{2\pi i}\int_{c-i\infty}^{c+i\infty}\left(\sum_{n=1}^{\infty} a_n\log(x/n)n^{-s}\right)\frac{x^s}{s}\,ds\\ &= \frac{1}{2\pi i}\int_{c-i\infty}^{c+i\infty}\left(\sum_{n=1}^{\infty} a_n\log(x/n)\left(\frac{x}{n}\right)^s\right)\frac{ds}{s}\\ &= -\frac{1}{2\pi i}\int_{c-i\infty}^{c+i\infty}\left(\sum_{n=1}^{\infty} a_n\log(x/n)\frac{1}{\log(x/n)}\left(\frac{x}{n}\right)^s\right)\frac{(-1)}{s^2}\,ds\\ &= \frac{1}{2\pi i}\int_{c-i\infty}^{c+i\infty} A(s)\frac{x^s}{s^2}\,ds\end{aligned}$$

yields another weighted version of the Perron formula. The boundary terms from the integration by parts tend to zero at infinity because $A(s)x^s/s \to 0$

for $s = c \pm iT$ and $T \to +\infty$. The improper integral in the weighted version

$$\sum_{n<x} a_n \log(x/n) = \frac{1}{2\pi i} \int_{c-i\infty}^{c+i\infty} A(s) \frac{x^s}{s^2}\, ds$$

of the Perron formula is also absolutely convergent.

The reason for the more favorable convergence properties of the weighted versions of the Perron formula is that the weights $x - n$ and $\log(x/n)$ have a gentle rather than an abrupt cut-off when n approaches x.

5.2. Bounds for Dirichlet L-functions

In typical applications of the Perron formula the integrand must be analytically continued to the left of the line of convergence of the Dirichlet series, and bounds for it established. The analytic continuations of $\zeta(s)$ and $L(s, \chi)$ to the whole complex plane are available by (the proofs of) their functional equations. But some of our applications of the method of contour integrals require only analytic continuation to a suitable half plane. In this section we will apply Euler-Maclaurin summation to obtain such analytic continuations, and establish some of the bounds that will be needed later.

To streamline the formulation of the various bounds, we introduce a special notation $\tau = |t| + 4$ connected with the imaginary part t of the complex variable $s = \sigma + it$.

Proposition 5.3. *The function $L(s, \chi)$ has an analytic continuation to the half plane $\sigma > 0$ and is holomorphic there, except at $s = 1$ if χ is principal. Moreover*

$$\left| L(s, \chi) - \frac{\phi(q)}{q} \frac{1}{s-1} \right| \le 2q\tau$$

for $\sigma \ge 1/4$, and the term involving $s - 1$ is missing unless χ is principal.

Proof. Note that

$$L(s, \chi) = \sum_{\substack{1 \le a \le q \\ \gcd(a,q)=1}} \chi(a) \sum_{k=0}^{\infty} (qk + a)^{-s}$$

and that

$$\sum_{k=0}^{\infty} (qk + a)^{-s} = \frac{a^{1-s}}{q(s-1)} + \frac{a^{-s}}{2} - qs \int_0^{\infty} S(u)(qu + a)^{-s-1}\, du$$

for $\sigma > 1$ by Euler-Maclaurin summation. The integral converges uniformly on compact sets in $\sigma > 0$ and is thus holomorphic there, and the other terms are holomorphic everywhere except for a simple pole at $s = 1$ with residue $1/q$. Moreover,

$$\left|\sum_{k=0}^{\infty}(qk+a)^{-s} - \frac{1}{q(s-1)}\right| \leq \left|\frac{a^{1-s}-1}{q(s-1)}\right| + \frac{a^{-\sigma}}{2} + \frac{|s|}{2\sigma}a^{-\sigma}$$

by bounding the integral trivially using $|S(u)| \leq 1/2$. If $|s-1| \geq 1$ then

$$\left|\frac{a^{1-s}-1}{q(s-1)}\right| \leq \frac{a^{1-\sigma}+1}{q} \leq 2,$$

while

$$\left|\frac{a^{1-s}-1}{q(s-1)}\right| \leq \frac{a^{|s-1|}-1}{q|s-1|} \leq \frac{a-1}{q} \leq 1$$

for $|s-1| \leq 1$ by expanding a^{1-s} into a power series in $s-1$ and applying the triangle inequality. Then

$$\left|\frac{a^{1-s}-1}{q(s-1)}\right| + \frac{a^{-\sigma}}{2} + \frac{|s|}{2\sigma}a^{-\sigma} \leq 2 + \frac{1}{2} + \frac{|s|}{2\sigma} \leq 2\tau$$

since $\sigma \geq 1/4$. The result follows by summing over a. □

The case of the Riemann zeta function is covered by the choice $q = 1$. The next result provides better estimates close to the line $\sigma = 1$.

Proposition 5.4. *The estimates*

$$L(s,\chi) = \frac{\phi(q)}{q}\frac{1}{s-1} + O(\log(q\tau))$$

and

$$L'(s,\chi) = -\frac{\phi(q)}{q}\frac{1}{(s-1)^2} + O(\log^2(q\tau))$$

hold for

$$1 - \frac{1}{\log(q\tau)} \leq \sigma \leq 2$$

uniformly. The term involving $s-1$ is missing unless χ is principal.

Proof. Write

$$L(s,\chi) = \sum_{n=1}^{N}\chi(n)n^{-s} + \sum_{n=N+1}^{\infty}\chi(n)n^{-s}$$

with N the largest integer satisfying $N \leq q\tau$, and note that

$$\left|\sum_{n=1}^{N}\chi(n)n^{-s}\right| \leq \sum_{n=1}^{N} n^{-(1-1/\log(q\tau))} \leq 1 + \int_1^{q\tau} u^{-(1-1/\log(q\tau))}\,du$$

$$\leq 1 + \frac{(q\tau)^{1/\log(q\tau)}}{1/\log(q\tau)} \ll \log(q\tau).$$

Furthermore

$$\sum_{n=N+1}^{\infty} \chi(n)n^{-s} = \sum_{\substack{N+1\le a\le N+q\\ \gcd(a,q)=1}} \chi(a)\sum_{k=0}^{\infty}(qk+a)^{-s}$$

where

$$\left|\sum_{k=0}^{\infty}(qk+a)^{-s} - \frac{1}{q(s-1)}\right| \le \left|\frac{a^{1-s}-1}{q(s-1)}\right| + \frac{a^{-\sigma}}{2} + \frac{|s|}{2\sigma}a^{-\sigma}$$

as in the proof of Proposition 5.3. Now treat the terms on the right-hand side in the same way as before, but divide into cases $|s-1| \ge 1/\log(q\tau)$ and $|s-1| \le 1/\log(q\tau)$ when bounding the first term, and use $1 - 1/\log(q\tau) \le \sigma \le 2$ and $q\tau \le a \le q\tau + q$. The estimate follows by summing over a.

Next

$$L'(s,\chi) = -\sum_{n=1}^{N}\chi(n)\log(n)n^{-s} - \sum_{n=N+1}^{\infty}\chi(n)\log(n)n^{-s},$$

with N as before. Then

$$\left|\sum_{n=1}^{N}\chi(n)\log(n)n^{-s}\right| \le \log(N)\sum_{n=1}^{N} n^{-(1-1/\log(q\tau))} \ll \log^2(q\tau)$$

as in the previous proof. Furthermore

$$\sum_{k=0}^{\infty}\log(qk+a)(qk+a)^{-s} = \int_0^{\infty}\log(qu+a)(qu+a)^{-s}\,du + \frac{\log(a)}{2}a^{-s}$$
$$+\int_0^{\infty} S(u)\left(\frac{q(qu+a)^{-s}}{qu+a} + \log(qu+a)(-s)q(qu+a)^{-s-1}\right)du$$

for $\sigma > 1$ by Euler-Maclaurin summation. Here

$$\begin{aligned}\int_0^{\infty}\log(qu+a)(qu+a)^{-s}\,du &= \log(qu+a)\frac{(qu+a)^{1-s}}{q(1-s)}\Bigg|_0^{\infty}\\ &\quad -\int_0^{\infty}\frac{q}{qu+a}\frac{(qu+a)^{1-s}}{q(1-s)}\,du\\ &= \log(a)\frac{a^{1-s}}{q(s-1)} + \frac{1}{s-1}\int_0^{\infty}(qu+a)^{-s}\,du\\ &= \log(a)\frac{a^{1-s}}{q(s-1)} + \frac{a^{1-s}}{q(s-1)^2}\\ &= \frac{1}{q(s-1)^2} + \frac{\log(a)(s-1)a^{1-s} + a^{1-s} - 1}{q(s-1)^2}\end{aligned}$$

by integration by parts. Thus

$$\left|\frac{\log(a)(s-1)a^{1-s}+a^{1-s}-1}{q(s-1)^2}\right| \le \log(a)\frac{a^{1-\sigma}}{q|s-1|}+\frac{a^{1-\sigma}-1}{q|s-1|^2}$$
$$\le \log(q\tau+q)\frac{(q\tau+q)^{1-(1-1\log(q\tau))}}{q/\log(q\tau)}+\frac{(q\tau+q)^{1-(1-1/\log(q\tau))}+1}{q/\log^2(q\tau)}$$
$$\ll \frac{\log^2(q\tau)}{q}$$

for $|s-1|\ge 1/\log(q\tau)$ as before, while for $|s-1|\le 1/\log(q\tau)$ we obtain

$$\left|\frac{\log(a)(s-1)a^{1-s}+a^{1-s}-1}{q(s-1)^2}\right| \le \frac{\log(a)|s-1|a^{1-\sigma}+a^{1-\sigma}-1}{q|s-1|^2}$$
$$\le \frac{\log(q\tau+q)(q\tau+q)^{1-(1-1/\log(q\tau))}/\log(q\tau)}{q/\log^2(q\tau)}$$
$$+\frac{(q\tau+q)^{1-(1-1/\log(q\tau))}-1}{q/\log^2(q\tau)} \ll \frac{\log^2(q\tau)}{q}$$

by expanding a^{1-s} into a power series in $s-1$ and applying the triangle inequality. The term $\log(a)a^{-s}/2$ causes no difficulty, and

$$\left|\int_0^\infty S(u)\left(\frac{q}{qu+a}(qu+a)^{-s}+\log(qu+a)(-s)q(qu+a)^{-s-1}\right)du\right|$$
$$\le \frac{q}{2}\int_0^\infty \left(1+|s|\log(qu+a)\right)(qu+a)^{-\sigma-1}\,du$$
$$\le \frac{q}{2}\frac{a^{-\sigma}}{q\sigma}+\frac{q}{2}|s|\left(\log(a)\frac{a^{-\sigma}}{q\sigma}+\frac{a^{-\sigma}}{q\sigma^2}\right) \ll \frac{\log(q\tau)}{q}$$

by an earlier calculation. The estimate follows by summing over a. □

Sometimes a bound for $L(s,\chi)$ is established first for primitive characters and afterwards extended to nonprincipal characters in general. Let $\chi\neq\chi_0$ be an imprimitive character modulo q induced by a primitive character χ^*. Then the relation

$$L(s,\chi)=\prod_p(1-\chi(p))^{-1}=\prod_{p\mid q}(1-\chi(p)p^{-s})^{-1}\prod_{p\nmid q}(1-\chi(p)p^{-s})^{-1}$$
$$=\prod_{p\nmid q}(1-\chi^*(p)p^{-s})^{-1}=L(s,\chi^*)\prod_{p\mid q}(1-\chi^*(p)p^{-s})$$

allows estimation of $L(s,\chi)$ in terms of $L(s,\chi^*)$.

5.3. Notes

Proposition 5.1 is due to Landau [**Lan05b**]. A special case of the Perron formula occurs in Riemann's paper [**Rie59**] on prime number theory. After a slightly weaker version had been established by Hadamard [**Had08**], the full result was proved by O. Perron [**Per08**].

Results like Propositions 5.3 and 5.4 were proved by Landau [**Lan03b**]. R. H. Mellin [**Mel00**] and Landau [**Lan05a**] gave similar but more precise bounds for the Riemann zeta function.

Exercises

(1) Calculate $\zeta(0)$.

(2) Show that

$$A(s) = \sum_{n=2}^{\infty} \frac{n^{-s}}{\sqrt{\log(n)}}$$

has a singularity at $s = 1$, but that this singularity is not a pole.

(3) a) Express

$$f(s) = \prod_{n=2}^{\infty} (1 - n^{-s})^{-1}$$

in terms of the Riemann zeta function.
b) Find the singularities of $f(s)$ in $\sigma > 0$ and determine their nature.

(4) a) Define

$$f(s) = 1 - \sum_{n=2}^{\infty} n^{-s}$$

on $\sigma > 1$. Show that $f(s)^{-1}$ has a formal Dirichlet series

$$f(s)^{-1} = \sum^{\infty} c_n n^{-s},$$

all of whose coefficients are nonnegative.
b) Determine the abscissa of convergence of the Dirichlet series in a). Show that the bound $c_n = O(n^{0.7})$ must be false.

(5) Show that

$$\frac{\zeta(2s)}{\zeta(s)} = s \int_1^{\infty} L(x) x^{-s} \frac{dx}{x}$$

for $\sigma > 1$. Here

$$L(x) = \sum_{n \leq x} \lambda(n)$$

is the summatory function of the Liouville function. Then show that no bound $L(x) = O(x^{1/2-\varepsilon})$ can hold with $\varepsilon > 0$.

(6) a) Calculate explicitly the Dirichlet series expansion of $\log(F_q)$ where

$$F_q(s) = \prod_{\chi \bmod q} L(s,\chi)$$

and q is a positive integer.
b) Show by contradiction that $L(1,\chi) \neq 0$ for each Dirichlet character χ modulo q.

(7) Show that

$$A(s) = \sum_{n=1}^{\infty} (-1)^{\omega(n)} n^{-s}$$

has an analytic continuation to the half plane $\sigma > 1/2$ and is holomorphic in a domain that contains the half plane $\sigma \geq 1$. Find the zeros of $A(s)$, and their multiplicities, in the half plane $\sigma > 1/2$.

(8) a) Expand $(1-2^{1-s})\zeta(s)$ and $(1-3^{1-s})\zeta(s)$ into Dirichlet series. Determine the abscissas of convergence and of absolute convergence.
b) Show that $\log(3)/\log(2)$ is an irrational number.
c) Extend $\zeta(s)$ to a holomorphic function in the half plane $\sigma > 0$ except for a simple pole at the point $s = 1$.
d) Show that $\zeta(\sigma) \neq 0$ for $\sigma > 0$. Whether this also holds for all Dirichlet L-functions with real characters is an important open problem.

(9) † Show that

$$\sum_{n \leq x} f(n) = \zeta(3/2)x^{1/2} + O(x^{1/3})$$

when $f(n)$ denotes the number of divisors $d|n$ so that d is a square and n/d is a cube.

(10) Show that for each positive integer k the sawtooth function $S(x)$ is the k-th derivative of a bounded function. Then show that the Riemann zeta function $\zeta(s)$ has an analytic continuation to the whole complex plane except $s = 1$.

(11) Express the sawtooth function $S(x)$ as a contour integral by means of the Perron formula.

(12) a) Find a contour integral representation of the smoothed summatory function

$$D_1(x) = \sum_{n \leq x} (x-n)d(n)$$

of the divisor function.
b) In Chapter 8 it is proved by the Phragmén-Lindelöf principle that

$$\zeta(s) = O(|s|^{1/2-\sigma/2+\delta})$$

for $0 \leq \sigma \leq 1$ and any $\delta > 0$. Use this to obtain an estimate for $D_1(x)$ that has an error term of the form $O(x^{1+\epsilon})$, for any $\epsilon > 0$.

(13) To establish an asymptotic estimate rigorously by the method of contour integrals requires in each case some auxiliary estimate. Without such estimates, we only obtain heuristic information.
a) Use the Perron formula and a shift of the contour of integration to the left across the singularity at $s = 1$ to predict in a formal way the asymptotic density of the sequence of integers that are not divisible by the cube of any prime.
b) Use the Perron formula to predict in a formal way the average number of solutions n of the equation $\phi(n) = m$ as $m \to +\infty$.
c) Let $f(n)$ be the sum of the block divisors of n and $F(x)$ its summatory function. Use the Perron formula to predict in a formal way the main term of an asymptotic estimate for $F(x)$. (A divisor $d|n$ is a block divisor if d and its complementary divisor n/d are coprime.)

(14) Show that

$$-\frac{\zeta'(\sigma)}{\zeta(\sigma)} \leq \frac{1}{\sigma - 1} + \frac{1}{e}$$

for $\sigma > 1$. (Hint: Euler-Maclaurin summation in the numerator and the denominator.)

(15) Show that $\zeta(s) = O(\log|t|)$ for $\sigma \geq 1$ and $|t| \geq 2$. (Hint: Euler-Maclaurin summation.)

(16) Calculate the mean value

$$\lim_{T\to+\infty} \frac{1}{2T} \int_{-T}^{T} \left| \sum_{n=1}^{\infty} a_n n^{-\sigma-it} \right|^2 dt$$

when the Dirichlet series converges absolutely (Hadamard).

Chapter 6

The Prime Number Theorem

6.1. A zero-free region

The Perron formula with $a_n = \Lambda(n)$ yields

$$\psi(x) = \frac{1}{2\pi i} \int_{c-i\infty}^{c+i\infty} \left(-\frac{\zeta'(s)}{\zeta(s)} \right) \frac{x^s}{s} \, ds$$

for any $c > 1$ and any $x > 0$ that is not an integer. Attempting to prove the Prime Number Theorem by the method of contour integrals brings up several issues. The first is that the integral is only conditionally convergent. This is not really troublesome, for we have already seen workarounds for the difficulty. It is far more serious that, though $\psi(x) \ll x$, the integrand in the formula above has modulus roughly of size x^c on part of every line $\sigma = c > 1$. Clearly there is substantial cancellation in the integral. But the behavior in detail of $-\zeta'(s)/\zeta(s)$ is very complicated on every line $\sigma = c > 1$, as one sees from the Dirichlet series. To account for this to the degree necessary for a nontrivial estimate would require impossible precision. We meet this issue by moving the path of integration to the left, in order to substantially reduce the modulus of x^s by making $\mathrm{Re}(s)$ smaller. For this device to have sufficient effect to allow us to bound the error term efficiently, at least part of the path of integration must be moved into the half plane $\sigma < 1$. This requires that the integrand be analytically continued to the left of the line $\sigma = 1$.

The function $-\zeta'(s)/\zeta(s)$ has an analytic continuation to the half plane $\sigma > 0$, with simple poles at $s = 1$ and at $s = \rho$ for every zero $\rho = \beta + i\gamma$

of $\zeta(s)$ in this half plane. (Here the notation γ is used in a sense different from the Euler-Mascheroni constant.) When we move part of the contour of integration out of the half plane $\sigma > 1$, we will move it across the simple pole at $s = 1$, and pick up a contribution to the integral by the Residue Theorem. This contribution yields the main term x in the form $\psi(x) = x + E(x)$ of the Prime Number Theorem with an error term. To bound the error term $E(x)$ efficiently by the Perron formula, it is necessary to know something about the size of $-\zeta'(s)/\zeta(s)$ to the left of the line $\sigma = 1$, and this requires information about the location of the zeros of $\zeta(s)$.

Preliminary to proving the Prime Number Theorem by the method of contour integrals, we establish a series of estimates for $\zeta(s)$ and closely related functions. The ultimate objective is to bound $-\zeta'(s)/\zeta(s)$ in a region slightly to the left of the line $\sigma = 1$. Having found an upper bound for $\zeta'(s)$ in a region to the left of $\sigma = 1$, we seek a lower bound for $\zeta(s)$ in order to obtain an upper bound for $-\zeta'(s)/\zeta(s)$. But nontrivial lower bounds to the left of $\sigma = 1$ imply information about the location of the zeros of $\zeta(s)$, which is hard to come by. The discovery that $\zeta(s)$ has no zeros on the line $\sigma = 1$ was essential to the first proofs of the Prime Number Theorem, due to Hadamard and de la Vallée Poussin.

It is by no means obvious that $\zeta(s) \neq 0$ on the line $\sigma = 1$, for the Euler product is not valid on this line, though it is valid immediately to the right of the line. We shall give the usual textbook proof whose detailed arrangement is due to Mertens, though the underlying idea of relating the behavior of $\zeta(s)$ at $s = \sigma + it$ to its behavior at $s = \sigma$ and at $s = \sigma + 2it$ was evolved by Hadamard and by de la Vallée Poussin independently.

We start from the trigonometric inequality

$$3 + 4\cos(\theta) + \cos(2\theta) = 3 + 4\cos(\theta) + 2\cos^2(\theta) - 1 = 2(1 + \cos(\theta))^2 \geq 0.$$

The Euler product formula yields

$$\begin{aligned}\log|\zeta(s)| = \operatorname{Re}\log(\zeta(s)) &= \operatorname{Re}\left(\sum_p \sum_{k=1}^{\infty} \frac{p^{-ks}}{k}\right) \\ &= \sum_{k=1}^{\infty} \frac{1}{k} \sum_p p^{-k\sigma} \cos(kt\log(p))\end{aligned}$$

for $\sigma > 1$. Now

$$\begin{aligned}&\log(|\zeta(\sigma)|^3 |\zeta(\sigma + it)|^4 |\zeta(\sigma + 2it)|) \\ &= \sum_{k=1}^{\infty} \frac{1}{k} \sum_p p^{-k\sigma}(3 + 4\cos(kt\log(p))) + \cos(2kt\log(p))) \geq 0\end{aligned}$$

for $\sigma > 1$. Clearly $\left|\zeta(\sigma)^3\zeta(\sigma+it)^4\zeta(\sigma+2it)\right| \geq 1$ on $\sigma > 1$. But since $\zeta(s)$ has a simple pole at $s = 1$, this inequality shows that if $\zeta(s)$ has a zero at $s = 1 + it$, then it must have a pole at $s = 1 + 2it$, which is false. Thus $\zeta(s)$ has no zeros on the line $\sigma = 1$.

There exist analytic proofs of the Prime Number Theorem in which the only nontrivial fact required about the zeta function is that $\zeta(1+it) \neq 0$. In particular these proofs dispense completely with estimates such as those in Proposition 5.4. On the other hand, it is not very difficult to deduce from the PNT that $\zeta(1+it) \neq 0$, so we may think of the Prime Number Theorem as equivalent to the absence of zeros of $\zeta(s)$ on the line $\sigma = 1$. But to prove the Prime Number Theorem with some bound for the error term, we need more information than this about the zeros.

Proposition 6.1. *There is some constant $\eta > 0$ so that the region*

$$1 - \frac{\eta}{\log^9(\tau)} \leq \sigma \leq 2$$

is free of zeros of $\zeta(s)$. Moreover, the bounds

$$\frac{1}{\zeta(s)} = O(\log^7(\tau))$$

and

$$-\frac{\zeta'(s)}{\zeta(s)} = \frac{1}{s-1} + O(\log^9(\tau))$$

hold, uniformly in σ, in the same region.

Proof. There is some open disk Δ around $s = 1$ that is free of zeros of $\zeta(s)$, and in which $1/\zeta(s)$ is bounded. It is clear that it will be sufficient to suppose that $s \notin \Delta$. In particular, we may use the bounds from Proposition 5.4 in the form $\zeta(s) = O(\log(\tau))$ and $\zeta'(s) = O(\log^2(\tau))$. Then the inequality

$$|\zeta(\sigma+it)| \geq |\zeta(\sigma)|^{-3/4}|\zeta(\sigma+2it)|^{-1/4} \geq A(\sigma-1)^{3/4}\log^{-1/4}(\tau)$$

holds in $1 < \sigma \leq 2$ for some constant $A > 0$. Now

$$\zeta(1+it) = \zeta(\sigma+it) + \int_{\sigma+it}^{1+it} \zeta'(u+it)\,du$$

yields

$$\begin{aligned} |\zeta(1+it)| &\geq |\zeta(\sigma+it)| - \int_{\sigma+it}^{1+it} |\zeta'(u+it)|\,|du| \\ &\geq A(\sigma-1)^{3/4}\log^{-1/4}(\tau) - B(\sigma-1)\log^2(\tau) \end{aligned}$$

for some positive constant B. Choose $\sigma - 1 = C\log^{-9}(\tau)$ for C a small positive parameter. Then

$$\begin{aligned}|\zeta(1+it)| &\geq AC^{3/4}\log^{-27/4}(\tau)\log^{-1/4}(\tau) - BC\log^{-9}(\tau)\log^2(\tau)\\ &= (AC^{3/4} - BC)\log^{-7}(\tau) \geq D\log^{-7}(\tau)\end{aligned}$$

for some constant $D > 0$, by choosing C small enough.

Thus far, we have used information on $\zeta(s)$ in the half plane $\sigma > 1$ to obtain a lower bound for $\zeta(s)$ on $\sigma = 1$. Next we use the information that we have gathered concerning $\zeta(s)$ on $\sigma = 1$ to obtain a lower bound in a narrow region near $\sigma = 1$.

The inequality

$$\begin{aligned}|\zeta(\sigma+it)| &\geq |\zeta(1+it)| - \left|\int_{1+it}^{\sigma+it} \zeta'(u+it)\,du\right|\\ &\geq D\log^{-7}(\tau) - B|1-\sigma|\log^2(\tau)\end{aligned}$$

holds in the region $1 - \log^{-1}(\tau) \leq \sigma \leq 2$. We note that the region $1 - \eta\log^{-9}(\tau) \leq \sigma \leq 1 + \eta\log^{-9}(\tau)$ is contained in the previous region if $0 < \eta \leq 1$, say. Choosing η so small that the last inequality for σ implies that

$$B|1-\sigma|\log^2(\tau) \leq \frac{1}{2}D\log^{-7}(\tau),$$

we obtain the lower bound

$$|\zeta(\sigma+it)| \geq \frac{D}{2}\log^{-7}(\tau) > 0$$

in the region $1-\eta\log^{-9}(\tau) \leq \sigma \leq 1+\eta\log^{-9}(\tau)$, choosing $\eta = \min(1, D/2B)$. In particular there are no zeros of $\zeta(s)$ in this region. The inequality

$$|\zeta(\sigma+it)| \geq A(\sigma-1)^{3/4}\log^{-1/4}(\tau)$$

implies that $|\zeta(\sigma+it)| \geq A\eta^{3/4}\log^{-7}(\tau)$ holds in the region $1+\eta\log^{-9}(\tau) \leq \sigma \leq 2$, so we finally obtain the bound $1/\zeta(s) = O(\log^7(\tau))$, uniformly in σ, in the region $1 - \eta\log^{-9}(\tau) \leq \sigma \leq 2$.

It remains to bound $-\zeta'(s)/\zeta(s)$ in the same region. The function

$$F(s) = -\frac{\zeta'(s)}{\zeta(s)} - \frac{1}{s-1}$$

is bounded on Δ, while $(s-1)^{-1}$ is bounded outside Δ. Now

$$\left|-\frac{\zeta'(s)}{\zeta(s)}\right| = |\zeta'(s)|\left|\frac{1}{\zeta(s)}\right| \leq O(\log^2(\tau))O(\log^7(\tau)) = O(\log^9(\tau))$$

holds within the region $1 - \eta\log^{-9}(\tau) \leq \sigma \leq 2$ less Δ. □

6.2. A proof of the PNT

Proposition 6.2 (Prime Number Theorem). *There exists some constant $c > 0$ such that*

$$\psi(x) = x + O\Big(x\, e^{-c \log^{1/10}(x)}\Big)$$

as $x \to +\infty$.

Proof. The auxiliary function

$$\psi_1(x) = \sum_{n \le x} (x-n)\Lambda(n)$$

is more easily handled by the Perron formula than $\psi(x)$ itself. We shall first find an estimate for $\psi_1(x)$, and afterwards it will prove straightforward to use this to establish the Prime Number Theorem.

The inequalities

$$\begin{aligned}\left| x\sum_{n<x} \Lambda(n) - \frac{1}{2\pi i}\int_{2-iT}^{2+iT} \left(-\frac{\zeta'(s)}{\zeta(s)}\right) \frac{x^{s+1}}{s}\, ds \right| \\ \ll x^3 \sum_{n=1}^{\infty} \frac{\Lambda(n) n^{-2}}{\max(1, T|\log(x/n)|)}\end{aligned}$$

and

$$\begin{aligned}\left| \sum_{n<x} n\Lambda(n) - \frac{1}{2\pi i}\int_{3-iT}^{3+iT} \left(-\frac{\zeta'(s-1)}{\zeta(s-1)}\right) \frac{x^{s}}{s}\, ds \right| \\ \ll x^3 \sum_{n=1}^{\infty} \frac{n\Lambda(n) n^{-3}}{\max(1, T|\log(x/n)|)}\end{aligned}$$

follow from Proposition 5.2. Change the variable from s to $s+1$ in the integral in the second inequality, and take the difference. Then

$$\left| \psi_1(x) - \frac{1}{2\pi i}\int_{2-iT}^{2+iT} \left(-\frac{\zeta'(s)}{\zeta(s)}\right) \frac{x^{s+1}}{s(s+1)}\, ds \right| \ll x^3 \sum_{n=1}^{\infty} \frac{\Lambda(n) n^{-2}}{\max(1, T|\log(x/n)|)}$$

for any $T > 0$ and any $x > 0$ that is not an integer.

Let x be of the form $x = [x] + 1/2$ and choose $T = x^4$. Then

$$\begin{aligned}\left| \psi_1(x) - \frac{1}{2\pi i}\int_{2-iT}^{2+iT} \left(-\frac{\zeta'(s)}{\zeta(s)}\right) \frac{x^{s+1}}{s(s+1)}\, ds \right| &\ll x^3 \sum_{n=1}^{\infty} \frac{\Lambda(n) n^{-2}}{\max(1, T|\log(x/n)|)} \\ &\ll \frac{x^3}{Tx^{-1}} = 1\end{aligned}$$

because $|\log(x/n)| \gg x^{-1}$. Applying the Residue Theorem on the rectangle with corners $2+iT, 1/2+iT, 1/2-iT, 2-iT$, we see that

$$\begin{aligned}\frac{1}{2\pi i}\int_{2-iT}^{2+iT}\frac{1}{s-1}\frac{x^{s+1}}{s(s+1)}\,ds &= \frac{x^2}{2} - \frac{1}{2\pi i}\int_{2+iT}^{1/2+iT}\frac{1}{s-1}\frac{x^{s+1}}{s(s+1)}\,ds \\ &- \frac{1}{2\pi i}\int_{1/2+iT}^{1/2-iT}\frac{1}{s-1}\frac{x^{s+1}}{s(s+1)}\,ds - \frac{1}{2\pi i}\int_{1/2-iT}^{2-iT}\frac{1}{s-1}\frac{x^{s+1}}{s(s+1)}\,ds,\end{aligned}$$

where

$$\left|\int_{1/2\pm iT}^{2\pm iT}\frac{1}{s-1}\frac{x^{s+1}}{s(s+1)}\,ds\right| \ll \frac{x^3}{T^3} \ll 1$$

and

$$\left|\int_{1/2+iT}^{1/2-iT}\frac{1}{s-1}\frac{x^{s+1}}{s(s+1)}\,ds\right| \ll x^{3/2}\int_{1/2-iT}^{1/2+iT}\frac{|ds|}{|s^3-s|} \ll x^{3/2},$$

because the last integral converges as $T \to +\infty$.

The function $F(s) = -\zeta'(s)/\zeta(s) - 1/(s-1)$ is holomorphic on a domain that contains the half plane $\sigma \geq 1$. It follows from Cauchy's theorem applied on the rectangle with corners $2+iT, 1+iT, 1-iT, 2-iT$ that

$$\begin{aligned}\frac{1}{2\pi i}\int_{2-iT}^{2+iT}F(s)\frac{x^{s+1}}{s(s+1)}\,ds &= \frac{1}{2\pi i}\int_{1-iT}^{1+iT}F(s)\frac{x^{s+1}}{s(s+1)}\,ds \\ &+ \frac{1}{2\pi i}\int_{1+iT}^{2+iT}F(s)\frac{x^{s+1}}{s(s+1)}\,ds - \frac{1}{2\pi i}\int_{1-iT}^{2-iT}F(s)\frac{x^{s+1}}{s(s+1)}\,ds,\end{aligned}$$

where

$$\left|\int_{1\pm iT}^{2\pm iT}F(s)\frac{x^{s+1}}{s(s+1)}\,ds\right| \ll x^3\frac{\log^9(T+4)}{T^2} \ll 1$$

by Proposition 6.1. Combining the various inequalities yields

$$\psi_1(x) = \frac{x^2}{2} + \frac{1}{2\pi i}\int_{1-iT}^{1+iT}F(s)\frac{x^{s+1}}{s(s+1)}\,ds + O(x^{3/2}).$$

We now introduce a parameter U with $0 < U < T$ and put $b(U) = \eta\log^{-9}(U+4)$. Replace the line segment from $1-iT$ to $1+iT$ by the polygonal path starting at $1-iT$ with corners at $1-iU, 1-b(U)-iU, 1-b(U)+iU, 1+iU$ and ending at $1+iT$, using Cauchy's theorem. See Figure 3 on page 176. Now

$$\int_{1-iT}^{1+iT} F(s)\frac{x^{s+1}}{s(s+1)}\,ds = \int_{1-iT}^{1-iU} F(s)\frac{x^{s+1}}{s(s+1)}\,ds$$
$$+\int_{1-iU}^{1-b(U)-iU} F(s)\frac{x^{s+1}}{s(s+1)}\,ds + \int_{1-b(U)-iU}^{1-b(U)+iU} F(s)\frac{x^{s+1}}{s(s+1)}\,ds$$
$$+\int_{1-b(U)+iU}^{1+iU} F(s)\frac{x^{s+1}}{s(s+1)}\,ds + \int_{1+iU}^{1+iT} F(s)\frac{x^{s+1}}{s(s+1)}\,ds$$

where

$$\left|\int_{1\pm iT}^{1\pm iU} F(s)\frac{x^{s+1}}{s(s+1)}\,ds\right| \ll x^2\int_U^T \frac{\log^9(t+4)}{t^2}\,dt \ll \frac{x^2}{U^{1/2}},$$

because $\log^9(t+4)\ll x^{1/2}$. Furthermore

$$\left|\int_{1\pm iU}^{1-b(U)\pm iU} F(s)\frac{x^{s+1}}{s(s+1)}\,ds\right| \ll x^2\frac{\log^9(U+4)}{U^2} \ll \frac{x^2}{U^{3/2}}$$

and

$$\left|\int_{1-b(U)-iU}^{1-b(U)+iU} F(s)\frac{x^{s+1}}{s(s+1)}\,ds\right|$$
$$\ll x^{2-b(U)}\int_{1-b(U)-iU}^{1-b(U)+iU}\frac{\log^9(|\mathrm{Im}(s)|+4)}{|s^2+s|}\,|ds| \ll x^{2-b(U)},$$

because the last integral converges as $U\to+\infty$. These bounds also follow by Proposition 6.1.

The estimate

$$\psi_1(x) = \frac{x^2}{2} + O\Big(x^{2-\eta\log^{-9}(U+4)}\Big) + O(x^2U^{-1/2}) + O(x^{3/2})$$

is valid for all x and U sufficiently large by the preceding calculations. Choosing $U = e^{2\log^{1/10}(x)} - 4$ yields $x^2U^{-1/2}\ll x^2e^{-\log^{1/10}(x)}$ and

$$x^{2-\eta\log^{-9}(U+4)} = x^2\exp\Big(-\log(x)\eta\log^{-9}(e^{2\log^{1/10}}(x))\Big)$$
$$= x^2e^{-\eta 2^{-9}\log(x)\log^{-9/10}(x)} = x^2e^{-2^{-9}\eta\log^{1/10}(x)}.$$

Then

$$\psi_1(x) = \frac{x^2}{2} + O\Big(x^2e^{-2c\log^{1/10}(x)}\Big)$$

with $2c = \min(1, 2^{-9}\eta)$.

The formula

$$\psi_1(x) = \int_1^x \psi(u)\,du$$

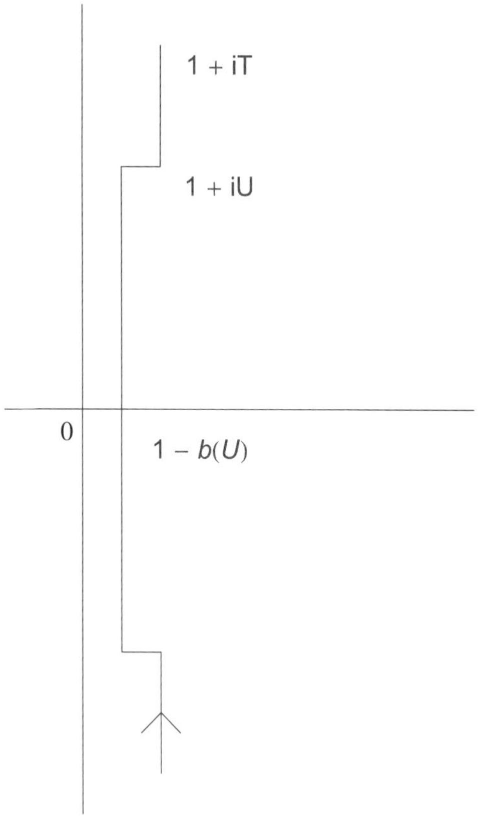

Figure 3. A contour in the proof of the PNT

holds by partial summation. The above estimate gives $|\psi_1(x) - x^2/2| \le \omega^2(x) = C^2x^2e^{-2c\log^{1/10}(x)}$ for $x \ge 1$, where C is some positive constant. We may choose $c > 0$ as small as we please, and we choose it so small that $\omega(x)$ is monotone increasing on $[1, \infty)$. Now

$$\psi(x) \ge \frac{1}{h}\int_{x-h}^{x} \psi(u)\, du = \frac{\psi_1(x) - \psi_1(x-h)}{h}$$

since $\psi(x)$ is monotone increasing. But

$$\begin{aligned}\psi_1(x) - \psi_1(x-h) &\ge \frac{x^2}{2} - \omega^2(x) - \frac{(x-h)^2}{2} - \omega^2(x-h)\\ &\ge xh - \frac{h^2}{2} - 2\omega^2(x),\end{aligned}$$

and so

$$\psi(x) \ge x - \frac{h}{2} - \frac{2\omega^2(x)}{h}.$$

Choosing $h = 2\omega(x)$ yields

$$\psi(x) \geq x - 2\omega(x) = x - 2Cxe^{-c\log^{1/10}(x)}.$$

The same argument applied on $[x, x+h]$ gives an upper bound. $\square$

6.3. Notes

The approach to the Prime Number Theorem in this chapter is due to Landau [**Lan03a**]. For our purposes the special merit of this particular proof is that it yields a useful error bound with modest effort and using no complex analysis beyond the residue calculus.

Proposition 6.1 is a consequence of stronger estimates found by de la Vallée Poussin in 1899 [**dlVP99**], but was singled out and given a simple proof by Landau [**Lan03a**]. The technique for bounding $\zeta(s)$ from below on the line $\sigma = 1$ is due to Hadamard [**Had96**] and de la Vallée Poussin [**dlVP96**] independently, and was streamlined by de la Vallée Poussin and by Mertens [**Mer98**]. Other techniques have since been discovered, but none that yield a better lower bound on the actual line $\sigma = 1$. Since the work of de la Vallée Poussin, the lower bound on $\sigma = 1$ has been combined with better upper bounds on the modulus of the zeta function to yield larger zero-free regions to the left of the line. See the second edition of *The Theory of the Riemann Zeta-Function* [**Tit39**] by E. C. Titchmarsh and revised by D. R. Heath-Brown, especially Chapters III and V and the associated end-of-chapter notes.

The first proofs of the Prime Number Theorem in the form of the asymptotic relation $\pi(x) \sim \operatorname{li}(x)$ by Hadamard [**Had96**] and de la Vallée Poussin [**dlVP96**] appeared in 1896 and relied heavily on the ideas of Riemann [**Rie59**]. These proofs were difficult and made essential use of complex analysis, especially the Hadamard factorization theorem for entire functions that had appeared [**Had93**] in 1893. But a long process of discovering less demanding proofs of the PNT began soon afterwards. Landau was a pioneer in this endeavor, and published a treatise *Handbuch der Lehre von der Verteilung der Primzahlen* [**Lan74**] in 1909. This became the bible of the subject for decades; see in particular the remarks on page 89 of *Littlewood's Miscellany* [**Lit86**].

It had been appreciated at least since the work of T. J. Stieltjes in the 1880s that zero-free regions for $\zeta(s)$ to the left of $\sigma = 1$ could yield bounds for the error term in the Prime Number Theorem. In 1899 de la Vallée Poussin [**dlVP99**] proved the PNT in the stronger form

$$\pi(x) = \operatorname{li}(x) + O\Big(xe^{-C\sqrt{\log(x)}}\Big)$$

with some constant $C > 0$ by means of a zero-free region $\sigma > 1 - c/\log(\tau)$. In the early 1920s Hardy and Littlewood found an improved bound for the modulus of the zeta function by means of the same method of Weyl that we encountered in Section 4.4. In 1922 Littlewood [**Lit22**] used this bound to establish a zero-free region $\sigma > 1 - c\log\log(\tau)/\log(\tau)$ which led to an improved bound

$$\pi(x) = \operatorname{li}(x) + O\Big(xe^{-C\sqrt{\log(x)\log\log(x)}}\Big)$$

for the error term in the Prime Number Theorem. Littlewood did not publish his proof, apparently because it was very complicated, and a simpler proof of this result was published by Landau [**Lan24b**].

Further improved bounds for the error term in the Prime Number Theorem have conformed to the pattern set by Littlewood's work; improved bounds for exponential sums leading to improved bounds for the modulus of the zeta function leading to better zero-free regions leading to better error bounds. Although there is now a somewhat broader range of techniques available to establish zero-free regions, none have as yet proved more efficacious than the combination of the original idea of Hadamard and de la Vallée Poussin with the most powerful technique available to bound exponential sums.

The current best bound is

$$\pi(x) = \mathrm{li}(x) + O\Big(xe^{-c\log(x)^{3/5}\log\log(x)^{-1/5}}\Big),$$

due independently to N. M. Korobov [**Kor58**] and I. M. Vinogradov [**Vin58**] in 1958. This is based on Vinogradov's technique for bounding Weyl sums. The bound had various precursors, also based on the Vinogradov technique, due to N. G. Chudakov, L. K. Hua, N. M. Korobov, T. Tatuzawa, and E. C. Titchmarsh. See page 193 of *Multiplicative Number Theory I. Classical Theory* by Montgomery and Vaughan for details of these contributions.

During the interwar years it was discovered that Fourier theory could substitute for complex analysis when proving the Prime Number Theorem. Both groups of proofs, the ones relying mainly on Fourier analysis as well as the ones using complex analysis, hinged on the fact that the Riemann zeta function has no zeros on the line $\sigma = 1$.

As we have already seen, many proofs in analytic number theory require neither complex analysis nor Fourier theory. Such proofs are said to be *elementary*. Techniques of combinatorics, elementary number theory, inequalities, and linear algebra are reckoned as elementary. Calculus is also generally regarded as elementary, though it does of course involve limiting processes.

For several decades after 1896 no elementary proof of the Prime Number Theorem was found, and all known proofs depended on the same crucial fact from complex analysis. So it was generally doubted whether the PNT had an elementary proof. But in 1948 Erdös [**Erd49**] and Selberg [**Sel49**], partly independently, discovered elementary proofs of the Prime Number Theorem. This was a wholly unexpected development. Both Erdös and Selberg relied for their proofs on a variant of the formula

$$\psi(x)\log(x) + \sum_{n\le x}\Lambda(n)\psi\left(\frac{x}{n}\right) = 2x\log(x) + O(x),$$

found by Selberg. Though the Selberg formula is not difficult to establish, the proofs of the Prime Number Theorem by Erdös and by Selberg were rather intricate. In analytic number theory, "elementary" is not a synonym for "easy". The discovery of the first elementary proofs of the PNT led to considerable activity up to the 1970s, partly to find simpler proofs and partly to obtain error bounds. Though no easy proof was discovered, various modifications and simplifications were achieved, and good error bounds were also established. The better elementary error bounds currently known lie between our Proposition 6.2 and de la Vallée Poussin's error bound in strength. These better elementary error bounds are quite arduous to prove, yet they are still far from matching the Korobov-Vinogradov bound. There is a survey of these developments up to 1982 in the paper [**Dia82**] by H. G. Diamond.

In 1984 H. Daboussi [**Dab84**] found the first elementary proof of the Prime Number Theorem that does not rely on some analogue of the Selberg formula. Instead he bases his proof on psixyology, which is the study of the set of positive integers less than or equal to x with all their prime factors less than or equal to y. There is an exposition of this proof in the appealing book *The Prime Numbers and their Distribution* [**TF00**] by G. Tenenbaum and M. Mendès France. A different elementary proof of the PNT that does not rely on the Selberg formula was discovered by A. J. Hildebrand [**Hil86**] in 1988.

Exercises

(1) Show that
$$\psi(x) - \psi\left(x - \frac{x}{\log^m(x)}\right) \sim \frac{x}{\log^m(x)}$$
as $x \to +\infty$, for any positive integer m.

(2) † Prove the relations
$$\pi(x) = \mathrm{li}(x) + \frac{\vartheta(x) - x}{\log(x)} + \int_2^x \frac{\vartheta(t) - t}{t\log^2(t)}\,dt + \frac{2}{\log(2)}$$
$$\vartheta(x) = x + (\pi(x) - \mathrm{li}(x))\log(x) - \int_2^x \frac{\pi(t) - \mathrm{li}(t)}{t}\,dt - 2$$
by means of partial summation and integration by parts.

(3) a) Establish the formal Dirichlet series identity
$$\frac{\zeta(s)\zeta(s-a)\zeta(s-b)\zeta(s-a-b)}{\zeta(2s-a-b)} = \sum_{n=1}^{\infty} \sigma_a(n)\sigma_b(n)n^{-s}$$
where
$$\sigma_a(n) = \sum_{d|n} d^a$$
and a and b are arbitrary complex constants (Ramanujan.)
b) Specialize to $a = ic$ and $b = -ic$ with c a positive real number in part a). Denote the resulting function by $A(s)$ and show that if $\zeta(s)$ has a zero on the line $\sigma = 1$, then the Dirichlet series of $A(s)$ must converge for $\sigma > 0$.
c) Show that $A(1/2) = 0$ and conclude that $\zeta(s)$ has no zeros on the line $\sigma = 1$ (Ingham.)

(4) This exercise requires an epsilon of Fourier analysis, specifically that

$$\int_{\mathbb{R}} f(t)e^{itu}\,dt \to 0$$

as $|u| \to \infty$ if f is integrable on $\mathbb{R}$. This is the *Riemann-Lebesgue Lemma.*

a) Show that

$$\sum_{n\le x}(x-n)(1-\Lambda(n)) = \frac{1}{2\pi i}\int_{c-i\infty}^{c+i\infty}\left(\zeta(s)+\frac{\zeta'(s)}{\zeta(s)}\right)\frac{x^{s+1}}{s(s+1)}\,ds$$

for $c > 1$.

b) Show that

$$f(t) = \left(\zeta(1+it)+\frac{\zeta'(1+it)}{\zeta(1+it)}\right)\frac{1}{(1+it)(2+it)}$$

is continuous and integrable on $\mathbb{R}$.

c) Show that

$$\sum_{n\le x}(x-n)(1-\Lambda(n)) = \frac{x^2}{2\pi}\int_{-\infty}^{\infty} f(t)e^{it\log(x)}\,dt$$

and conclude that

$$\sum_{n\le x}(x-n)(1-\Lambda(n)) = o(x^2)$$

as $x \to +\infty$.

d) By peeking ahead to the end of the proof of Proposition 6.2 and making small modifications, prove the Prime Number Theorem in the form $\psi(x) \sim x$.

(5) Show that

$$\pi(x) = \mathrm{li}(x) + O\Big(xe^{-c\log^{1/10}(x)}\Big)$$

for some constant $c > 0$.

(6) Show that

$$\liminf_{k\to+\infty}\frac{d_k}{\log(p_k)} \le 1 \le \limsup_{k\to+\infty}\frac{d_k}{\log(p_k)},$$

where $d_k = p_{k+1} - p_k$.

(7) a) By splitting the interval of integration into two subintervals, show that

$$\int_2^x \frac{dt}{\log^n(t)} = O\left(\frac{x}{\log^n(x)}\right)$$

for any positive integer n.

b) By means of integration by parts, show that

$$\int_2^x \frac{dt}{\log^n(t)} = \frac{x}{\log^n(x)} - \frac{2}{\log^n(2)} + n\int_2^x \frac{dt}{\log^{n+1}(t)}$$

for any positive integer n.

c) Show that

$$\mathrm{li}(x) = \frac{x}{\log(x)} + \cdots + (n-1)!\frac{x}{\log^n(x)} + O\left(\frac{x}{\log^{n+1}(x)}\right)$$

for any positive integer n.

d) Show that, if $\pi(x) = x/(\log(x) - A(x))$,, then $A(x) \to 1$ as $x \to +\infty$.

(8) Show that there is some constant c such that

$$\sum_{n\le x} \frac{\Lambda(n)}{n} = \log(x) + c + O_m\left(\frac{1}{\log^m(x)}\right)$$

for each positive integer m.

(9) a) Define

$$M_1(x) = \sum_{n\le x}(x-n)\mu(n)$$

and show that

$$M_1(x) \ll x^2 e^{-2c\log^{1/10}(x)}$$

for some $c > 0$.

b) Show that

$$\frac{M_1(x) - M_1(x-h)}{h} - 1 - \frac{h}{2} \le M(x) \le \frac{M_1(x) - M_1(x-h)}{h} + 1 + \frac{h}{2}$$

for any $h > 0$.

c) Prove that $M(x) \ll x\exp(-c\log^{1/10}(x))$ for some $c > 0$.

d) † Prove that

$$Q(x) = \frac{6}{\pi^2}x + O\left(\sqrt{x}e^{-c'\log^{1/10}(x)}\right)$$

for some $c' > 0$. (Hint: The hyperbola method.)

(10) † Show that if

$$\psi(x;q,a) = \sum_{\substack{n\le x\\ n\equiv a \bmod q}} \Lambda(n)$$

then

$$\psi(x;q,a) = \frac{x}{\phi(q)} + O\left(x\,e^{-c\log^{1/10}(x)}\right)$$

for fixed q and a with $\gcd(q,a) = 1$. This is a modest version of the prime number theorem for arithmetic progressions. Note that there is no uniformity with respect to q, so we cannot allow q to grow with x.

(11) Show that $(1, x]$ contains more primes than $(x, 2x]$ for all x sufficiently large. (Stated by F. J. E. Lionnet in 1872 and established by E. Landau in 1901.)

(12) † Find an asymptotic estimate for the summatory function of $\Lambda * \log$.

Chapter 7

The Siegel-Walfisz Theorem

7.1. Zero-free regions for L-functions

Our objective in this chapter is a prime number theorem for arithmetic progressions $n \equiv a \pmod q$. To make the result versatile it is necessary to pay careful attention to the dependence of the error term on the modulus q.

For a and q coprime we shall estimate the summatory function

$$\psi(x; q, a) = \sum_{\substack{n \leq x \\ n \equiv a \bmod q}} \Lambda(n) = \sum_{n \leq x} \Lambda(n) \mathrm{I}_{q,a}(n)$$

of the coefficients of the Dirichlet series

$$\Psi(s; q, a) = \sum_{n=1}^{\infty} \Lambda(n) \mathrm{I}_{q,a}(n) n^{-s},$$

where $\mathrm{I}_{q,a}$ is the indicator function of the arithmetic progression $n \equiv a \pmod q$. This was harmonically analyzed in terms of Dirichlet characters back in Section 3.6. Thus

$$\begin{aligned}\Psi(s; q, a) &= \sum_{n=1}^{\infty} \Lambda(n) n^{-s} \frac{1}{\phi(q)} \sum_{\chi \bmod q} \overline{\chi(a)} \chi(n) \\ &= \frac{1}{\phi(q)} \sum_{\chi \bmod q} \overline{\chi(a)} \sum_{n=1}^{\infty} \Lambda(n) \chi(n) n^{-s} = \frac{-1}{\phi(q)} \sum_{\chi \bmod q} \overline{\chi(a)} \frac{L'(s, \chi)}{L(s, \chi)},\end{aligned}$$

as one sees by logarithmically differentiating the Euler product of $L(s,\chi)$. Now apply the truncated Perron formula of Chapter 5 to obtain

$$\left|\psi(x;q,a) - \frac{1}{2\pi i}\int_{c-iT}^{c+iT} \Psi(s;q,a)\frac{x^s}{s}\,ds\right| \le 2x^c \sum_{n=1}^{\infty} \frac{|\Lambda(n)\mathrm{I}_{q,a}(n)|n^{-c}}{\max(1, T|\log(x/n)|)}$$

for $x > 1$ not an integer, $c > 1$ and $T > 0$. In Chapter 6 we used the truncated Perron formula with smoothing to establish the PNT with an error term. This time we shall apply it in a different style, making use of the nice inequality $|\Lambda(n)I_{q,a(n)}| \le \log(n)$ for the coefficients on the right-hand side.

As one might expect, information about the zeros of $L(s,\chi)$ is required to apply the Residue Theorem to move the contour of integration sufficiently far to the left. It is moreover necessary to bound the logarithmic derivative L'/L on the new contour. There are several ways to do this. The classical method uses the Hadamard factorization theorem. There is another method, in its original form due to Landau, which has the advantage of only using information about those zeros of $L(s,\chi)$ that lie close to s. This local method does not require the Hadamard factorization theorem, and is the one that we shall use.

Proposition 7.1. *Let $f(z)$ be meromorphic on a domain that contains the closed disk $|z-a| \le R$ and have no zero or pole at $z = a$ or on the circle $|z-a| = R$. Let $z_k = a + r_k \exp(i\theta_k)$ be the zeros and poles of $f(z)$ in the disk. Then*

$$\mathrm{Re}\frac{f'(a)}{f(a)} = \sum_k n_k \left(\frac{r_k}{R^2} - \frac{1}{r_k}\right)\cos(\theta_k) + \frac{1}{R}\int_0^{2\pi} \cos(\theta)\log|f(a+Re^{i\theta})|\frac{d\theta}{\pi},$$

where n_k denotes the multiplicity of z_k viewed as a zero, thus negative if z_k is a pole.

Proof. Assume without loss of generality that $a = 0$. The integral

$$I = \frac{1}{2\pi i}\oint_{|z|=R} \left(\frac{1}{z} - \frac{z}{R^2}\right)\frac{f'(z)}{f(z)}\,dz$$

may be evaluated as

$$I = \frac{f'(0)}{f(0)} + \sum_k n_k \left(\frac{1}{z_k} - \frac{z_k}{R^2}\right)$$

by residue calculus. But integration by parts yields

$$I = \left(\frac{1}{z} - \frac{z}{R^2}\right) \frac{\log(f(z))}{2\pi i}\Bigg|_{R\exp(2\pi i 0^+)}^{R\exp(2\pi i 0^-)} + \oint_{|z|=R} \left(\frac{1}{z^2} + \frac{1}{R^2}\right) \log(f(z)) \frac{dz}{2\pi i},$$

$$= 0 - 0 + \frac{1}{\pi R} \int_0^{2\pi} \cos(\theta) \log(f(Re^{i\theta}))\, d\theta,$$

where the path of integration starts and ends at $z = R$ on the positive real axis, and runs in the counterclockwise direction around the circle. Taking the real part of both expressions for I and setting them equal yields the desired identity. □

In the present state of knowledge, what one can prove about the zeros of $L(s, \chi)$ depends significantly on whether χ is a real or complex character.

Proposition 7.2. *There is an absolute constant $c_1 > 0$ such that for any complex character χ modulo q the region*

$$\sigma \geq 1 - \frac{c_1}{\log(q\tau)}$$

is free of zeros of $L(s, \chi)$.

Proof. Put $\theta_{n,t} = \arg(\chi(n)) + t\log(n)$ for n coprime with q. Then

$$3\,\mathrm{Re}\left(-\frac{L'(\sigma, \chi_0)}{L(\sigma, \chi_0)}\right) + 4\,\mathrm{Re}\left(-\frac{L'(\sigma + it, \chi)}{L(\sigma + it, \chi)}\right) + \mathrm{Re}\left(-\frac{L'(\sigma + 2it, \chi^2)}{L(\sigma + 2it, \chi^2)}\right)$$

$$= \sum_{n=1}^{\infty} \chi_0(n) n^{-\sigma} \Lambda(n)(3 + 4\cos(\theta_{n,t}) + \cos(2\theta_{n,t})) \geq 0$$

for $\sigma > 1$, by logarithmic differentiation of the Euler product for the L-function, and the trigonometric inequality

$$3 + 4\cos(\theta) + \cos(2\theta) = 3 + 4\cos(\theta) + 2\cos^2(\theta) - 1 = 2(1 + \cos(\theta))^2 \geq 0.$$

Let $\sigma > 1$ and $t \in \mathbb{R}$, and choose $R > 0$ so that none of the L-functions modulo q have any zeros on the circle $|z - \sigma - it| = R$. Then

$$\mathrm{Re}\left(-\frac{L'(\sigma + it, \chi)}{L(\sigma + it, \chi)}\right) = \sum_k n_k \left(\frac{1}{r_k} - \frac{r_k}{R^2}\right) \cos(\theta_k)$$

$$+ \frac{1}{\pi R} \int_0^{2\pi} (-\cos(\theta)) \log|L(\sigma + it + Re^{i\theta}, \chi)|\, d\theta,$$

by applying Proposition 7.1 to $L(z, \chi)$ on the disk $|z - \sigma - it| \leq R$. Here $\sigma + it + r_k \exp(i\theta_k)$ is any zero (or pole) of this L-function in the disk and n_k the corresponding multiplicity. To obtain the information required from this identity, it is necessary to bound the integral term from above. Thus for $\pi/2 \leq \theta \leq 3\pi/2$ it is necessary to bound $|L(\sigma + it + Re^{i\theta}, \chi)|$ from above, while for the other values of θ it must be bounded from below. To bound $L(s,\chi)$ from above we will use Proposition 5.3. To bound $L(s,\chi)$ from below, first note that

$$\left|\zeta(\sigma)-\frac{1}{\sigma-1}-\frac{1}{2}\right|=\left|-\sigma\int_1^\infty S(u)u^{-\sigma-1}\right|\leq\sigma\int_1^\infty\frac{1}{2}u^{-\sigma-1}\,du=\frac{1}{2}$$

by Euler-Maclaurin summation applied to the Dirichlet series of $\zeta(\sigma)$. Then

$$\left|\frac{1}{L(s,\chi)}\right|=\prod_p|1-\chi(p)p^{-s}|\leq\prod_p(1+p^{-\sigma})\leq\zeta(\sigma)\leq\frac{1}{\sigma-1}+1,$$

so $|L(s,\chi)|\geq(\sigma-1)/\sigma$ for $\sigma>1$.

The conclusion is that for χ a nonprincipal character

$$\operatorname{Re}\left(-\frac{L'(\sigma+it,\chi)}{L(\sigma+it,\chi)}\right)\leq\sum_k n_k\left(\frac{1}{r_k}-\frac{r_k}{R^2}\right)\cos(\theta_k)+\frac{\log(2q\tau)}{R}+\frac{1}{R}\log\left(\frac{\sigma}{\sigma-1}\right)$$

if R is small enough that the disk $|z-\sigma-it|\leq R$ is contained in the half plane $\operatorname{Re}(z)\geq 1/4$. While for the principal character

$$\operatorname{Re}\left(-\frac{L'(\sigma+it,\chi_0)}{L(\sigma+it,\chi_0)}\right)\leq\sum_k n_k\left(\frac{1}{r_k}-\frac{r_k}{R^2}\right)\cos(\theta_k)+\frac{1}{R}\log\left(\frac{\sigma}{\sigma-1}\right)+\frac{1}{R}\log\left(2q\tau+\frac{1}{|-R+\sigma+it-1|}\right)$$

under the same conditions.

It is advantageous to choose R not too small, and R may be chosen larger than $\sigma-1/2$. Fix

$$\sigma_*=1+\frac{1}{K\log(q\tau)}$$

with $K\geq 1$ an undetermined constant. Then

$$\operatorname{Re}\left(-\frac{L'(\sigma_*,\chi_0)}{L(\sigma_*,\chi_0)}\right)\leq\sum_k n_k\left(\frac{1}{r_k}-\frac{r_k}{R^2}\right)\cos(\theta_k)+2\log\left(1+K\log(q\tau)\right)+2\log\left(2+2q\tau\right),$$

where $z_k=r_k\exp(i\theta_k)$ are the zeros and poles of $L(\sigma_*+z,\chi_0)$ in the disk $|z|<R$, and n_k the associated multiplicities. There is a simple pole of $L(s,\chi_0)$ at $s=1$, say for $k=1$, with $n_1=-1$, $\theta_1=\pi$ and $r_1=\sigma-1$. Any zeros z_k of $L(\sigma_*+z,\chi_0)$ have $\pi/2<\theta_k<3\pi/2$, while $1/r_k-r_k/R^2>0$ in the disk. Thus the inequality still holds after throwing away the terms in the sum corresponding to possible zeros (actually none in this particular disk). Now

$$\begin{aligned}\mathrm{Re}\biggl(-\frac{L'(\sigma_*,\chi_0)}{L(\sigma_*,\chi_0)}\biggr) &\le \frac{1}{r_1} - \frac{r_1}{R^2} + 2\log\left(1+K\log(q\tau)\right) + 2\log\left(2+2q\tau\right)\\ &\le K\log(q\tau) + 2\log\left(1+K\log(q\tau)\right) + 4\log(q\tau)\end{aligned}$$

since $R \ge 1/2$, while $q \ge 1$ and $\tau \ge 4$.

Next

$$\mathrm{Re}\biggl(-\frac{L'(\sigma_*+2it,\chi^2)}{L(\sigma_*+2it,\chi^2)}\biggr) \le 2\log\left(1+K\log(q\tau)\right) + 4\log(q\tau),$$

because the whole sum on the right-hand side may be thrown away. All the terms correspond to zeros since $\chi^2 \ne \chi_0$, thus $n_k > 0$, and these zeros lie in the left-hand half of the disk, thus $\cos(\theta_k) < 0$, hence all the terms are negative.

Furthermore

$$\begin{aligned}\mathrm{Re}\biggl(-\frac{L'(\sigma_*+it,\chi)}{L(\sigma_*+it,\chi)}\biggr) &\le \sum_k n_k\biggl(\frac{1}{r_k} - \frac{r_k}{R^2}\biggr)\cos(\theta_k)\\ &\quad + \frac{\log(2q\tau)}{R} + \frac{1}{R}\log\biggl(\frac{\sigma_*}{\sigma_*-1}\biggr)\end{aligned}$$

where we assume that z_1 is a zero on the real axis, corresponding to a zero of $L(s,\chi)$ whose imaginary part equals t. Then $\cos(\theta_1) = -1$, and we throw away all terms except the one corresponding to z_1. Now

$$\begin{aligned}\mathrm{Re}\biggl(-\frac{L'(\sigma_*+it,\chi)}{L(\sigma_*+it,\chi)}\biggr) &\le -\frac{1}{r_1} + \frac{r_1}{R^2} + 2\log(2q\tau) + 2\log\left(1+K\log(q\tau)\right)\\ &\le -\frac{1}{r_1} + 6\log(q\tau) + 2\log\left(1+K\log(q\tau)\right)\end{aligned}$$

since $R \ge 1/2$, while $r_1 \le R$ and thus $r_1/R^2 \le 2 \le 2\log(q\tau)$.

Finally multiply the inequality at σ_* by 3, the inequality at σ_*+it by 4 and the inequality at σ_*+2it by 1 and add, obtaining

$$\begin{aligned}0 &\le -3\,\mathrm{Re}\frac{L'(\sigma_*,\chi_0)}{L(\sigma_*,\chi_0)} - 4\,\mathrm{Re}\frac{L'(\sigma_*+it,\chi)}{L(\sigma_*+it,\chi)} - \mathrm{Re}\frac{L'(\sigma_*+2it,\chi^2)}{L(\sigma_*+2it,\chi^2)}\\ &\le 3K\log(q\tau) - \frac{4}{r_1} + 40\log(q\tau) + 16\log(K\log(q\tau)+1)\\ &\le 3Kq\tau - \frac{4}{r_1} + 40\log(q\tau) + 16\log(K\log(q\tau)+K)\\ &\le 3Kq\tau - \frac{4}{r_1} + 40\log(q\tau) + 16\log(K) + 16\log(q\tau+1)\\ &\le 3Kq\tau - \frac{4}{r_1} + 72\log(q\tau) + 4\log(K)\log(q\tau),\end{aligned}$$

since $q\tau \geq 4$ and $K \geq 1$. Then

$$r_1 - (\sigma_* - 1) \geq \frac{4}{3K\log(q\tau) + 72\log(q\tau) + 4\log(K)\log(q\tau)} - \frac{1}{K\log(q\tau)}$$
$$= \frac{1}{K\log(q\tau)}\left(\frac{4}{3 + \frac{72}{K} + \frac{4\log(K)}{K}} - 1\right),$$

and because $4/3 > 1$ the result follows by choosing K sufficiently large. □

When χ is real, this argument has to be modified, for then the bound on the logarithmic derivative of $L(s, \chi^2)$ is much weaker when $|t|$ is very small.

Proposition 7.3. *There is an absolute constant $c_2 > 0$ such that for any quadratic character χ modulo q the region*

$$\sigma \geq 1 - \frac{c_2}{\log(q\tau)}$$

contains at most one zero of $L(s, \chi)$. If this zero exists, it is real and simple.

Proof. If χ is quadratic the same mode of reasoning as for the previous result goes through, but now the bound obtained is

$$r_1 - (\sigma_* - 1) \geq \frac{1}{K\log(q\tau)}\left(\frac{4}{3 + \frac{1}{|1+itK\log(q\tau)|} + \frac{72}{K} + \frac{4\log(K)}{K}} - 1\right)$$

because $\chi^2 = \chi_0$, and the disk $|s - \sigma_* - it| \leq R$ will contain the pole at $s = 1$ when $|t|$ is small enough. At the outset, K should be at least so large that $72/K + 4\log(K)/K \leq 1/4$, say. If $|t| \geq 2/(K\log(q\tau))$, then

$$|1 + itK\log(q\tau)| \geq |t|K\log(q\tau) \geq \frac{2}{K\log(q\tau)}K\log(q\tau) = 2,$$

and thus

$$r_1 - (\sigma_* - 1) \geq \frac{1}{15K\log(q\tau)}$$

follows.

By symmetry with respect to the real axis, it will now be enough to show that there is at most one zero of $L(s, \chi)$ in the region $\mathcal{R}$ determined by

$$0 \leq t \leq \frac{2}{K\log(q\tau)}, \quad 1 - \frac{1}{15K\log(q\tau)} \leq \sigma \leq 1.$$

Choosing R as before, with $R \geq 1/2$, the region $\mathcal{R}$ is contained in the disk $|s - \sigma_* - it| \leq R$ when K is sufficiently large, so if there are two zeros in $\mathcal{R}$, both will lie in this disk. This preliminary reasoning allows us to assume two zeros in a single disk, which is necessary in order to apply the local method with Proposition 7.1.

So assume that s_1 and s_2 are zeros of $L(s,\chi)$ in $\mathcal{R}$, and apply Proposition 7.1 with $a = \sigma_* + it$ where t is the imaginary part of s_1. Then

$$\begin{aligned}\mathrm{Re}\left(-\frac{L'(\sigma_*+it,\chi)}{L(\sigma_*+it,\chi)}\right) &\leq \sum_k n_k\left(\frac{1}{r_k}-\frac{r_k}{R^2}\right)\cos(\theta_k)\\ &\quad + 4\log(q\tau) + 2\log(K\log(q\tau)+1),\end{aligned}$$

where $s_1 = a + z_1$ and $s_2 = a + z_2$. Now an inequality

$$0 \leq 4K\log(q\tau) - \frac{4}{r_1} + \frac{4}{r_2}\cos(\theta_2) + 80\log(q\tau) + 4\log(K)\log(q\tau)$$

is obtained in the same way as before.

By considering a suitable triangle one obtains $\cos(\theta_2) \leq -2/\sqrt{5}$. See Figure 4 on page 190 for the way that the minimal value of θ_2 arises. For the sake of clarity, and to avoid giving a false impression that $\mathcal{R}$ is a rectangle, the height of the triangle is exaggerated in the figure. The bound

$$\begin{aligned}r_2 &\leq \max_{s\in\mathcal{R}}|s-\sigma_*-it|\\ &\leq \frac{1}{K\log(q\tau)} + \frac{2}{K\log(4q)} + \frac{1}{15K\log(4q)} \leq \frac{4}{K\log(q\tau)}\end{aligned}$$

follows, and implies that

$$\begin{aligned}\frac{4}{r_1} &\leq 4K\log(q\tau) - 4\frac{K\log(q\tau)}{4}\frac{2}{\sqrt{5}} + 80\log(q\tau) + 4\log(K)\log(q\tau)\\ &\leq \left(4-\frac{2}{\sqrt{5}}\right)K\log(q\tau) + 80\log(q\tau) + 4\log(K)\log(q\tau),\end{aligned}$$

thus

$$r_1 - (\sigma_* - 1) \geq \frac{1}{K\log(q\tau)}\left(\frac{4}{4-\frac{2}{\sqrt{5}}+\frac{80}{K}+\frac{4\log(K)}{K}} - 1\right).$$

Choosing K sufficiently large shows that the desired bound is also valid in the range $|t| \leq 2/(K\log(q\tau))$, with at most one exception.

A multiple zero is a limiting case of two distinct zeros, so a single zero in the region

$$\sigma \geq 1 - \frac{c_2}{\log(q\tau)}$$

must be simple, and it must lie on the real axis, otherwise there would be two distinct ones by symmetry, since $L(s,\chi)$ is real on the real axis when χ is real. □

This chapter is focused on reaching the Siegel-Walfisz theorem quickly and with few prerequisites. There are interesting topics around the Prime Number Theorem in arithmetic progressions that are neglected here. See *Multiplicative Number Theory I. Classical Theory* by Hugh L. Montgomery

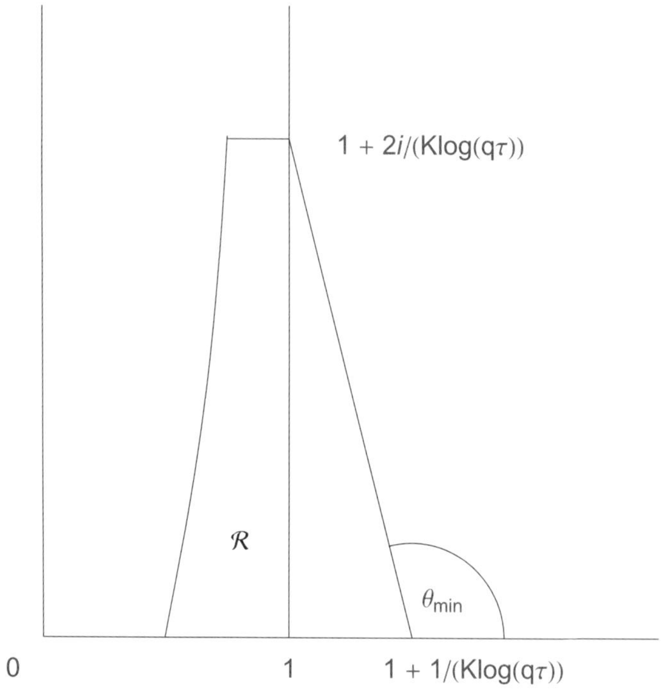

Figure 4. The smallest attainable angle

and Robert C. Vaughan for a more expansive treatment. Their book contains an outstanding collection of exercises of diverse levels of difficulty. *Multiplicative Number Theory* by Harold Davenport is a classic of analytic number theory, emphasizing the distribution of primes in arithmetic progressions. Davenport applies Hadamard factorization to prove the Siegel-Walfisz theorem, rather than the local method applied here.

7.2. An idea of Landau

The following result dates to 1918 and is due to Landau. The proof starts with an important idea that Siegel later used to prove his lower bound for $L(1,\chi)$.

Proposition 7.4. *There is an absolute constant $c_3 > 0$ such that for any two distinct quadratic characters χ_1, χ_2 modulo q_1, q_2 respectively with $\chi_1\chi_2$ not principal, the bound*

$$\min(\beta_1, \beta_2) < 1 - \frac{c_3}{\log(q_1q_2)}$$

holds for any real zeros β_1, β_2 of $L(s,\chi_1), L(s,\chi_2)$ respectively.

Proof. The inequality

$$-\frac{\zeta'(\sigma)}{\zeta(\sigma)} - \frac{L'(\sigma,\chi_1)}{L(\sigma,\chi_1)} - \frac{L'(\sigma,\chi_2)}{L(\sigma,\chi_2)} - \frac{L'(\sigma,\chi_1\chi_2)}{L(\sigma,\chi_1\chi_2)}$$
$$= \sum_{n=1}^{\infty}(1+\chi_1(n))(1+\chi_2(n))\Lambda(n)n^{-\sigma} \geq 0$$

is obtained by logarithmic differentiation of Euler products.

The bound

$$-\frac{\zeta'(\sigma)}{\zeta(\sigma)} \leq \frac{1}{\sigma-1} + \frac{1}{e}$$

may be established by Euler-Maclaurin summation in the numerator and denominator. The other terms will be bounded from above by means of the inequality

$$\mathrm{Re}\left(-\frac{L'(\sigma+it,\chi)}{L(\sigma+it,\chi)}\right) \leq \sum_k n_k\left(\frac{1}{r_k} - \frac{r_k}{R^2}\right)\cos(\theta_k) + \frac{\log(2q\tau)}{R} + \frac{1}{R}\log\left(\frac{\sigma}{\sigma-1}\right)$$

from the proof of Proposition 7.2. Fix

$$\sigma_* = 1 + \frac{1}{K\log(q_1q_2)}$$

with some constant $K \geq 1$, and R arbitrarily close to $\sigma_* - 1/2$. Applying the inequality with $t = 0$ and $\chi = \chi_1$ yields

$$\begin{aligned}\mathrm{Re}\left(-\frac{L'(\sigma_*,\chi_1)}{L(\sigma_*,\chi_1)}\right) &\leq \sum_k n_k\left(\frac{1}{r_k} - \frac{r_k}{R^2}\right)\cos(\theta_k) \\ &\quad + \frac{\log(2q\tau)}{R} + \frac{1}{R}\log\left(\frac{\sigma_*}{\sigma_*-1}\right) \\ &\leq -\frac{1}{\sigma_*-\beta_1} + 2 + 2\log(4q_1) + 2\log(K\log(q_1q_2)+1),\end{aligned}$$

by omitting all terms in the sum except the one corresponding to β_1. Of course an analogous bound holds for the character χ_2, and the bound

$$\mathrm{Re}\left(-\frac{L'(\sigma_*,\chi_1\chi_2)}{L(\sigma_*,\chi_1\chi_2)}\right) \leq 2\log(4q_1q_2) + 2\log(K\log(q_1q_2)+1)$$

holds by throwing the whole sum away. Moreover

$$\log(K\log(q_1q_2)+1) \leq \log(K) + \log(\log(q_1q_2)+1) \leq \log(K) + \log(q_1q_2),$$

and thus

$$K\log(q_1q_2) - \frac{1}{\sigma_*-\beta_1} - \frac{1}{\sigma_*-\beta_2} + 10\log(q_1q_2) + 6\log(K) + 13 \geq 0$$

by adding up the bounds. This yields

$$\frac{1}{\sigma_* - \beta_1} + \frac{1}{\sigma_* - \beta_2} \leq K\log(q_1q_2) + 17\log(q_1q_2) + 3\log(K)\log(q_1q_2).$$

Assume without loss of generality that $\beta_1 \leq \beta_2$. Then

$$\frac{2}{\sigma_* - \beta_1} \leq K\log(q_1q_2) + 17\log(q_1q_2) + 3\log(K)\log(q_1q_2),$$

and so

$$\begin{aligned}\beta_1 &\leq \sigma_* - \frac{2}{K\log(q_1q_2) + 17\log(q_1q_2) + 3\log(K)\log(q_1q_2)}\\ &= 1 - \frac{1}{K\log(q_1q_2)}\left(\frac{2}{1+\frac{17}{K}+\frac{3\log(K)}{K}} - 1\right).\end{aligned}$$

The desired bound follows by choosing K sufficiently large. □

The following result will prove useful when we establish the prime number theorem for arithmetic progressions.

Proposition 7.5. *There is an absolute constant $c_4 > 0$ such that there is a real zero*

$$\beta \geq 1 - \frac{c_4}{\log(q)}$$

of $L(s,\chi)$ for at most one character χ modulo q.

Proof. To exclude complex characters by Proposition 7.2, it is enough to choose $c_4 \leq c_1$. If χ_1, χ_2 is any pair of distinct primitive quadratic characters modulo q, then

$$\min(\beta_1, \beta_2) < 1 - \frac{c_3}{\log(\mathfrak{f}_1\mathfrak{f}_2)} \leq 1 - \frac{c_3}{\log(q^2)} = 1 - \frac{c_3}{2\log(q)}$$

for β_1, β_2 positive zeros of $L(s,\chi_1), L(s,\chi_2)$ respectively, and $\mathfrak{f}_1, \mathfrak{f}_2$ the conductors of χ_1, χ_2 respectively. Thus $c_4 = \min(c_1, c_3/2)$ will work. □

A zero of $L(s,\chi)$, with χ a Dirichlet character modulo q, in the region

$$\sigma > 1 - \frac{c}{\log(q\tau)},$$

for a sufficiently small $c > 0$, is called a *Landau-Siegel zero,* or just an *exceptional zero.* The choice of c is a pragmatic matter; we choose

$$c = \frac{1}{3}\min(1, c_1, c_2, c_4),$$

where c_1, c_2, c_4 are the constants in Propositions 7.2, 7.3, 7.5 respectively. In particular an exceptional zero is necessarily real and simple, the associated character is quadratic, and to any modulus there is at most one character associated with an exceptional zero. This associated character is called an *exceptional character* and its conductor an *exceptional conductor.* No exceptional conductor is known, and the existence or nonexistence of Landau-Siegel zeros is an outstanding problem in analytic number theory. But Landau proved that exceptional conductors, if they exist at all, are exceedingly scarce.

Proposition 7.6. *If* $\mathfrak{f}_1 < \mathfrak{f}_2$ *are exceptional conductors, then* $\mathfrak{f}_1^2 < \mathfrak{f}_2$.

Proof. Let χ_1 and χ_2 be the primitive quadratic characters associated to $\mathfrak{f}_1$ and $\mathfrak{f}_2$ respectively, and β_1 and β_2 the associated Landau-Siegel zeros. Then

$$1 - \frac{c}{\log(\mathfrak{f}_1)} < \min(\beta_1, \beta_2) < 1 - \frac{c_3}{\log(\mathfrak{f}_1\mathfrak{f}_2)} < 1 - \frac{3c}{\log(\mathfrak{f}_1\mathfrak{f}_2)},$$

and thus $\mathfrak{f}_1^2 < \mathfrak{f}_2$ by unraveling the inequality. □

That the definition of Landau-Siegel zeros involves a constant that we are free to choose may be disconcerting. But actually it leads to an important insight about exceptional zeros: In essence, either there are infinitely many Landau-Siegel zeros or else none. For we can choose the constant in the definition small enough to exclude any given finite set of zeros.

7.3. The theorem of Siegel

We were unable to exclude the possibility of an exceptional zero close to $s = 1$ for a Dirichlet L-function. The existence of an exceptional zero for a real character modulo q would profoundly influence the distribution of the primes in the arithmetic progressions with modulus q. As they would contradict the analogue for Dirichlet L-functions of the Riemann Hypothesis for the Riemann zeta function, the general expectation is that exceptional zeros do not exist. The following theorem of Siegel limits how close to $s = 1$ exceptional zeros may be found, and leads to a strong form of the PNT for arithmetic progressions.

Proposition 7.7 (Siegel's theorem). *For every* $\varepsilon > 0$ *there exists a constant* $C(\varepsilon) > 0$ *such that*

$$L(1, \chi) > C(\varepsilon) q^{-\varepsilon}$$

for any quadratic character χ *modulo* q.

An important remark about this result is that it is a pure existence statement: For $\varepsilon < 1/2$ it is the mere existence of $C(\varepsilon)$ that is asserted, and the method of proof does not allow us to assign any definite value to this constant. So though the theorem of Siegel will later prove adequate to draw an important theoretical conclusion about the class numbers of imaginary quadratic fields, we shall find it inadequate to obtain numerical bounds for class numbers. A result that asserts the existence of constants without allowing any definite values to be assigned to them is called *noneffective.* If definite values may be assigned to the constants, the result is called *effective.* An effective result is what is required for computational work. There do exist effective analogues of the theorem of Siegel, but they are weaker.

We now prove the theorem of Siegel by a method due to D. M. Goldfeld. At the appropriate place in the proof we indicate the reason why it leads to a noneffective result.

Proof. Suppose that χ and χ_1 are primitive quadratic characters to distinct moduli q, q_1 respectively. The Dirichlet series of

$$F(s) = \zeta(s)L(s,\chi_1)L(s,\chi)L(s,\chi_1\chi)$$

has nonnegative coefficients a_n since

$$\begin{aligned}\log(F(s)) &= \sum_p \sum_{k=1}^{\infty} \frac{1}{k}\Big(1+\chi_1(p^k)+\chi(p^k)+\chi_1(p^k)\chi(p^k)\Big)p^{-ks}\\ &= \sum_p \sum_{k=1}^{\infty} \frac{1}{k}\Big(1+\chi_1(p^k)\Big)\Big(1+\chi(p^k)\Big)p^{-ks},\end{aligned}$$

and exponentiating a Dirichlet series with nonnegative coefficients yields a Dirichlet series with nonnegative coefficients. Also note that $a_1 = 1$ from the Euler product for $F(s)$. Modifying the Perron Formula appropriately, we obtain

$$\sum_{n<x}(x-n)^5 a_n n^{-\beta} = \frac{1}{2\pi i}\int_{2-i\infty}^{2+i\infty} \frac{F(s+\beta)x^{s+5}}{s(s+1)(s+2)(s+3)(s+4)(s+5)}\,ds$$

for $1/2 < \beta < 1$, and thus

$$\frac{1}{2\pi i}\int_{2-i\infty}^{2+i\infty} F(s+\beta)\frac{x^s}{P(s)}\,ds \gg 1$$

where $P(s)$ is the polynomial in the denominator. We are going to shift the contour of integration to the line from $-\beta/2 - i\infty$ to $-\beta/2 + i\infty$. Note that

$$F(s) = O\big(q_1^2 q^2 \tau^4\big)$$

for $\sigma \geq 1/4$, except near $s = 1$, by Proposition 5.3. Since $\operatorname{Im}(s+\beta) \geq -\beta/2+\beta = \beta/2 \geq 1/4$ because $\beta > 1/2$, the integrals over the horizontal line segments tend to zero as $T \to +\infty$. The integrand of the line integral has poles at $s = 0$ and at $s = 1-\beta$. The first pole is simple and the residue there is $F(\beta)/120$. Since χ_1 and χ are primitive characters to distinct moduli, the product $\chi_1\chi$ is a nonprincipal character, so $F(s)$ has a simple pole at $s = 1$ with residue $\lambda = L(1,\chi_1)L(1,\chi)L(1,\chi_1\chi)$. Then the second pole of the integrand is also simple and the residue there is $\lambda x^{1-\beta}/P(1-\beta)$. Now

$$\lambda\frac{x^{1-\beta}}{P(1-\beta)}+\frac{F(\beta)}{2}+\frac{1}{2\pi i}\int_{-\beta/2-i\infty}^{-\beta/2+i\infty}\frac{F(s+\beta)x^s}{P(s)}\,ds\gg 1$$

by the Residue Theorem. Taking the absolute value in the integral, we obtain

$$\lambda\frac{x^{1-\beta}}{P(1-\beta)}+\frac{F(\beta)}{2}+O\Big(q_1^2q^2x^{-\beta/2}\Big)\gg 1.$$

The integrand decreases like τ^{-2} on the line $\sigma=-\beta/2$ and $P(s)\gg 1$ there.

We now distinguish two cases depending on a parameter δ with $0<\delta<1/2$. In the first case, there are no zeros of any L-function of any real primitive character in the interval $(1-\delta,1)$. In this case we choose some arbitrary real primitive character χ_1 modulo some $q_1\geq 3$. In the other case there exists a real primitive character χ_1 with modulus $q_1\geq 3$ such that $L(\beta,\chi_1)=0$ for some β with $1-\delta<\beta<1$. It is here that the noneffectiveness enters into the proof. We have no way to estimate the modulus q_1, since we know nothing about χ_1 beyond its mere existence in the second case.

In the first case, observe that $L(1,\chi_1)L(1,\chi)L(1,\chi_1\chi)$ is positive by the Euler product. The formula

$$(1-2^{1-\sigma})\zeta(\sigma)=1-2^{-\sigma}+3^{-\sigma}-\cdots$$

implies that $\zeta(\sigma)<0$ for $0<\sigma<1$. Thus $F(\beta)\leq 0$ for any β with $1-\delta<\beta<1$. In the second case, $F(\beta)=0$, and so $F(\beta)\leq 0$ for some such β in both cases. Now fix the character χ_1 and the point β in $(1-\delta,1)$ with $F(\beta)\leq 0$ for the rest of the proof. Next

$$\lambda\frac{x^{1-\beta}}{P(1-\beta)}+O\Big(q_1^2q^2x^{-\beta/2}\Big)\gg 1$$

by $F(\beta)\leq 0$, and choosing some x so large that

$$q_1^2q^2x^{-\beta/2}=c>0$$

with c a sufficiently small constant, we can obtain

$$\lambda x^{1-\beta}\gg 1.$$

Furthermore

$$\lambda=L(1,\chi_1)L(1,\chi)L(1,\chi_1\chi)\ll L(1,\chi)\log^2(q_1q)$$

by Proposition 5.4. Now

$$L(1,\chi)\log^2(q_1q)\gg x^{\beta-1}$$

while

$$x=(c^{-1}q_1^2q^2)^{2/\beta}$$

so

$$L(1,\chi)\log^2(q) \gg q^{-\frac{8(1-\beta)}{\beta}}$$

by assuming $q > q_1$, as we may. But by choosing δ sufficiently small, we can force $8(1-\beta)/\beta < \varepsilon/2$ for any $\varepsilon > 0$, and the power of $\log(q)$ is dominated by $q^{\varepsilon/2}$, so $L(1,\chi) \gg q^{-\varepsilon}$.

We now remove the restriction that χ be primitive. Suppose that $\chi \neq \chi_0$ is an imprimitive character modulo q induced by a primitive character χ_2 modulo q_2. Then

$$\begin{aligned} L(1,\chi) = L(1,\chi_2)\prod_{p|q}(1-\chi_2(p)p^{-1}) &\gg q_2^{-\epsilon/2}\prod_{p|q}(1-\chi_2(p)p^{-1}) \\ &\geq q^{-\epsilon/2}\prod_{p\leq q}(1-p^{-1}) \gg q^{-\epsilon/2}\frac{1}{\log(q)} \gg q^{-\epsilon} \end{aligned}$$

by the Mertens formula. □

This proof is a beautiful application of the method of contour integrals. The argument about choosing x large enough is a valuable simplification due to Friedlander.

7.4. The Borel-Carathéodory lemma

A *domain* is an open connected set. The Maximum Principle of complex analysis states that if f is a nonconstant holomorphic function on a domain $D \subseteq \mathbb{C}$, then $|f|$ cannot attain a local maximum at any point of D. There are many ways to see this, and perhaps the simplest is by means of power series. If $s_0 \in D$ then

$$f(s) = f(s_0) + \frac{f^{(n)}(s_0)}{n!}(s-s_0)^n + \cdots$$

in some neighborhood of s_0, for some positive integer n. Here $f^{(n)}(s_0)$ is the first nonzero derivative of f at s_0. Such a nonzero derivative must exist since f was assumed nonconstant. Let $s = s_0 + re^{i\theta}$ and choose θ so that $\arg(f^{(n)}(s_0)e^{in\theta}) = \arg(f(s_0))$, and choose $r > 0$ so small that the power series converges. If $f(s_0) = 0$ it does not matter what choice is made for θ, while if $f(s_0) \neq 0$ the choice made ensures that the first two nonzero terms of the power series pull in the same direction. Then

$$|f(s)| = |f(s_0)| + \frac{|f^{(n)}(s_0)|}{n!}r^n + O(r^{n+1}),$$

and choosing r sufficiently small, $|f(s)| > |f(s_0)|$. The analogue of the Maximum Principle for local minima fails, since a holomorphic function in a domain may have a zero there.

A very useful boundary version of the Maximum Principle states that if the domain D is bounded and $|f| \leq M$ in $\mathcal{N} \cap D$ where $\mathcal{N}$ is an open neighborhood of the boundary ∂D of D, then $|f| \leq M$ in the whole of D. For otherwise $|f|$ must attain a global maximum value larger than M on the closure of the complement in D of $\mathcal{N} \cap D$, since this closure is compact, and $|f|$ is continuous on D. If f is a continuous function on the closure of a bounded domain D and holomorphic in D, we may simply assert that $|f|$ attains its global maximum value on the boundary of D.

We use the Maximum Principle to establish an important inequality between the absolute value and the real part of a holomorphic function on a disk. This inequality will also prove useful later when establishing the Hadamard factorization theorem in Chapter 8.

Proposition 7.8 (Borel-Carathéodory lemma). *Suppose that $f(z)$ is holomorphic on a domain containing $|z-a| \leq R$ and put*

$$A(r) = \max_{|z-a|=r} \mathrm{Re}(f(z)).$$

Then

$$|f(z)| \leq \frac{2|z-a|}{R-|z-a|} A(R) + \frac{R+|z-a|}{R-|z-a|} |f(a)|$$

and

$$|f^{(n)}(z)| \leq \frac{2^{n+2} n! R}{(R-|z-a|)^{n+1}} (A(R) + |f(a)|)$$

for $0 \leq |z-a| < R$.

Proof. Assume without loss of generality that $a = 0$. We may also assume that $f(z)$ is nonconstant, for otherwise the desired inequalities hold trivially. Applying the maximum principle to $\exp(f(z))$ one sees that $A(r)$ is strictly increasing. If we assume that $f(0) = 0$ then

$$\phi(z) = \frac{f(z)}{z(2A(R) - f(z))}$$

is holomorphic on $|z| \leq R$, because $\mathrm{Re}(f(z)) < 2A(R)$ there. Now

$$\begin{aligned} |\phi(z)|^2 &= \frac{f(z)}{z(2A(R) - f(z))} \frac{\overline{f(z)}}{\overline{z}(2A(R) - \overline{f(z)})} \\ &= \frac{|f(z|^2}{|z|^2((2A(R) - \mathrm{Re}(f(z)))^2 + |f(z)|^2 - \mathrm{Re}(f(z))^2)} \leq \frac{1}{R^2} \end{aligned}$$

on $|z| = R$, since $|2A(R) - \mathrm{Re}(f(z))| \geq |\mathrm{Re}(f(z)|$ there. Then the maximum principle implies that $|\phi(z)| \leq 1/R$ in $|z| \leq R$. Thus

$$\left| \frac{2A(R)}{f(z)} - 1 \right| \geq \frac{R}{|z|}$$

there, and so

$$\frac{2A(R)}{|f(z)|} \geq \frac{R}{|z|} - 1$$

on the same disk. Then the first of the two inequalities follows in the case where $f(0) = 0$. Applying it with $f(z) - f(0)$ instead of $f(z)$ yields

$$|f(z) - f(0)| \leq \frac{2|z|}{R - |z|}(A(R) - \operatorname{Re}(f(0)))$$

if $f(0) \neq 0$ and so

$$\begin{aligned} |f(z)| &\leq \frac{2|z|}{R - |z|}A(R) + |f(0)| - \frac{2|z|}{R - |z|}\operatorname{Re}(f(0)) \\ &\leq \frac{2|z|}{R - |z|}A(R) + |f(0)| + \frac{2|z|}{R - |z|}|f(0)| \\ &= \frac{2|z|}{R - |z|}A(R) + \frac{R + |z|}{R - |z|}|f(0)| \end{aligned}$$

holds generally.

Let z be a fixed point with $|z| < R$ and denote by C the circular path of integration $|w - z| = \rho = (R - |z|)/2$ traversed in the positive direction. Then

$$\begin{aligned} |f^{(n)}(z)| &= \left| \frac{n!}{2\pi i} \oint_C \frac{f(w) - f(0)}{(w - z)^{n+1}} dw \right| \leq \frac{n!}{2\pi} \oint_C \frac{|f(w) - f(0)|}{|w - z|^{n+1}} |dw| \\ &\leq \frac{n!}{2\pi} \frac{2(|z| + \rho)}{R - |z| - \rho}(A(R) - \operatorname{Re}(f(0))) \frac{2\pi\rho}{\rho^{n+1}} \\ &\leq \frac{2^{n+2} n! R}{(R - |z|)^{n+1}}(A(R) + |f(0)|) \end{aligned}$$

by Cauchy's integral formula for derivatives. □

7.5. The PNT for arithmetic progressions

Using Proposition 7.1 we have obtained the information about the zeros of $L(s, \chi)$ that we need in order to prove the Siegel-Walfisz theorem. But we also need to bound $L'(s, \chi)/L(s, \chi)$, for which purpose we apply the Borel-Carathéodory lemma. It is in fact possible to use the Borel-Carathéodory lemma for both purposes.

Proposition 7.9. *There is an absolute constant $c_5 > 0$ so that in the region*

$$\sigma \geq 1 - \frac{c_5}{\log(q\tau)}$$

a bound

$$-\frac{L'(s, \chi)}{L(s, \chi)} = \frac{1}{s - 1} - \frac{1}{s - \beta} + O(\log(q\tau))$$

holds uniformly in q. *Here the term* $1/(s-1)$ *is absent unless* χ *is principal, and the term* $1/(s-\beta)$ *is absent unless* β *is the exceptional zero associated to an exceptional character* χ.

Proof. Assume χ is neither principal nor exceptional and choose

$$s_0 = \sigma_0 + it_0 = 1 + \frac{c}{\log(q(|t_0|+4))} + it_0$$

where $c > 0$ is the constant in the definition of Landau-Siegel zeros. Let $R \geq 1/2$ be such that the circle $|s - s_0| = R$ lies in the half plane $\sigma \geq 1/4$ and such that no zero of $L(s,\chi)$ lies on it. Define a function

$$f(s) = \frac{L(s,\chi)}{\prod(s-\rho)},$$

where the product in the denominator is taken over all zeros ρ of $L(s,\chi)$ lying in the disk $|s - s_0| \leq R/2$, counted with multiplicity. Then

$$\left|\frac{f(s)}{f(s_0)}\right| = \frac{|L(s,\chi)|}{|L(s_0,\chi)|}\prod \frac{|s_0-\rho|}{|s-\rho|} \leq \frac{|L(s,\chi)|}{|L(s_0,\chi)|} \leq \frac{2q\tau}{\frac{\sigma_0-1}{\sigma_0}} \ll (q\tau)^2$$

on $|s - s_0| = R$. For the distance from any point s on the circle $|s - s_0| = R$ to any ρ in $|\rho - s_0| \leq R/2$ is at least as large as the distance from ρ to s_0.

There is a holomorphic branch $g(s)$ of $\log(f(s)/f(s_0))$, with $g(s_0) = 0$, on an open disk containing $|s - s_0| \leq R/2$, since $f(s)/f(s_0)$ has no zeros there. Moreover

$$\left|\frac{f(s)}{f(s_0)}\right| \ll (q\tau)^2$$

in $|s - s_0| \leq R$ by the Maximum Principle, so $\mathrm{Re}(g(s)) \ll \log(q\tau)$ in $|s-s_0| \leq R/2$. Then

$$\begin{aligned}|g'(s)| &\leq \frac{8\frac{R}{2}}{\left(\frac{R}{2} - |s-s_0|\right)^2}\left(\max_{|s-s_0|=R/2} \mathrm{Re}(g(s)) + |g(s_0)|\right)\\ &\leq \frac{64}{R}\max_{|s-s_0|=R/2} \mathrm{Re}(g(s)) \ll \log(q\tau)\end{aligned}$$

in $|s - s_0| \leq R/4$ by the Borel-Carathéodory lemma, since $g(s_0) = 0$ and $R \geq 1/2$.

Now

$$\left|\frac{f'(s)}{f(s)}\right| \ll \log(q\tau),$$

and thus

$$\mathrm{Re}\left(-\frac{L'(s,\chi)}{L(s,\chi)}\right) = -\sum_{\rho}\frac{1}{s-\rho} + O(\log(q\tau))$$

for $|s-s_0| \le R/4$. Let $s=\sigma+it$ and $\rho=\beta+i\gamma$ and note that $\mathrm{Re}(s-\rho)=\sigma-\beta$ to see that $\mathrm{Re}(1/(s-\rho))>0$ if $\sigma>\beta$. Furthermore $|\rho-s_0|\le R/2\le 1/4$, so $|\gamma-t_0|\le 1/4$. If

$$\sigma \ge \sigma_0 - r_0 = \sigma_0 - \frac{3c}{2}\frac{1}{\log(q(|t_0|+4))}$$

then

$$\begin{aligned}\sigma-\beta &\ge 1+\frac{c}{\log(q(|t_0|+4))}-\frac{3c}{2}\frac{1}{\log(q(|t_0|+4))}-\left(1-\frac{c}{\log(q(|\gamma|+4))}\right)\\ &\ge \frac{c}{\log(q(|t_0|+\frac{1}{4}+4))}-\frac{c}{2}\frac{1}{\log(q(|t_0|+4))}>0,\end{aligned}$$

and so

$$\mathrm{Re}\left(-\frac{L'(s,\chi)}{L(s,\chi)}\right) \ll \log(q\tau),$$

since the sum is negative.

Now

$$\begin{aligned}\left|\frac{L'(s,\chi)}{L(s,\chi)}\right| &\le \frac{2|s-s_0|}{r_0-|s-s_0|}\max_{|s-s_0|=r_0}\mathrm{Re}\left(-\frac{L'(s,\chi)}{L(s,\chi)}\right)\\ &\quad+\frac{r_0+|s-s_0|}{r_0-|s-s_0|}\left|\frac{L'(s_0,\chi)}{L(s_0,\chi)}\right|\\ &\le 2\max_{|s-s_0|=r_0}\mathrm{Re}\left(-\frac{L'(s,\chi)}{L(s,\chi)}\right)+3\left|\frac{\zeta'(s_0)}{\zeta(s_0)}\right| \ll \log(q\tau)\end{aligned}$$

in $|s-s_0|\le r_0/2$ by the Borel-Carathéodory lemma applied on the disk $|s-s_0|\le r_0$. But for $s=\sigma+it_0$ we see that

$$\sigma \ge 1-\frac{c}{4}\frac{1}{\log(q(|t_0|+4))}$$

implies that $|s-s_0|\le r_0/2$ and so we may choose any $c_5>0$ with $c_5\le c/4$.

The case when χ is principal or exceptional is handled in the same way, but using one of

$$f(s)=\frac{L(s,\chi_0)(s-1)}{\prod(s-\rho)} \quad \text{or} \quad f(s)=\frac{L(s,\chi)/(s-\beta)}{\prod(s-\rho)},$$

with the exceptional zero not occurring in the product. In both cases straigthforward modifications must be made when bounding $f(s)$ by means of Proposition 5.3. □

We now establish a preliminary estimate that makes the influence of the exceptional zero explicit. This can be useful on its own, for example, when considering the difference $\psi(x;q,a)-\psi(y;q,a)$ for y close to x.

Proposition 7.10. *There is a constant $C_0 > 0$ so that if a and q are coprime then*

$$\psi(x;q,a) = \frac{x}{\phi(q)} - \frac{\chi(a)}{\phi(q)}\frac{x^\beta}{\beta} + O\Big(xe^{-C_0\sqrt{\log(x)}}\Big)$$

holds uniformly in $q \le x$ if χ is an exceptional character modulo q, and β the associated Landau-Siegel zero. If there is no exceptional character modulo q, the estimate holds without the term containing χ and β.

Proof. The bound

$$\left|\psi(x;q,a) - \frac{1}{2\pi i}\int_{c-iT}^{c+iT} \Psi(s;q,a)\frac{x^s}{s}\,ds\right| \le 2x^c \sum_{n=1}^{\infty} \frac{\log(n)n^{-c}}{\max(1,T|\log(x/n)|)}$$

holds by 5.1.2. If $n \le x/e$ or $n > ex$, then $|\log(x/n)| \ge 1$ so

$$\begin{aligned}2x^c\left(\sum_{n\le x/e} + \sum_{n>ex}\right)\frac{\log(n)n^{-c}}{\max(1,T|\log(x/n)|)} &\le \frac{2x^c}{T}\sum_{n=1}^{\infty}\log(n)n^{-c}\\ &= \frac{2x^c}{T}(-\zeta(c)) \ll \frac{x^c}{T(c-1)^2}.\end{aligned}$$

Assume that $x = [x] + 1/2$, and consider the range $x/e < n \le x$. Then

$$\log\left(\frac{x}{n}\right) = -\log\left(1 - \frac{x-n}{n}\right) \ge \frac{x-n}{x}$$

and so

$$\begin{aligned}2x^c \sum_{x/e<n\le x} \frac{\log(n)n^{-c}}{\max(1,T|\log(x/n)|)} &\le \frac{2x^c}{T} \sum_{x/e<n\le x} \frac{\log(x)(x/e)^{-c}}{\frac{x-n}{x}}\\ &= \frac{2e^c x\log(x)}{T} \sum_{x/e<n\le x} \frac{1}{x-n}\\ &\ll \frac{x\log^2(x)}{T}\end{aligned}$$

for $c \le 2$, say, and

$$2x^c \sum_{x<n\le ex} \frac{\log(n)n^{-c}}{\max(1,T|\log(x/n)|)} \ll \frac{x\log^2(x)}{T}$$

in a similar way. Choosing $c = 1 + 1/\log(x)$ yields

$$\psi(x;q,a) = \frac{1}{2\pi i}\int_{c-iT}^{c+iT} \Psi(s;q,a)\frac{x^s}{s}\,ds + O\left(\frac{x\log^2(x)}{T}\right),$$

where

$$\begin{aligned}\Psi(s;q,a) &= -\frac{1}{\phi(q)}\sum_{\chi \bmod q}\overline{\chi(a)}\frac{L'(s,\chi)}{L(s,\chi)} \\ &= \frac{1}{\phi(q)}\frac{1}{(s-1)} - \frac{\chi(a)}{\phi(q)}\frac{1}{s-\beta} + O(\log(q\tau))\end{aligned}$$

by Proposition 7.9. The term with $1/(s-\beta)$ is missing unless there is an exceptional character χ modulo q, and $\Psi(s;q,a)$ is holomorphic on the domain

$$\sigma > 1 - \frac{c_5}{\log(q\tau)}$$

apart from the point $s=1$ and possibly a point $s=\beta$.

Applying the Residue Theorem on the rectangle with corners $c+iT, 1/2+iT, 1/2-iT, c-iT$, we see that

$$\begin{aligned}\frac{1}{2\pi i}\int_{c-iT}^{c+iT}\left(\frac{1}{\phi(q)}\frac{1}{s-1} - \frac{\chi(a)}{\phi(q)}\frac{1}{s-\beta}\right)\frac{x^s}{s}\,ds &= \frac{x}{\phi(q)} - \frac{\chi(a)}{\phi(q)}\frac{x^\beta}{\beta} \\ &- \frac{1}{2\pi i}\int_{c+iT}^{1/2+iT}\left(\frac{1}{\phi(q)}\frac{1}{s-1} - \frac{\chi(a)}{\phi(q)}\frac{1}{s-\beta}\right)\frac{x^s}{s}\,ds \\ &- \frac{1}{2\pi i}\int_{1/2+iT}^{1/2-iT}\left(\frac{1}{\phi(q)}\frac{1}{s-1} - \frac{\chi(a)}{\phi(q)}\frac{1}{s-\beta}\right)\frac{x^s}{s}\,ds \\ &- \frac{1}{2\pi i}\int_{1/2-iT}^{c-iT}\left(\frac{1}{\phi(q)}\frac{1}{s-1} - \frac{\chi(a)}{\phi(q)}\frac{1}{s-\beta}\right)\frac{x^s}{s}\,ds\end{aligned}$$

where

$$\left|\int_{1/2\pm iT}^{c\pm iT}\left(\frac{1}{\phi(q)}\frac{1}{s-1} - \frac{\chi(a)}{\phi(q)}\frac{1}{s-\beta}\right)\frac{x^s}{s}\,ds\right| \ll \frac{x^c}{T^2} \ll \frac{x}{T^2}$$

and

$$\begin{aligned}\left|\int_{1/2+iT}^{1/2-iT}\left(\frac{1}{\phi(q)}\frac{1}{s-1} - \frac{\chi(a)}{\phi(q)}\frac{1}{s-\beta}\right)\frac{x^s}{s}\,ds\right| &\ll x^{1/2}\int_{1/2-iT}^{1/2+iT}\frac{|ds|}{|s^2-\beta s|} \\ &\ll x^{1/2},\end{aligned}$$

because the last integral converges as $T\to+\infty$. The function

$$F(s;q,a) = \Psi(s;q,a) - \frac{1}{\phi(q)}\frac{1}{s-1} + \frac{\chi(a)}{\phi(q)}\frac{1}{s-\beta}$$

is holomorphic on the domain

$$\sigma > 1 - \frac{c_5}{\log(q\tau)},$$

and satisfies the bound $F(s;q,a) \ll \log(q\tau)$ there. It follows from Cauchy's theorem applied on the rectangle with corners $c+iT, b+iT, b-iT, b-iT$ that

$$\begin{aligned}\frac{1}{2\pi i}\int_{c-iT}^{c+iT} F(s;q,a)\frac{x^s}{s}\,ds &= \frac{1}{2\pi i}\int_{b-iT}^{b+iT} F(s;q,a)\frac{x^s}{s}\,ds \\ &+ \frac{1}{2\pi i}\int_{b+iT}^{c+iT} F(s;q,a)\frac{x^s}{s}\,ds - \frac{1}{2\pi i}\int_{b-iT}^{c-iT} F(s;q,a)\frac{x^s}{s}\,ds\end{aligned}$$

if this rectangle is contained in the above domain. Here

$$\left|\int_{b\pm iT}^{c\pm iT} F(s;q,a)\frac{x^s}{s}\,ds\right| \ll \frac{x\log(q(T+4))}{T} \ll \frac{x\log(x)+x\log(T+4)}{T}$$

since $q \le x$. Now choose

$$T = e^{B\sqrt{\log(x)}} - 4$$

with $B > 0$ undetermined, and x sufficiently large in terms of B so that $T \ge 1$, say. Combining the various error terms then yields

$$\begin{aligned}\psi(x;q,a) &= \frac{x}{\phi(q)} - \frac{\chi(a)}{\phi(q)}\frac{x^\beta}{\beta} \\ &+ \frac{1}{2\pi i}\int_{b-iT}^{b+iT} F(s;q,a)\frac{x^s}{s}\,ds + O\Big(x\log^2(x)e^{-B\sqrt{\log(x)}}\Big) \\ &= \frac{x}{\phi(q)} - \frac{\chi(a)}{\phi(q)}\frac{x^\beta}{\beta} \\ &+ O\Big(\log(qT)x^b\log(T)\Big) + O\Big(x\log^2(x)e^{-B\sqrt{\log(x)}}\Big).\end{aligned}$$

Using $1 \le q \le x$, and choosing

$$b = 1 - \frac{c_5}{2\log(q(T+4))} = 1 - \frac{c_5}{2\log\left(qe^{B\sqrt{\log(x)}}\right)}$$

by Proposition 7.9, we see that

$$\begin{aligned}\psi(x;q,a) &= \frac{x}{\phi(q)} - \frac{\chi(a)}{\phi(q)}\frac{x^\beta}{\beta} \\ &+ O\Big(\log^{3/2}(x)xe^{-\frac{c_5}{2B}\sqrt{\log(x)}}\log(T)\Big) + O\Big(x\log^2(x)e^{-B\sqrt{\log(x)}}\Big).\end{aligned}$$

Choosing $C_0 = B/2$ where B is the positive solution of the equation $B = c_5/(2B)$ completes the proof. □

The constants in the various bounds in the proof are independent of q. In fact $1 \le q \le x$ is the only information about the size of q that is appealed to. Taking $q = 1$ and noting that there is no exceptional zero, de la Vallée Poussin's error term in the Prime Number Theorem is obtained as a corollary.

Proposition 7.11. *There is a constant $C_0 > 0$ so that*

$$\psi(x) = x + O\Big(x\, e^{-C_0\sqrt{\log(x)}}\Big)$$

as $x \to +\infty$.

If there is no exceptional character modulo q then

$$\psi(x; q, a) = \frac{x}{\phi(q)} + O\Big(x\, e^{-C_0\sqrt{\log(x)}}\Big)$$

uniformly in q with a and q coprime. For a fixed large x, as q grows bigger the subset of the arithmetic progression $n \equiv a \pmod{q}$ lying in $1 \le n \le x$ becomes sparser. So it seems natural that $\psi(x; q, a)$ should be harder to bound well for q large, at least from below. Thus we ask how large q can be taken in the estimate above and still have the main term dominate the error term. Since $\phi(q)$ can be as large as $q - 1$, we can take $q \ll \exp(C\sqrt{\log(x)}$ with any $C < C_0$, and have an asymptotic estimate. It is the possible existence of exceptional zeros that prevents us from taking q as large as this for all moduli. We can allow q to grow with x in the general case, but only considerably more slowly.

Proposition 7.12 (Siegel-Walfisz theorem). *For arbitrary $A > 0$ there is a constant $C_A > 0$ such that*

$$\psi(x; q, a) = \frac{x}{\phi(q)} + O\Big(x\, e^{-C_A\sqrt{\log(x)}}\Big)$$

uniformly in coprime q and a with $q \ll \log^A(x)$.

Note how much smaller we must take q compared to the situation when there is no exceptional zero modulo q. Though the constant C_A in the Siegel-Walfisz theorem is uniform in q, it cannot be computed for A large. This is because we use Siegel's theorem, which is ineffective. It should also be mentioned that in 1935 A. Page obtained a version of this result valid for small A, which does not depend on Siegel's theorem and is effective. Consequently the result is sometimes called the Page-Siegel-Walfisz theorem.

Proof. Suppose that β is an exceptional zero of $L(s, \chi)$ where χ is a primitive character modulo q. Then

$$L(1, \chi) = |L(1, \chi) - L(\beta, \chi)| = \left|\int_\beta^1 L'(\sigma, \chi)\, d\sigma\right| \le |1 - \beta| \log^2(4q)$$

by Proposition 5.4, and

$$1 - \beta \ge \frac{L(1, \chi)}{\log^2(4q)} \ge \frac{C(\epsilon) q^{-\epsilon}}{\log^2(4q)} \ge D(\epsilon) q^{-2\epsilon}$$

by Siegel's theorem. Now

$$\left|\frac{\chi(a)}{\phi(q)}\frac{x^\beta}{\beta}\right| \ll x^\beta \le x^{1-D(\epsilon)q^{-2\epsilon}} \le x\, e^{-D(\epsilon)\log(x)^{1-2\epsilon A}},$$

and so by Proposition 7.10 it is sufficient to take $\epsilon = 1/(6A)$, say. □

7.6. Notes

The local method of Landau is in [**Lan24a**]. Proposition 7.1 is from the paper [**HB92**] of Heath-Brown on the Linnik constant. This is the smallest constant L such that every arithmetic progression $n \equiv a \pmod q$ with a and q coprime contains a prime $p \ll_\epsilon q^{L+\epsilon}$, for arbitrary $\epsilon > 0$. The approach of Heath-Brown yielded better numerical constants for zero-free regions than were known before, which allowed him to obtain the bound $L \le 5.5$. Proposition 7.2 is due to T. H. Gronwall [**Gro13**], and Proposition 7.3 to A. Page [**Pag35**] (real zeros) and E. C. Titchmarsh [**Tit30**] (complex zeros).

Proposition 7.4 and 7.6 are due to Landau [**Lan18b**], Proposition 7.5 to Page [**Pag35**].

Proposition 7.7 is due to Siegel [**Sie35**], after work of Landau [**Lan18b, Lan35**]. The proof given here is due to D. M. Goldfeld [**Gol74**]. The best effective analogue of Proposition 7.7 has been obtained by a method due to Goldfeld [**Gol76**]. The method relies on the existence of a suitable elliptic curve with an L-function having a high order zero at $s = 1/2$. The existence of the requisite elliptic curve was established by B. Gross and D. Zagier [**GZ83**]. J. Oesterlé [**Oes85**] used the method to find an explicit effective bound.

Proposition 7.8 is due to E. Borel [**Bor99**].

Proposition 7.9 is due to Landau [**Lan08**], and the method of proof also to him [**Lan24a**]. Proposition 7.10 is due to Page [**Pag35**], and Proposition 7.11 to de la Vallée Poussin [**dlVP99**]. Proposition 7.12 is due to Walfisz [**Wal36**].

Exercises

(1) Show that if it were possible to prove that $\psi(x;q,a) \le (2-\epsilon)x/\phi(q)$ uniformly in $q \le x^\theta$ for some constants $\epsilon, \theta > 0$ and all x sufficiently large, then one could exclude the existence of Landau-Siegel zeros merely by adopting a sufficiently strong definition of such zeros. A related inequality for $\pi(x;q,a)$ known as the Brun-Titchmarsh lemma holds uniformly in the range $q \le x$ and can be proved by sieve methods. Unfortunately the best constant obtained in the Brun-Titchmarsh lemma thus far is 2 rather than $2-\epsilon$, and this just fails to exclude Landau-Siegel zeros.

(2) The usual way to organize the proof of the Siegel-Walfisz theorem is in terms of the summatory functions

$$\psi(x,\chi) = \sum_{n\leq x} \chi(n)\Lambda(n).$$

Formulate the estimates obtained in Section 7.5 for $\psi(x,\chi)$ with χ principal, exceptional, and neither principal nor exceptional.

(3) Show that the sequence of exceptional conductors in increasing order will grow at least as fast as the doubly exponential sequence

$$Q^{2^{n-1}}$$

with Q fixed but arbitrarily large, if there are exceptional conductors.

(4) Find an asymptotic estimate for

$$\sum_{\substack{p\leq x \\ p\equiv a \bmod q}} \frac{\Lambda(n)}{n},$$

with a good error term.

(5) Deduce a Siegel-Walfisz type theorem for the counting function

$$\pi(x;q,a) = \sum_{\substack{p\leq x \\ p\equiv a \bmod q}} 1$$

of the primes in an arithmetic progression.

(6) In Section 9.3 there is a lower bound for $L(1,\chi)$ for χ a quadratic character that is effective, though weaker than Siegel's theorem. Use it to determine a range of values of A for which the Siegel-Walfisz theorem can be made effective.

(7) The class P consists of all holomorphic functions $f : \mathbb{D} \to \mathbb{C}$ from the *unit disk* $\mathbb{D} = \{z \in \mathbb{C} \mid |z| < 1\}$, satisfying $f(0) = 1$ and $\mathrm{Re}(f(z)) > 0$ on $\mathbb{D}$. Use the Borel-Carathéodory lemma to show that $|f(z)| \leq (1 + |z|)/(1 - |z|)$ on $\mathbb{D}$ for all $f \in P$.

The functions in P are those that have an integral representation

$$f(z) = \int_{|\eta|=1} \frac{1+\eta z}{1-\eta z} d\mu$$

with μ some probability measure on the unit circle. This is called the *Herglotz representation.* Use this to deduce the above bound for functions in the class P.

(8) Apply the method of proof of Proposition 7.2 to $\Psi(s;q,1)$ and obtain a bound for the exceptional zero. Explain why this bound is much poorer than the one obtained from the theorem of Siegel.

(9) † a) Find an estimate for the smoothed summatory function

$$\psi_1(x;q,a) = \sum_{\substack{n \leq x \\ n \equiv a \bmod q}} (x-n)\Lambda(n)$$

analogous to Proposition 7.10.

b) Deduce the Siegel-Walfisz theorem from part a).

(10) Show that

$$\psi(x;q,a) - \psi(x;q,b) = O\Big(xe^{-C_0\sqrt{\log(x)}}\Big)$$

for any a and b coprime with q, uniformly in q, unless there is a Landau-Siegel zero for the modulus q. In the latter case we must impose an extra condition on a and b to draw the same conclusion. What is it?

Chapter 8

Mainly Analysis

8.1. The Poisson summation formula

The subgroup $\mathbb{Z}$ of the abelian group $\mathbb{R}$ under addition acts on $\mathbb{R}$ by $x \mapsto k+x$. The functions $f : \mathbb{R} \to \mathbb{C}$ invariant under the action are the ones that are periodic with period 1. The trigonometric basis functions

$$e(nx) = e^{2\pi i n x}, \qquad n \in \mathbb{Z}$$

are periodic with period 1, and we endeavor to express functions f as above as linear combinations of these harmonics. The calculation

$$\int_0^1 \overline{e(mx)}\, e(nx)\, dx = \int_0^1 e^{2\pi i(n-m)x}\, dx = \begin{cases} 1 & \text{if } n-m=0, \\ 0 & \text{if } n-m \neq 0, \end{cases}$$

shows that the exponentials $e(nx)$ form an orthonormal set. The coefficients of finite trigonometric sums

$$t(x) = \sum_n c_n\, e(nx)$$

may be recovered by the integration

$$\int_0^1 t(x)\, e(-mx)\, dx = \sum_n c_n \int_0^1 \overline{e(mx)}\, e(nx)\, dx = c_m$$

using the orthonormality.

The integrals

$$\hat{f}_n = \int_0^1 f(x)\, e(-nx)\, dx$$

define the *Fourier coefficients* $\hat{f}_n$ of f when f is integrable on $[0,1]$. The infinite series

$$\sum_{n=-\infty}^{\infty} \hat{f}_n \, e(nx)$$

is called the *Fourier series* of f. We note that the Fourier series of a finite trigonometric sum $t(x)$ equals $t(x)$. An example due to A. N. Kolmogorov shows that for sufficiently general integrable functions f the sequence of partial sums

$$\sigma_N(f;x) = \sum_{n=-N}^{N} \hat{f}_n \, e(nx)$$

may fail to converge anywhere.

The calculation

$$\begin{aligned}
&\int_0^1 \left| f(x) - \sum_{n=-N}^{N} \hat{f}_n \, e(nx) \right|^2 dx \\
&= \int_0^1 \left(f(x) - \sum_{m=-N}^{N} \hat{f}_m \, e(mx) \right) \overline{\left(f(x) - \sum_{n=-N}^{N} \hat{f}_n \, e(nx) \right)} \, dx \\
&= \int_0^1 |f(x)|^2 dx - \sum_{n=-N}^{N} |\hat{f}_n|^2
\end{aligned}$$

shows that

$$\sum_{n=-\infty}^{\infty} |\hat{f}_n|^2 \le \int_0^1 |f(x)|^2 \, dx.$$

This is known as *Bessel's inequality*, and is considered in greater detail in the L^2-theory of Fourier series.

The sawtooth function $S(x)$ is periodic with period 1, and we investigate its Fourier series. Let $z = re^{i\theta}$ with $r \le 1$, but $z \ne 1$. Summation of a finite geometric series implies that

$$\frac{e^{i\theta}}{1 - re^{i\theta}} = e^{i\theta} + re^{2i\theta} + \cdots + r^{N-1} e^{iN\theta} + \frac{r^N e^{i(N+1)\theta}}{1 - re^{i\theta}}$$

for $1 \ne z = re^{i\theta}$. Integrating with respect to r from 0 to 1 we obtain

$$\int_0^1 \frac{e^{i\theta}}{1 - re^{i\theta}} \, dr = \sum_{n=1}^{N} \frac{e^{in\theta}}{n} + \int_0^1 \frac{r^N e^{i(N+1)\theta}}{1 - re^{i\theta}} \, dr.$$

Here

$$\int_0^1 \frac{e^{i\theta}}{1 - re^{i\theta}} \, dr = \int_0^{\exp(i\theta)} \frac{dz}{1 - z} = -\log(1 - e^{i\theta})$$

so

$$\int_0^1 \frac{e^{i\theta}}{1-re^{i\theta}}\,dr - \int_0^1 \frac{e^{-i\theta}}{1-re^{-i\theta}}\,dr = \log\left(\frac{1-e^{-i\theta}}{1-e^{i\theta}}\right) = i(\pi-\theta),$$

which is valid for $0<\theta<2\pi$. Furthermore one sees that

$$\left|\int_0^1 \frac{r^N e^{i(N+1)\theta}}{1-re^{i\theta}}\,dr\right| \le \int_0^1 \frac{r^N}{\sin(\theta/2)}\,dr = \frac{1}{(N+1)\sin(\theta/2)}$$

by considering the distance from the point $z=1$ to the ray $z=re^{i\theta}$. Thus

$$\sum_{0<|n|\le N} \frac{e^{in\theta}}{n} \to i(\pi-\theta)$$

uniformly on $[\varepsilon, 2\pi-\varepsilon]$ for each $\varepsilon>0$ as $N\to+\infty$, and so

$$S(x) = -\sum_{0\ne n\in\mathbb{Z}} \frac{e^{2\pi inx}}{2\pi in}$$

for all $x\notin\mathbb{Z}$ by the periodicity. We now have a trigonometric series converging to $S(x)$ everywhere except at the integers, but it is not obvious that this is the Fourier series of $S(x)$. So we calculate the Fourier coefficients

$$\begin{aligned}\hat{S}_n &= \int_0^1 S(x)e(-nx)\,dx = \int_0^1\left(x-\frac{1}{2}\right)e^{-2\pi inx}\,dx,\\ &= \left(x-\frac{1}{2}\right)\frac{e^{-2\pi inx}}{-2\pi in}\Bigg|_0^1 - \int_0^1 \frac{e^{-2\pi inx}}{-2\pi in}\,dx = \begin{cases}-(2\pi in)^{-1} & \text{if } n\ne 0,\\ 0 & \text{if } n=0\end{cases}\end{aligned}$$

to make sure. Clearly the Fourier series of $S(x)$ converges to zero at all integer values of x, so we may redefine the sawtooth function to be zero at all its jump discontinuities, and then the Fourier series converges to $S(x)$ everywhere. This convergence is neither absolute nor uniform, though it is uniform off $(-\varepsilon,\varepsilon)+\mathbb{Z}$ for every $\varepsilon>0$.

One of the main themes of Fourier analysis is the investigation of wide classes of functions, of modes of convergence, and of techniques of summation for which the Fourier series of a function sums to that function. For such work the Lebesgue integral has proved indispensable. But our objective is the Poisson summation formula, which we shall require only for a fairly narrow class of functions, and this enables us to rely on the Riemann integral familiar from calculus.

We shall impose on our functions successively stronger conditions to simplify the analysis, while still ending up with a useful version of the Poisson summation formula. First we require absolute convergence of the Fourier series of f to $f(x)$ for all $x\in\mathbb{R}$. Absolute convergence is very useful, because it allows various manipulations of the series. Moreover an absolutely convergent Fourier series is uniformly convergent, by $|\hat{f}_n\, e(nx)| = |\hat{f}_n|$ and a result from advanced calculus often called the Weierstrass M-test. Since the terms of a Fourier series are continuous functions, the uniform convergence implies that the sum must be a continuous function. Continuous functions are Riemann integrable, thus the Riemann integral suffices to write down the Fourier coefficients of those functions in which we are interested. The Fourier series of a continuous function may not converge to the function everywhere, and may well fail to converge absolutely where it converges. So we must further restrict the class of functions under consideration.

Proposition 8.1. *The Fourier series of a periodic function with everywhere continuous first derivative converges absolutely to that function everywhere.*

Proof. Assume without loss of generality that f has period 1. We obtain a relation

$$\begin{aligned}\widehat{f'}_n &= \int_0^1 f'(x)e(-nx)\,dx = f(x)e(-nx)\Big|_{x=0}^{x=1} - \int_0^1 f(x)(-2\pi i n)e(-nx)\,dx \\ &= 2\pi i n \int_0^1 f(x)e(-nx)\,dx = 2\pi i n \hat{f}_n\end{aligned}$$

between the Fourier coefficients of f and f', since the boundary terms in the integration by parts cancel by periodicity. Now

$$\sum_{n=-\infty}^{\infty} n^2|\hat{f}_n|^2 = (2\pi)^{-2} \sum_{n=-\infty}^{\infty} |\widehat{f'}_n|^2 \le (2\pi)^{-2} \int_0^1 |f'(x)|^2\,dx < \infty$$

by Bessel's inequality, for f' is continuous. Then

$$\left(\sum_{0\neq n\in\mathbb{Z}} |\hat{f}_n|\right)^2 = \left(\sum_{0\neq n\in\mathbb{Z}} n^{-1}n|\hat{f}_n|\right)^2 \le \left(\sum_{0\neq n\in\mathbb{Z}} n^{-2}\right) \sum_{0\neq n\in\mathbb{Z}} n^2|\hat{f}_n|^2 < \infty$$

by the Cauchy-Schwarz inequality. Thus the Fourier series converges absolutely.

It remains to show that the Fourier series converges to f. Since we may replace $f(x)$ by an arbitrary translate $f(x+a)$, it will be enough to show convergence in a single point. The formula

$$f(0) = \frac{1}{2}f(0) + \frac{1}{2}f(1) = \int_0^1 f(u)\,du + \int_0^1 S(u)f'(u)\,du$$

follows from the Euler-Maclaurin summation formula by choosing $A = 0$ and $B = 1$ in Proposition 1.8. For arbitrary $\varepsilon > 0$ the interchange of sum and integral in

$$\int_{\varepsilon}^{1-\varepsilon} S(u)f'(u)\,du = \int_{\varepsilon}^{1-\varepsilon} \left(-\sum_{0\neq n\in\mathbb{Z}} \frac{e^{2\pi inu}}{2\pi in} \right) f'(u)\,du$$
$$= \sum_{0\neq m\in\mathbb{Z}} \frac{1}{2\pi im} \int_{\varepsilon}^{1-\varepsilon} f'(u)e(-mu)\,du$$

is valid by uniform convergence. Since f' is continuous, there exists a constant $M > 0$ such that $|f'| \leq M$, and so

$$\left| \int_{|u|\leq\varepsilon} S(u)f'(u)\,du \right| \leq \int_{|u|\leq\varepsilon} |S(u)||f'(u)|\,du \leq \frac{1}{2}M2\varepsilon \ll \varepsilon.$$

Bessel's inequality implies that

$$\sum_{m=-\infty}^{\infty} \left| \int_{|u|\leq\varepsilon} f'(u)e(-mu)\,du \right|^2 \leq \int_{|u|\leq\varepsilon} |f'(u)|^2\,du \leq M^2 2\varepsilon \ll \varepsilon,$$

since $f'(u)I_\varepsilon(u)$ is piecewise continuous. Here $I_\varepsilon(u)$ denotes the indicator function of $(-\varepsilon,\varepsilon)+\mathbb{Z}$. But then

$$\left| \sum_{0\neq m\in\mathbb{Z}} \frac{1}{2\pi im} \int_{|u|\leq\varepsilon} f'(u)e(-mu)\,du \right|^2$$
$$\leq \left(\sum_{0\neq m\in\mathbb{Z}} \frac{1}{4\pi^2m^2} \right) \sum_{m=-\infty}^{\infty} \left| \int_{|u|\leq\varepsilon} f'(u)e(-mu)\,du \right|^2 \ll \varepsilon$$

by the Cauchy-Schwarz inequality. Since

$$\int_0^1 f(u)\,du = \hat{f}_0$$

we obtain

$$f(0) - \sum_{n=-\infty}^{\infty} \hat{f}_n \ll \varepsilon + \varepsilon^{1/2}$$

with $\varepsilon > 0$ arbitrary, so the Fourier series of f converges to f at $x = 0$. □

The *Fourier transform* $\hat{f}$ of an integrable function $f : \mathbb{R} \to \mathbb{C}$ is defined as

$$\hat{f}(y) = \int_{-\infty}^{\infty} f(x)\,e(-xy)\,dx.$$

For our purposes, integrability is taken to mean integrable as an improper Riemann integral. The Fourier transform may be defined under much more general conditions on f, and there is an extensive theory whose simplest results parallel those of the Fourier analysis on finite groups of Section 3.5.

But this theory requires a background in analysis which we cannot presuppose here, and we will not pursue it. Instead we shall prove a version of the Poisson summation formula.

A function $f : \mathbb{R} \to \mathbb{C}$ will have a *periodization*

$$g(x) = \sum_{m=-\infty}^{\infty} f(x+m)$$

if $|f(x)|$ decreases rapidly enough to zero as $|x| \to \infty$. The periodization is periodic with period 1.

Proposition 8.2 (Poisson summation formula). *Suppose that $f : \mathbb{R} \to \mathbb{C}$ is a function with everywhere continuous first derivative and that*

$$\sum_{n=-\infty}^{\infty} h(x+m)$$

converges uniformly on compact sets for $h = f, f'$. Then

$$\sum_{n=-\infty}^{\infty} \hat{f}(n)\, e(nx) = \sum_{m=-\infty}^{\infty} f(x+m)$$

for all $x \in \mathbb{R}$.

Proof. The periodization

$$g(x) = \sum_{m=-\infty}^{\infty} f(x+m)$$

has everywhere continuous first derivative by the analogous assumption on f and the assumption of uniform convergence. Hence

$$g(x) = \sum_{n=-\infty}^{\infty} \hat{g}_{\mathbf{n}}\, e(nx)$$

everywhere by Proposition 8.1. Now

$$\begin{aligned} \hat{g}_n &= \int_0^1 \left(\sum_{m=-\infty}^{\infty} f(x+m) \right) e(-nx)\, dx \\ &= \sum_{m=-\infty}^{\infty} \int_0^1 f(x+m)\, e(-n(x+m))\, dx = \int_{-\infty}^{\infty} f(x)\, e(-nx)\, dx = \hat{f}(n) \end{aligned}$$

by uniform convergence again, and the fact that the intersection of distinct intervals $[m_1, m_1+1]$ and $[m_2, m_2+1]$ has length zero. □

When speaking of the Poisson summation formula, often the special case

$$\sum_{n=-\infty}^{\infty} \hat{f}(n) = \sum_{m=-\infty}^{\infty} f(m)$$

for $x = 0$ is meant. In number theory the Poisson formula is used to exploit the additive structure of the group $\mathbb{Z}$ by making particular choices for f.

From the one-variable Poisson summation formula, a several-variable version may be obtained inductively. For this purpose we note that

$$\hat{f}(\mathbf{y}) = \int_{\mathbb{R}^d} f(\mathbf{x})\, e(-\mathbf{x}\cdot\mathbf{y})\, d\mathbf{x}$$

defines the Fourier transform $\hat{f}$ of an integrable function $f : \mathbb{R}^d \to \mathbb{C}$ of several variables. Our convention on the meaning of integrability stays in force: The improper multiple Riemann integral should converge absolutely.

Proposition 8.3. *Suppose that $f : \mathbb{R}^d \to \mathbb{C}$ is a function with everywhere continuous partial derivatives up to d-th order and that*

$$\sum_{\mathbf{n}\in\mathbb{Z}^d} |h(\mathbf{x}+\mathbf{m})|$$

converges uniformly on compact sets for every partial derivative h of f up to d-th order. Then

$$\sum_{\mathbf{n}\in\mathbb{Z}^d} \hat{f}(\mathbf{n})\, e(\mathbf{n}\cdot\mathbf{x}) = \sum_{\mathbf{m}\in\mathbb{Z}^d} f(\mathbf{x}+\mathbf{m})$$

for all $\mathbf{x} \in \mathbb{R}^d$.

Proof. The conclusion holds for $d = 1$ by Proposition 8.2. Assume that it holds for an integer $d > 0$ and that $f : \mathbb{R}^{d+1} \to \mathbb{C}$ satisfies the above conditions.

The function $f(\mathbf{x}, x_{d+1})$ of the first d variables with x_{d+1} fixed satisfies the conditions of Proposition 8.3 and thus

$$\sum_{\mathbf{n}\in\mathbb{Z}^d} \hat{f}(\mathbf{n}, x_{d+1})\, e(\mathbf{n}\cdot\mathbf{x}) = \sum_{\mathbf{m}\in\mathbb{Z}^d} f(\mathbf{x}+\mathbf{m}, x_{d+1})$$

where the Fourier transform is taken with respect to the first d variables. Clearly for fixed $\mathbf{x}$ the function $u_{\mathbf{x}}(x_{d+1})$ of a single variable that equals either side of the above identity satisfies the conditions of Proposition 8.2.

Thus

$$
\begin{aligned}
\sum_{\mathbf{m}\in\mathbb{Z}^{d+1}} f(\mathbf{x}+\mathbf{m}) &= \sum_{m_{d+1}\in\mathbb{Z}} u_{\mathbf{x}}(x_{d+1}+m_{d+1}) \\
&= \sum_{n_{d+1}\in\mathbb{Z}} \hat{u}_{\mathbf{x}}(n_{d+1})e(n_{d+1}x_{d+1}) \\
&= \sum_{n_{d+1}\in\mathbb{Z}} \int_{\mathbb{R}} u_{\mathbf{x}}(y_{d+1})e(-n_{d+1}y_{d+1})\,dy_{d+1}\,e(n_{d+1}x_{d+1}) \\
&= \sum_{\mathbf{n}\in\mathbb{Z}} \sum_{n_{d+1}\in\mathbb{Z}} \int_{\mathbb{R}} \hat{f}(\mathbf{n}, y_{d+1})\,e(-n_{d+1}y_{d+1})\,dy_{d+1}\,e(\mathbf{n}{\cdot}\mathbf{x}+n_{d+1}x_{d+1}) \\
&= \sum_{\mathbf{n}\in\mathbb{Z}} \sum_{n_{d+1}\in\mathbb{Z}} \int_{\mathbb{R}}\int_{\mathbb{R}^d} f(\mathbf{y}, y_{d+1})e(-\mathbf{n}{\cdot}\mathbf{y})\,d\mathbf{y}e(-n_{d+1}y_{d+1})\,dy_{d+1} \\
&\quad\times e(\mathbf{n}{\cdot}\mathbf{x}+n_{d+1}x_{d+1}) \\
&= \sum_{\mathbf{n}\in\mathbb{Z}^{d+1}} \int_{\mathbb{R}^{d+1}} f(\mathbf{y})e(-\mathbf{n}{\cdot}\mathbf{y})\,d\mathbf{y}\,e(\mathbf{n}{\cdot}\mathbf{x}) = \sum_{\mathbf{n}\in\mathbb{Z}^{d+1}} \hat{f}(\mathbf{n})e(\mathbf{n}{\cdot}\mathbf{x})
\end{aligned}
$$

by absolute and uniform convergence. Here $\mathbf{m}, \mathbf{n}, \mathbf{x}, \mathbf{y}$ have d or $d+1$ components according to context. □

Fourier Analysis. An Introduction by Elías M. Stein and Rami Shakarchi is an exposition of Fourier theory requiring only a modest background in analysis, and has numerous interesting exercises. *Trigonometric Series* by Antoni Zygmund is a classic monograph on Fourier series. *Introduction to Fourier Analysis on Euclidean Spaces* by Elías M. Stein and Guido M. Weiss has an indepth treatment of Fourier theory in several variables.

8.2. Theta functions

The *theta function* $\vartheta : \mathbb{C}\times\mathbb{H}\to\mathbb{C}$ defined by

$$\vartheta(z,\tau) = \sum_{n=-\infty}^{\infty} e^{\pi i n^2\tau}e(nz)$$

was introduced by Jacobi for the purpose of expressing elliptic functions as ratios of entire functions. Here $\mathbb{H} = \{\tau\in\mathbb{C} \mid \mathrm{Im}(\tau) > 0\}$ is the *upper half plane*. The bound

$$\left|e^{\pi i n^2\tau}e(nz)\right| \le e^{-\pi n^2\mathrm{Im}(\tau)+2\pi|n||\mathrm{Im}(z)|}$$

implies that the series converges uniformly on compact sets $K\subset\mathbb{C}\times\mathbb{H}$, to a holomorphic function on $\mathbb{C}\times\mathbb{H}$.

The form of the series defining the theta function implies various transformation properties. The periodicity $\vartheta(z+1,\tau) = \vartheta(z,\tau)$ is obvious, and the identity

$$\begin{aligned}\vartheta(z+\tau,\tau) &= \sum_{n=-\infty}^{\infty} e^{\pi i n^2\tau} e(nz+n\tau) = e^{-2\pi i z} \sum_{n=-\infty}^{\infty} e^{\pi i n^2\tau + 2\pi i n\tau} e(nz+z) \\ &= e^{-\pi i\tau - 2\pi i z} \sum_{n=-\infty}^{\infty} e^{\pi i (n+1)^2\tau} e(nz+z) = e^{-\pi i\tau-2\pi i z}\vartheta(z,\tau)\end{aligned}$$

scarcely less so. We can determine the zeros of $\vartheta(z,\tau)$ by means of these two identities. First note that $\vartheta(z,\tau)$ is an even function of the variable z, so that

$$\vartheta\Big(\frac{1}{2}+\frac{\tau}{2},\tau\Big) = \vartheta\Big(\frac{1}{2}-\frac{\tau}{2}+\tau,\tau\Big) = -\vartheta\Big(\frac{1}{2}-\frac{\tau}{2},\tau\Big) = -\vartheta\Big(-\frac{1}{2}-\frac{\tau}{2},\tau\Big)$$

implies that $z = 1/2+\tau/2$ is a zero of $\vartheta(z,\tau)$. Let C be a contour traversed in the counterclockwise direction along a parallelogram with corners in the points $z_0, z_0+1, z_0+\tau, z_0+1+\tau$. Choose z_0 so that $\vartheta(z,\tau)$ has no zeros on C, and that $z = 1/2+\tau/2$ lies inside C. Let $C_1,\dots,C_4$ denote the sides of C taken in the counterclockwise direction, starting with the line segment from z_0 to z_0+1. The transformation formulas yield

$$\frac{\vartheta'(z+1,\tau)}{\vartheta(z+1,\tau)} = \frac{\vartheta'(z,\tau)}{\vartheta(z,\tau)}$$

and

$$\frac{\vartheta'(z+\tau,\tau)}{\vartheta(z+\tau,\tau)} = \frac{(e^{-\pi i\tau-2\pi i z})'}{e^{-\pi i\tau-2\pi i z}} + \frac{\vartheta'(z,\tau)}{\vartheta(z,\tau)} = -2\pi i + \frac{\vartheta'(z,\tau)}{\vartheta(z,\tau)}$$

so

$$\begin{aligned}\oint_C \frac{\vartheta'(z,\tau)}{\vartheta(z,\tau)}\,dz &= \Big(\int_{C_1}+\int_{C_2}+\int_{C_3}+\int_{C_4}\Big)\frac{\vartheta'(z,\tau)}{\vartheta(z,\tau)}\,dz \\ &= \Big(\int_{C_1}+\int_{C_2}-\int_{C_1}-\int_{C_2}\Big)\frac{\vartheta'(z,\tau)}{\vartheta(z,\tau)}\,dz \\ &\quad + \int_{C_3}(-2\pi i)\,dz = 2\pi i.\end{aligned}$$

But to each zero of a holomorphic function corresponds a simple pole of its logarithmic derivative at the same point, whose residue equals the multiplicity of the zero. This implies that $\vartheta(z,\tau)$ has only the zero $z = 1/2+\tau/2$ within C, and that this is a simple zero. In particular $\vartheta(z,\tau)$ is nonconstant in the variable z, and by the transformation properties, $1/2+\tau/2+\mathbb{Z}+\mathbb{Z}\tau$ constitutes its full set of zeros, all of which are simple.

The theta function also has transformation properties with respect to the variable τ. It is easy to see that

$$\vartheta(z,\tau+1)=\vartheta\left(z+\frac{1}{2},\tau\right)$$

by noting that n^2 is even or odd according as n is even or odd, respectively. In particular $\vartheta(z,\tau+2)=\vartheta(z,\tau)$. The theta function also has another transformation property with respect to τ, which lies deeper than the others. This is called the functional equation of the Jacobi theta function.

Before proving the functional equation, we must calculate a certain definite integral. Using a change to polar coordinates in

$$\left(\int_{-\infty}^{\infty} e^{\pi i u^2\tau}\,du\right)\left(\int_{-\infty}^{\infty} e^{\pi i v^2\tau}\,dv\right)=\int_{-\infty}^{\infty}\int_{-\infty}^{\infty} e^{\pi i(u^2+v^2)\tau}\,du\,dv$$
$$=\int_0^{2\pi}\int_0^{\infty} e^{\pi i r^2\tau}\,r dr\,d\theta=\pi\int_0^{\infty} e^{\pi i x\tau}\,dx=\frac{i}{\tau}$$

determines the square of the definite integral that we want. Since

$$\mathrm{Re}\left(\int_{-\infty}^{\infty} e^{\pi i u^2\tau}\,du\right)=2\int_0^{\infty} e^{-\pi u^2\mathrm{Im}(\tau)}\cos(\pi u^2\mathrm{Re}(\tau))\,du>0$$

by the monotone decrease of the exponential, we see that

$$\int_{-\infty}^{\infty} e^{\pi i u^2\tau}\,du=\sqrt{\frac{i}{\tau}}$$

with the square root lying in the right half plane.

Proposition 8.4 (Functional equation of the Jacobi theta function). *The identity*

$$\vartheta\left(\frac{z}{\tau},-\frac{1}{\tau}\right)=\sqrt{\frac{\tau}{i}}\,e^{\pi i z^2/\tau}\vartheta(z,\tau)$$

holds for $z\in\mathbb{C}$ and $\tau\in\mathbb{H}$, with the square root lying in the right half plane.

Proof. The periodization of $f(x)=e^{-\pi i x^2/\tau}$ is

$$\sum_{n=-\infty}^{\infty} e^{-\pi i(x+n)^2/\tau}=e^{-\pi i x^2/\tau}\sum_{n=-\infty}^{\infty} e^{-\pi i n^2/\tau}e\left(n\frac{x}{\tau}\right)=e^{-\pi i x^2/\tau}\vartheta\left(\frac{x}{\tau},-\frac{1}{\tau}\right)$$

and the series

$$\sum_{n\in\mathbb{Z}} f(x+n)\quad\text{and}\quad\sum_{n\in\mathbb{Z}} f'(x+n)$$

converge uniformly on compact sets for fixed $\tau \in \mathbb{H}$ because

$$\begin{aligned}|f'(x+n)| &= \frac{2\pi}{|\tau|}|(x+n)|e^{-\pi(x+n)^2\mathrm{Im}(-1/\tau)} \\ &\le \frac{2\pi}{|\tau|}(|n|+b)e^{\pi(2b|n|-n^2)\mathrm{Im}(-1/\tau)}\end{aligned}$$

for $x \in [-b, b]$, and the uniform convergence of the periodization of $f(x)$ follows from that of $f'(x)$ by integration.

Calculating the Fourier transform

$$\begin{aligned}\hat{f}(n) &= \int_{-\infty}^{\infty} e^{-\pi i x^2/\tau} e(-nx)\,dx = e^{\pi i n^2 \tau}\int_{-\infty}^{\infty} e^{\pi i (x+n\tau)^2/(-\tau)}\,dx \\ &= e^{\pi i n^2\tau}\int_{-\infty}^{\infty} e^{\pi i x^2/(-\tau)}\,dx = \sqrt{\frac{i}{-1/\tau}}\,e^{\pi i n^2/\tau} = \sqrt{\frac{\tau}{i}}\,e^{\pi i n^2/\tau}\end{aligned}$$

by Cauchy's theorem, the Poisson summation formula yields

$$\sqrt{\frac{\tau}{i}}\,\vartheta(x,\tau) = \sum_{n=-\infty}^{\infty}\hat{f}(n)e(nx) = \sum_{n=-\infty}^{\infty} f(x+n) = e^{-\pi i x^2/\tau}\vartheta\left(\frac{x}{\tau}, -\frac{1}{\tau}\right),$$

and so the functional equation is valid for all real z. But since $\vartheta(z,\tau)$ is holomorphic in $z \in \mathbb{C}$ for arbitrary τ, the functional equation is valid for all z. □

For $z = 0$ the functional equation of the Jacobi theta function simplifies to

$$\vartheta\left(0, -\frac{1}{\tau}\right) = \sqrt{\frac{\tau}{i}}\,\vartheta(0,\tau).$$

The holomorphic function $\vartheta(0,\tau)$ on $\mathbb{H}$ is called a *theta-constant.*

It is not hard to determine the classical Gauss sum

$$\tau_p = \sum_{m=1}^{p}\left(\frac{m}{p}\right)e(m/p) = \sum_{k=1}^{p} e(k^2/p)$$

up to sign. The determination of the sign is altogether more difficult, and we shall achieve this by means of the Jacobi theta function.

Proposition 8.5 (Landsberg-Schaar formula). *The identity*

$$\frac{1}{\sqrt{h}}\sum_{m=1}^{h} e\left(\frac{m^2 k}{h}\right) = \frac{e^{\pi i/4}}{\sqrt{2k}}\sum_{n=1}^{2k} e\left(-\frac{n^2 h}{4k}\right)$$

holds for any positive integers h and k.

Proof. The asymptotic behavior of $\vartheta(0,\tau)$ for $\tau = 2k/h + i\delta$ with $\delta \to 0^+$ may be determined in two different ways. First

$$\begin{aligned}\sum_{n=-\infty}^{\infty} e^{-\pi n^2\delta} e(n^2k/h) &= \sum_{a=1}^{h}\sum_{m=-\infty}^{\infty} e^{-\pi(hm+a)^2\delta} e((hm+a)^2k/h)\\ &= \sum_{a=1}^{h} e(a^2k/h)\sum_{m=-\infty}^{\infty} e^{-\pi h^2m^2\delta} e^{-2\pi hma\delta-\pi a^2\delta}\\ &= \sum_{a=1}^{h} e(a^2k/h)\sum_{m=-\infty}^{\infty} e^{-\pi h^2m^2\delta}\\ &\quad+\sum_{a=1}^{h} e(a^2k/h)\sum_{m=-\infty}^{\infty} e^{-\pi h^2m^2\delta}\\ &\quad\times\left(e^{-2\pi hma\delta-\pi a^2\delta}-1\right)\end{aligned}$$

where

$$\sum_{m=-\infty}^{\infty} e^{-\pi h^2m^2\delta} = h^{-1}\delta^{-1/2} + O(1)$$

by the functional equation. Next

$$\begin{aligned}&\left|\sum_{a=1}^{h} e(a^2k/h)\sum_{m=-\infty}^{\infty} e^{-\pi h^2m^2\delta}\left(e^{-2\pi hma\delta-\pi a^2\delta}-1\right)\right|\\ &\quad\le \sum_{a=1}^{h}\sum_{m=-\infty}^{\infty} e^{-\pi h^2m^2\delta}\left|e^{-2\pi hma\delta-\pi a^2\delta}-1\right|\\ &\quad\ll \sum_{a=1}^{h}\int_{-\infty}^{\infty} e^{-\pi h^2x^2\delta}\left|e^{-2\pi hxa\delta-\pi a^2\delta}-1\right| dx\\ &\quad= \delta^{-1/2}\sum_{a=1}^{h}\int_{-\infty}^{\infty} e^{-\pi h^2u^2}\left|e^{-2\pi hua\delta^{1/2}-\pi a^2\delta}-1\right| du\end{aligned}$$

for sufficiently small δ, say $0 < \delta < \delta_1$. For arbitrary $\varepsilon > 0$ choose b so that

$$\begin{aligned}&\sum_{a=1}^{h}\int_{|u|\ge b} e^{-\pi h^2u^2}\left|e^{-2\pi hua\delta^{1/2}-\pi a^2\delta}-1\right| du\\ &\quad\le \sum_{a=1}^{h}\int_{|u|\ge b} e^{-\pi h^2u^2+2\pi h|u|a\delta^{1/2}} du < \frac{\varepsilon}{2}\end{aligned}$$

for $0 < \delta < \delta_1$. Now

$$\lim_{\delta\to 0^+} \sum_{a=1}^{h} \int_{-b}^{b} e^{-\pi h^2 u^2} \left| e^{-2\pi h u a \delta^{1/2} - \pi a^2 \delta} - 1 \right| du = 0$$

for arbitrary b, passing the limit under the integral sign by uniform convergence. Note that it is not valid in general to pass a limit inside an *improper* integral even if the limit is uniform. Having fixed b, choose δ_2 with $0 < \delta_2 < \delta_1$ such that

$$\sum_{a=1}^{h} \int_{-b}^{b} e^{-\pi h^2 u^2} \left| e^{-2\pi h u a \delta^{1/2} - \pi a^2 \delta} - 1 \right| du < \frac{\varepsilon}{2}$$

for $0 < \delta < \delta_2$. Then

$$\left| \sum_{a=1}^{h} e(a^2 k/h) \sum_{m=-\infty}^{\infty} e^{-\pi h^2 m^2 \delta} \left(e^{-2\pi h m a \delta - \pi a^2 \delta} - 1 \right) \right| < \varepsilon$$

for $0 < \delta < \delta_2$. These considerations yield

$$\sum_{n=-\infty}^{\infty} e^{-\pi n^2 \delta} e(n^2 k/h) \sim \delta^{-1/2} \frac{1}{h} \sum_{a=1}^{h} e(a^2 k/h)$$

as $\delta \to 0^+$.

Next apply the functional equation by

$$\begin{aligned} \sum_{n=-\infty}^{\infty} e^{-\pi n^2 \delta} e(n^2 k/h) &= \frac{1}{\sqrt{\delta - 2ik/h}} \sum_{n=-\infty}^{\infty} e^{-\frac{\pi n^2}{\delta - 2ik/h}} \\ &= \frac{1}{\sqrt{\delta - 2ik/h}} \sum_{n=-\infty}^{\infty} e^{-\frac{\pi n^2 \delta}{\delta^2 + 4k^2/h^2}} \\ &\quad \times e\left(-\frac{n^2 k/h}{\delta^2 + 4k^2/h^2} \right), \end{aligned}$$

with

$$\frac{1}{\sqrt{\delta - 2ik/h}} = \frac{1}{\sqrt{-2ik/h}} + O(\delta)$$

and

$$e^{-\frac{\pi n^2 \delta}{\delta^2 + 4k^2/h^2}} e\left(-\frac{n^2 k/h}{\delta^2 + 4k^2/h^2} \right) = e^{-\frac{\pi n^2 \delta h^2}{4k^2}} e\left(-\frac{n^2 h}{4k} \right) e^{O(n^2 \delta^2)},$$

so that

$$\sum_{n=-\infty}^{\infty} e^{-\pi n^2\delta} e(n^2k/h) \sim \frac{1}{\sqrt{-2ik/h}}\left(\frac{\delta}{4k^2/h^2}\right)^{-1/2}\frac{1}{4k}\sum_{a=1}^{4k} e(-a^2h/4k)$$

$$= \delta^{-1/2}\frac{e^{\pi i/4}}{\sqrt{h}}\frac{1}{\sqrt{2k}}\sum_{a=1}^{2k} e(-a^2h/4k)\,.$$

as $\delta \to 0^+$ by the asymptotic estimate above. Note that the mapping $a \mapsto 2k - a$ preserves each term of the sum from $a = 1$ to $4k$. Comparison of the two asymptotic relations yields the Landsberg-Schaar identity. □

Now the classical Gauss sum is seen to be

$$\tau_p = \sqrt{p}\,\frac{e^{\pi i/4}}{2}\sum_{n=1}^{2} e\left(-\frac{n^2 p}{4}\right) = \begin{cases} \sqrt{p} & \text{if } p \equiv 1 \pmod 4, \\ i\sqrt{p} & \text{if } p \equiv 3 \pmod 4, \end{cases}$$

by the Landsberg-Schaar formula. Gauss kept a research journal, in which he states that it took him four years to determine the sign.

The determination of quadratic Gauss sums implies the Law of Quadratic Reciprocity. Let χ_1 denote the Legendre symbol to the modulus p and χ_2 to the modulus q, where p and q are distinct odd primes. On the one hand

$$\tau(\chi_1, q)\tau(\chi_2, p) = \left(\frac{q}{p}\right)\tau_p\left(\frac{p}{q}\right)\tau_q = \left(\frac{q}{p}\right)\left(\frac{p}{q}\right)\sqrt{pq}\, i^{\frac{p-1}{2}+\frac{q-1}{2}}$$

by Proposition 3.21 and the determination of the classical Gauss sum. On the other hand

$$\tau(\chi_1, q)\tau(\chi_2, p) = \sum_{m=1}^{p} e(m^2q/p)\sum_{n=1}^{q} e(n^2p/q) = \sum_{m=1}^{p}\sum_{n=1}^{q} e\left(\frac{m^2q^2 + n^2p^2}{pq}\right)$$

$$= \sum_{m=1}^{p}\sum_{n=1}^{q} e\left(\frac{(mq+np)^2}{pq}\right) = \sum_{N=1}^{pq} e(N^2/pq)$$

$$= \sqrt{pq}\,\frac{e^{\pi i/4}}{\sqrt{2}}(1 + e(-pq/4))$$

by the Landsberg-Schaar formula. Comparison yields the Law of Quadratic Reciprocity.

It may be worth noting here that the calculation of the classical Gauss sum shows that every quadratic field is a subfield of a cyclotomic field. For it expresses the square root of any odd prime as a sum of roots of unity, and the remaining cases -1 and 2 are clear.

Elliptic Functions by K. Chandrasekharan covers theta functions and their connections to elliptic functions, with many arithmetical applications.

Tata Lectures on Theta I by David Mumford also presents applications of theta functions to number theory. Volumes II and III of this work deal with theta functions in relation to questions of algebraic geometry and nonlinear partial differential equations that have a different flavor.

There are many arithmetical applications of Fourier analysis that are unrelated to theta functions. The collection *La fonction zêta* edited by Nicole Berline and Claude Sabbah contains an exposition by Jean-Benoît Bost of a proof of the Prime Number Theorem due to Jean-Pierre Kahane that relies on the Fourier transform. Both *Analytic Number Theory* by Henryk Iwaniec and Emmanuel Kowalski and *Ten Lectures on the Interface Between Analytic Number Theory and Harmonic Analysis* by Hugh L. Montgomery contain much material on harmonic analysis related to number theory.

8.3. The gamma function

One of the most important of the special functions of classical analysis is the gamma function $\Gamma(s)$. Following K. T. W. Weierstrass we define it by means of an infinite product

$$\frac{1}{\Gamma(s)} = se^{\gamma s} \prod_{n=1}^{\infty} \left(1 + \frac{s}{n}\right) e^{-s/n}$$

converging uniformly on compact sets to an entire function by the estimate $(1+z)e^{-z} = 1 + O(|z|^2)$ as $z \to 0$. From the product one sees that $\Gamma(s)$ has simple poles at $s = 0, -1, -2, \dots$ and no other singularities. Moreover the gamma function clearly has no zeros.

There is an extensive apparatus of identities for $\Gamma(s)$, of which we shall establish only a few that will prove important later. The calculations

$$\begin{aligned}
\frac{1}{\Gamma(s+1)} &= (s+1)e^{\gamma(s+1)} \prod_{n=1}^{\infty} \left(1 + \frac{s+1}{n}\right) e^{-(s+1)/n} \\
&= (s+1)e^{\gamma(s+1)} \prod_{n=1}^{\infty} \frac{n+1+s}{n+1} \left(1 + \frac{1}{n}\right) e^{-1/n} e^{-s/(n+1) - s/n(n+1)} \\
&= (s+1)e^{\gamma(s+1)} \prod_{n=1}^{\infty} e^{-s/n(n+1)} \prod_{n=1}^{\infty} \left(1 + \frac{1}{n}\right) e^{-1/n} \\
&\quad \times \prod_{n=1}^{\infty} \left(1 + \frac{s}{n+1}\right) e^{-s/(n+1)} \\
&= (s+1)e^{\gamma(s+1)} e^{-s} \frac{e^{-\gamma}}{\Gamma(1)} \frac{1}{(1+s)e^{-s}} \frac{1}{se^{\gamma s}} \frac{1}{\Gamma(s)} = \frac{1}{\Gamma(1) s \Gamma(s)}
\end{aligned}$$

and

$$\prod_{n=1}^{N}\left(1+\frac{1}{n}\right)e^{-1/n}=\frac{2}{1}\frac{3}{2}\cdots\frac{N+1}{N}e^{-1-1/2-\cdots-1/N}$$
$$=(N+1)e^{-\log(N)-\gamma-O(1/N)}\to e^{-\gamma}$$

as $N \to +\infty$ establish that $\Gamma(1) = 1$, and also the so-called *functional equation*

$$\Gamma(s+1)=s\,\Gamma(s)$$

of the gamma function. Calculating the values of the gamma function at the positive integers iteratively by the functional equation, one sees that $\Gamma(n) = (n-1)!$. Apart from an inessential shift of the independent variable, the gamma function extends the factorial to the complex plane.

The function $e^{2\pi is}\Gamma(s)$ also extends the factorial, and it is a meromorphic function satisfying the same functional equation as $\Gamma(s)$. In fact the gamma function is far from uniquely determined by the requirements mentioned. But we do obtain a very useful uniqueness statement by imposing a boundedness condition.

Proposition 8.6 (Wielandt uniqueness theorem)**.** *Suppose that F is a holomorphic function on* $\mathrm{Re}(s) > 0$ *that is bounded in the strip* $1 \le \sigma \le 2$. *If* $F(1) = 1$ *and* $F(s+1) = sF(s)$ *then* $F(s) = \Gamma(s)$ *everywhere.*

Proof. If $1 \le \sigma \le 2$ then

$$|s|\geq\sigma\quad\text{and}\quad|e^{\gamma s}|=e^{\gamma\sigma}\quad\text{and}\quad\left|\left(1+\frac{s}{n}\right)e^{-s/n}\right|\geq\left(1+\frac{\sigma}{n}\right)e^{-\sigma/n},$$

so $|\Gamma(s)| \le \Gamma(\sigma)$ in the vertical strip by its definition.

Clearly F also has an analytic continuation to the complex plane by the functional equation, with at worst poles at $s = 0, -1, -2, \ldots$. The difference $G(s) = F(s) - \Gamma(s)$ satisfies the same functional equation, is holomorphic everywhere except possibly for poles at $s = 0, -1, -2, \ldots$, and is also bounded in the strip. But in addition, $G(1) = F(1)-\Gamma(1) = 1-1 = 0$. Now

$$\lim_{s\to0}sG(s)=\lim_{s\to0}G(s+1)=0,$$

and so G is holomorphic at $s = 0$ by the Riemann removable singularities theorem. But then the functional equation implies that G is an entire function.

The formula $G(s) = G(s+1)/s$ now implies that G is bounded on the strip $0 \le \sigma \le 1$. Then the entire function $H(s) = G(s)G(1-s)$ is also bounded on this strip, and

$$H(s+1)=G(s+1)G(1-(s+1))=\frac{G(s)}{s}(-s)G(1-s)=-H(s).$$

This implies that $H(s+2) = H(s)$, so H is a bounded entire function, thus constant by Liouville's theorem. Then H and thus G is identically zero, since $H(0) = G(0)G(1) = 0$. □

Proposition 8.7 (Euler integral formula). *The integral representation*

$$\Gamma(s) = \int_0^\infty e^{-u} u^{s-1}\, du$$

holds in the half plane $\sigma > 0$.

Proof. Define a function

$$F(s) = \int_0^\infty e^{-u} u^{s-1}\, du$$

for $\sigma > 0$. The integral converges uniformly on compact subsets of $\sigma > 0$, so F is holomorphic there. Moreover

$$|F(s)| = \left| \int_0^\infty e^{-u} u^{s-1}\, du \right| \leq \int_0^\infty e^{-u} u^{\sigma-1}\, du = F(\sigma),$$

so F is bounded in the strip $1 \leq \sigma \leq 2$.

The integration by parts

$$\int_0^\infty e^{-u} u^s\, du = -e^{-u} u^s \Big|_0^\infty - \int_0^\infty (-e^{-u}) s u^{s-1}\, du = s \int_0^\infty e^{-u} u^{s-1}\, du$$

establishes the functional equation $F(s+1) = s\,F(s)$ for $\sigma > 0$. Since

$$F(1) = \int_0^\infty e^{-u}\, du = 1$$

the Wielandt uniqueness theorem implies the claim. □

The next result is closely related to the famous factorization

$$\sin(z) = z \prod_{n=1}^\infty \left(1 - \frac{z^2}{\pi^2 n^2}\right)$$

of the sine, and both are due to Euler.

Proposition 8.8 (Reflection formula). *The identity*

$$\Gamma(s)\Gamma(1-s) = \frac{\pi}{\sin(\pi s)}$$

holds everywhere.

Proof. We are going to establish the identity for $s = \sigma$ with $0 < \sigma < 1$. This will be sufficient by analytic continuation.

First

$$\begin{aligned}\Gamma(\sigma)\Gamma(1-\sigma) &= \int_0^\infty e^{-u}u^{\sigma-1}\,du \int_0^\infty e^{-v}v^{-\sigma}\,dv \\ &= \int_0^\infty \int_0^\infty e^{-u-v}\left(\frac{u}{v}\right)^\sigma \frac{du}{u}\,dv \\ &= \int_0^\infty \int_0^\infty e^{-vx-v}x^{\sigma-1}\,dx\,dv \\ &= \int_0^\infty x^{\sigma-1}\int_0^\infty e^{-vx-v}\,dv\,dx = \int_0^\infty \frac{x^{\sigma-1}}{x+1}\,dx\end{aligned}$$

by the substitution $u = vx$ and an interchange of the order of integration.

The last integral will be evaluated by residue calculus. Let $C_{r,R}$ with $0 < r < 1 < R$ denote the path of integration that begins at $x = r$ on the positive real axis, runs along the real axis to $x = R$, follows the circle $|z| = R$ counterclockwise to $x = R$, runs down the real axis to $x = r$, and follows the circle $|z| = r$ clockwise to $x = r$. Let $z^{\sigma-1}/(z+1)$ denote the single-valued meromorphic branch on the complex plane slit along the positive real axis whose boundary values from above on the positive real axis are $x^{\sigma-1}/(x+1)$. This function is holomorphic on and inside $C_{r,R}$ with the exception of a simple pole at $z = -1$, where the residue is $-e^{\pi i\sigma}$. Then

$$\oint_{C_{r,R}} \frac{z^{\sigma-1}}{z+1}\,dz = -2\pi i e^{\pi i\sigma}$$

by the Residue Theorem. We note that

$$\left|\int_{|z|=R} \frac{z^{\sigma-1}}{z+1}\,dz\right| \le \int_{|z|=R} \frac{|z|^{\sigma-1}}{|z+1|}|dz| \le \frac{R^{\sigma-1}}{R-1}2\pi R \to 0$$

as $R \to +\infty$, and similarly with the integral over $|z| = r$ as $r \to 0^+$. Then

$$\int_0^\infty \frac{x^{\sigma-1}}{x+1}\,dx + \int_\infty^0 \frac{e^{2\pi i(\sigma-1)}x^{\sigma-1}}{x+1}\,dx = -2\pi i e^{\pi i\sigma},$$

and solving for the integral yields the desired formula when $0 < \sigma < 1$. □

Note that $\Gamma(1/2) = \sqrt{\pi}$ by substituting $s = 1/2$ in the reflection formula, since the gamma function is positive on the positive real axis. The next result is due to Legendre.

Proposition 8.9 (Duplication formula). *The identity*

$$\Gamma(s)\Gamma\left(s + \frac{1}{2}\right) = \sqrt{\pi}\,2^{1-2s}\Gamma(2s)$$

holds everywhere.

Proof. Put

$$F(s) = \frac{\Gamma\left(\frac{s}{2}\right)\Gamma\left(\frac{s}{2}+\frac{1}{2}\right)}{\sqrt{\pi}\,2^{1-s}}$$

and note that $F(1) = 1$ because $\Gamma(1/2) = \sqrt{\pi}$, and that $F(s)$ is holomorphic in $\sigma > 0$. Moreover

$$F(s+1) = \frac{\Gamma\left(\frac{s}{2}+\frac{1}{2}\right)\Gamma\left(\frac{s}{2}+1\right)}{\sqrt{\pi}\,2^{1-s-1}} = \frac{\Gamma\left(\frac{s}{2}+\frac{1}{2}\right)\frac{s}{2}\Gamma\left(\frac{s}{2}\right)}{\sqrt{\pi}\,2^{1-s}2^{-1}} = sF(s),$$

while the argument at the beginning of the proof of the Wielandt uniqueness theorem implies that the gamma function is bounded in any fixed closed strip in the half plane $\sigma > 0$, so $F(s)$ is bounded in the strip $1 \le \sigma \le 2$. □

Further theory of the gamma function may be pursued in *The Gamma Function* by Emil Artin. *Classical Topics in Complex Function Theory* by Reinhold Remmert also has an excellent exposition. *A Course of Modern Analysis* by E. T. Whittaker and G. N. Watson has an extensive treatment of the apparatus of gamma function identities, and is a very accessible source of information on the special functions of classical analysis. Moreover, anyone who wishes to sharpen his or her manipulative skills in analysis will find there many challenging problems taken from nineteenth century Cambridge exams.

8.4. The functional equation of $\zeta(s)$

In Chapter 5 the Riemann zeta function was analytically continued to $\sigma > 0$ and a weak bound was proved for it. More precise information may be obtained by means of a functional equation satisfied by $\zeta(s)$. We shall give Riemann's second proof of this functional equation, using the functional equation of the Jacobi theta function. From this proof it is clear that the functional equation is a consequence of the additive structure of the integers, just as the Euler product formula is a consequence of their multiplicative structure.

Integral transforms of the type

$$(\mathcal{M}f)(s) = \int_0^\infty f(u)u^{s-1}\,du = \int_{\mathbb{R}^+} f(u)u^s\frac{du}{u}$$

are called *Mellin transforms.* The integral is sometimes written with du/u as above to display the invariance property $d(au)/(au) = du/u$, emphasizing that the Mellin transform may naturally be viewed as a transform on the multiplicative group $\mathbb{R}^+$.

The Mellin transform representation of the gamma function enables us to represent the Riemann zeta function by means of a Mellin transform. The change of variable $u = \pi n^2 v$ in the integral

$$\Gamma\Big(\frac{s}{2}\Big) = \int_0^\infty e^{-u} u^{s/2-1}\, du = \int_0^\infty e^{-\pi n^2 v} \pi^{s/2-1} n^{s-2} v^{s/2-1} \pi n^2\, dv$$

shows that

$$\pi^{-s/2}\Gamma\Big(\frac{s}{2}\Big) n^{-s} = \int_0^\infty e^{-\pi n^2 u} u^{s/2-1}\, du,$$

and so

$$\pi^{-s/2}\Gamma\Big(\frac{s}{2}\Big)\zeta(s) = \sum_{n=1}^\infty \int_0^\infty e^{-\pi n^2 u} u^{s/2-1}\, du$$

for $\sigma > 1$. To proceed further we would like to pass the summation sign under the integral sign, but the series

$$\sum_{n=1}^\infty e^{-\pi n^2 u}$$

does not converge uniformly on $[0, \infty)$. Moreover, it is not true that uniform convergence is sufficient to allow interchange of the sum and integral when the integral is over an infinite interval. We proceed in the following way, which is also adequate for similar cases encountered later. Let $0 < a < b < \infty$ and note that the above series converges uniformly on $[a, b]$. Then

$$\begin{aligned}\sum_{n=1}^\infty \int_a^b e^{-\pi n^2 u} u^{s/2-1}\, du &= \int_a^b \left(\sum_{n=1}^\infty e^{-\pi n^2 u}\right) u^{s/2-1}\, du \\ &= \int_a^b \frac{\vartheta(0, iu) - 1}{2} u^{s/2-1}\, du,\end{aligned}$$

by uniform convergence, while

$$\begin{aligned}&\left|\sum_{n=1}^\infty \int_0^\infty e^{-\pi n^2 u} u^{s/2-1}\, du - \sum_{n=1}^\infty \int_a^b e^{-\pi n^2 u} u^{s/2-1}\, du\right| \\ &\qquad \le \sum_{n=1}^\infty \left(\int_0^a + \int_b^\infty\right) e^{-\pi n^2 u} u^{\sigma/2-1}\, du.\end{aligned}$$

Now

$$\begin{aligned}\sum_{n=1}^N \int_0^a e^{-\pi n^2 u} u^{\sigma/2-1}\, du &= \int_0^a \left(\sum_{n=1}^N e^{-\pi n^2 u}\right) u^{\sigma/2-1}\, du \\ &\le \int_0^a \frac{\vartheta(0, iu) - 1}{2} u^{\sigma/2-1}\, du,\end{aligned}$$

and thus

$$\sum_{n=1}^\infty \int_0^a e^{-\pi n^2 u} u^{\sigma/2-1}\, du \le \int_0^a \frac{\vartheta(0, iu) - 1}{2} u^{\sigma/2-1}\, du \to 0$$

as $a \to 0^+$, and similarly for the integral over $[b, \infty)$. Finally

$$\begin{aligned}\pi^{-s/2}\Gamma\Big(\frac{s}{2}\Big)\,\zeta(s) &= \sum_{n=1}^{\infty}\int_0^{\infty} e^{-\pi n^2 u} u^{s/2-1}\,du \\ &= \int_a^b \Big(\sum_{n=1}^{\infty} e^{-\pi n^2 u}\Big)\, u^{s/2-1}\,du \\ &= \int_0^{\infty} \frac{\vartheta(0,iu)-1}{2}\, u^{s/2-1}\,du.\end{aligned}$$

What makes this simple argument work is absolute convergence of the improper integral.

Proposition 8.10 (Functional equation of the zeta function)**.** *The Riemann zeta function has an analytic continuation to the whole complex plane as a meromorphic function and satisfies the functional equation*

$$\pi^{-(1-s)/2}\Gamma\Big(\frac{1-s}{2}\Big)\zeta(1-s) = \pi^{-s/2}\Gamma\Big(\frac{s}{2}\Big)\zeta(s).$$

It is holomorphic except for a simple pole at $s = 1$.

Proof. We already know that

$$\pi^{-s/2}\Gamma\Big(\frac{s}{2}\Big)\,\zeta(s) = \int_0^{\infty} \frac{\vartheta(0,iu)-1}{2}\, u^{s/2-1}\,du,$$

for $\sigma > 1$. Then

$$\begin{aligned}\pi^{-s/2}\Gamma\Big(\frac{s}{2}\Big)\zeta(s) &= \Big(\int_0^1 + \int_1^{\infty}\Big)\,\frac{\vartheta(0,iu)-1}{2}\, u^{s/2-1}\,du \\ &= \int_{\infty}^1 \frac{\vartheta(0,i/v)-1}{2} v^{-s/2+1}(-v^{-2})\,dv + \int_1^{\infty}\frac{\vartheta(0,iu)-1}{2}\, u^{s/2-1}\,du \\ &= \int_1^{\infty} \frac{v^{1/2}\vartheta(0,iv)-v^{1/2}}{2}\, v^{-s/2-1}\,dv + \int_1^{\infty}\frac{v^{1/2}-1}{2}\, v^{-s/2-1}\,dv \\ &\quad + \int_1^{\infty}\frac{\vartheta(0,iu)-1}{2}\, u^{s/2-1}\,du \\ &= \frac{1}{s(s-1)} + \frac{1}{2}\int_1^{\infty} (\vartheta(0,iu)-1)\Big(u^{s/2}+u^{(1-s)/2}\Big)\frac{du}{u}\end{aligned}$$

where $\vartheta(0,iu) - 1$ decays exponentially as $u \to +\infty$, so the last integral converges to an entire function invariant under the mapping $s \mapsto 1-s$.

The formula obtained implies that $\zeta(s)$ cannot have singularities except at $s = 0$ and $s = 1$, because $\Gamma(s/2)$ has no zeros. Furthermore $\zeta(s)$ must have a simple pole at $s = 1$ since $\Gamma(s/2)$ has no pole at this point. But then $\zeta(s)$ must be holomorphic at $s = 0$, because $\Gamma(s/2)$ has a simple pole there. □

The function $Z(s) = \pi^{-s/2}\Gamma(s/2)\zeta(s)$ is called the *completed Riemann zeta function.* The notation for the completed zeta function has never been standardized; see the Notes for various notations in use. Clearly the functional equation takes the form $Z(1-s) = Z(s)$ for the completed Riemann zeta function. Since $\zeta(s)$ is holomorphic in $\sigma > 0$ except for a simple pole at $s = 1$, and $\Gamma(s/2)$ has no singularities in $\sigma > 0$ and is never zero, $Z(s)$ is holomorphic in $\sigma > 0$ except for a simple pole at $s = 1$. Then by the symmetry around the *critical line* $\sigma = 1/2$, it is also holomorphic in $\sigma < 1$ except for a simple pole at $s = 0$. Moreover, $Z(s)$ has no zeros in the half plane $\sigma > 1$, for the gamma function it has no zeros whatsoever, and we already know that the zeta function has none in $\sigma > 1$. Clearly $Z(s)$ does not have any zeros in $\sigma < 0$ either, again by the symmetry property, so all its zeros lie in the *critical strip* $0 \le \sigma \le 1$. Since $Z(s)$ is real on the real axis, the zeros lie symmetrically about both the real axis and the critical line.

As $Z(s)$ is holomorphic and nonzero on $\sigma < 0$ the zeros of $\zeta(s)$ in this half plane occur precisely where $\Gamma(s/2)$ has a pole. The poles are simple poles at $s = -2, -4, -6, \ldots$, so $\zeta(s)$ has simple zeros at these points. These zeros are called the *trivial zeros,* and are the only zeros outside the critical strip. The zeros inside the critical strip are of great importance for the distribution of primes. Riemann conjectured that they all lie on the critical line $\sigma = 1/2$. This is the famous Riemann Hypothesis.

The functional equation of the zeta function is often written in an unsymmetrical form. We may pass from the symmetrical form to one of the unsymmetrical forms by a calculation

$$\begin{aligned}\zeta(1-s) &= \frac{\pi^{-s/2}\Gamma\left(\frac{s}{2}\right)}{\pi^{-(1-s)/2}\Gamma\left(\frac{1-s}{2}\right)}\zeta(s) = \pi^{1/2}\pi^{-s}\frac{\Gamma\left(\frac{s}{2}\right)}{\Gamma\left(1-\frac{s+1}{2}\right)}\zeta(s) \\ &= \pi^{-1/2}\pi^{-s}\Gamma\left(\frac{s}{2}\right)\Gamma\left(\frac{s}{2}+\frac{1}{2}\right)\cos\left(\frac{\pi s}{2}\right)\zeta(s) \\ &= 2(2\pi)^{-s}\Gamma(s)\cos\left(\frac{\pi s}{2}\right)\zeta(s)\end{aligned}$$

using the reflection formula and the duplication formula. This form of the functional equation is the one obtained from Riemann's first proof, using residue calculus.

Riemann introduced an entire function $\xi(s) = s(s-1)Z(s)/2$ also satisfying a symmetric functional equation $\xi(1-s) = \xi(s)$. Clearly the zeros of $\xi(s)$ coincide with the zeros of $Z(s)$, and with the same multiplicities. The symmetry of $\xi(s)$ with respect to the critical line motivated Riemann to introduce a new function $\Xi(z) = \xi(1/2 + iz)$ by a change of variable. The functional equation takes its simplest form for this function, just stating that $\Xi(z)$ is an even function.

The standard reference for the analytic theory of the Riemann zeta function is the second edition of *The Theory of the Riemann Zeta-function* by E. C. Titchmarsh, revised and with end-of-chapter notes by D. R. Heath-Brown. There is also a comprehensive treatise *The Riemann Zeta-Function* by Aleksandar Ivić. *An Introduction to the Theory of the Riemann Zeta-Function* by S. J. Patterson is strong on explanations and motivation, and has many exercises. It goes more deeply into the theory of the Riemann zeta function than this book, but does not cover as broad a canvas as the treatises of Ivić and Titchmarsh.

8.5. ⋆ The functional equation of $L(s,\chi)$

There is a nearly, but not quite, symmetric form of functional equation for Dirichlet L-functions. For imprimitive characters χ the Euler product of $L(s,\chi)$ has factors with zeros on the imaginary axis, and these factors also complicate the functional equation. So we restrict the Dirichlet characters to be primitive. Dirichlet characters are termed *even* if $\chi(-1)=1$ and *odd* if $\chi(-1)=-1$. Because $\chi(-1)^2=\chi((-1)^2)=\chi(1)=1$, these are the only two possibilities for $\chi(-1)$. It turns out that the completed L-function takes slightly different forms in the two cases. Introducing a gamma factor

$$\gamma(s,\chi) \overset{\text{def}}{=} \left(\frac{\pi}{q}\right)^{-\frac{s+v}{2}} \Gamma\left(\frac{s+v}{2}\right), \qquad v=v(\chi)=\frac{1-\chi(-1)}{2},$$

the completed L-function is $Z(s,\chi) \overset{\text{def}}{=} \gamma(s,\chi)L(s,\chi)$.

Proposition 8.11 (Functional equation of Dirichlet L-functions). *The completed Dirichlet L-function of a primitive character χ modulo $q\geq 3$ has an analytic continuation to the whole complex plane as an entire function and satisfies the functional equation*

$$Z(s,\chi)=W(\chi)Z(1-s,\overline{\chi}), \qquad W(\chi)=\frac{\tau(\chi)}{i^v\sqrt{q}}.$$

Proof. The integral representation of the gamma function yields

$$\left(\frac{\pi}{q}\right)^{-\frac{s+v}{2}} \Gamma\left(\frac{s+v}{2}\right) n^{-s} = \int_0^\infty n^v e^{-\pi n^2 u/q} u^{\frac{s+v}{2}-1}\, du$$

as in Riemann's second proof of the functional equation of the zeta function. Then

$$Z(s,\chi)=\frac{1}{2}\int_0^\infty \left(\sum_{n=-\infty}^{\infty} \chi(n) n^v e^{-\pi n^2 u/q}\right) u^{\frac{s+v}{2}-1}\, du,$$

noting that $\chi(0)=0$ since χ is not the principal character modulo 1. Discussion of the convergence of this integral will be deferred until we know a little more about the integrand.

The series under the integral sign can be expressed in terms of theta functions due to the periodicity of χ. Introducing the notation

$$\vartheta(\chi, z, \tau) = \sum_{n=-\infty}^{\infty} \chi(n) e^{\pi i n^2 \tau} e(nz)$$

we have

$$\sum_{n=-\infty}^{\infty} \chi(n) n^v e^{-\pi n^2 u/q} = (2\pi i)^{-v} \left. \frac{d^{(v)}}{dz^{(v)}} \vartheta(\chi, z, iu/q) \right|_{z=0}.$$

Now

$$\begin{aligned}
\vartheta(\chi, z, iu/q) &= \sum_{r=1}^{q} \sum_{m=-\infty}^{\infty} \chi(qm+r) e^{-\pi(mq+r)^2 u/q} e((mq+r)z) \\
&= \sum_{r=1}^{q} \chi(r) e^{-\pi r^2 u/q} e(rz/q) \sum_{m=-\infty}^{\infty} e^{-\pi m^2 qu} e(mq(z + riu/q)) \\
&= \sum_{r=1}^{q} \chi(r) e^{-\pi r^2 u/q} e(rz/q) \vartheta(qz + riu, qiu).
\end{aligned}$$

Next

$$\begin{aligned}
\vartheta(qz + riu, qiu) &= \sqrt{\frac{i}{qiu}} e^{-\pi i(qz+riu)^2/(qiu)} \vartheta\left(\frac{qz + riu}{qiu}, -\frac{1}{qiu} \right) \\
&= \frac{1}{\sqrt{qu}} e^{-\pi(z+riu)^2/(qu)} \vartheta\left(\frac{qz + riu}{qiu}, -\frac{1}{qiu} \right)
\end{aligned}$$

by the functional equation of the Jacobi theta function. Thus

$$\begin{aligned}
\vartheta(\chi, z, iu/q) &= \sum_{r=1}^{q} \chi(r) e^{-\pi r^2 u/q} e(rz/q) \frac{e^{-\pi(z+riu)^2/(qu)}}{\sqrt{qu}} \\
&\quad \times \vartheta\left(\frac{qz + riu}{qiu}, -\frac{1}{qiu} \right) \\
&= \frac{e^{-\pi z^2/(qu)}}{\sqrt{qu}} \sum_{r=1}^{q} \chi(r) \vartheta\left(\frac{qz + riu}{qiu}, -\frac{1}{qiu} \right) \\
&= \frac{e^{-\pi z^2/(qu)}}{\sqrt{qu}} \sum_{r=1}^{q} \chi(r) \sum_{n=-\infty}^{\infty} e^{-\pi i n^2/(qiu)} e\left(n \frac{qz + riu}{qiu} \right),
\end{aligned}$$

and so

$$\begin{aligned}
\vartheta(\chi, z, iu/q) &= \frac{e^{-\pi z^2/(qu)}}{\sqrt{qu}} \sum_{n=-\infty}^{\infty} e^{\pi i n^2/(-qiu)} e\left(n\frac{z}{iu}\right) \sum_{r=1}^{q} \chi(r) e\left(\frac{nr}{q}\right) \\
&= \frac{e^{-\pi z^2/(qu)}}{\sqrt{qu}} \sum_{n=-\infty}^{\infty} e^{\pi i n^2/(-qiu)} e\left(n\frac{z}{iu}\right) \tau(\chi, n) \\
&= \tau(\chi)\frac{e^{-\pi z^2/(qu)}}{\sqrt{qu}} \sum_{n=-\infty}^{\infty} e^{\pi i n^2/(-qiu)} e\left(n\frac{z}{iu}\right) \overline{\chi(n)} \\
&= \tau(\chi)\frac{e^{-\pi z^2/(qu)}}{\sqrt{qu}} \vartheta\left(\overline{\chi}, \frac{z}{iu}, -\frac{1}{qiu}\right),
\end{aligned}$$

by Proposition 3.23, since χ is primitive. Differentiating once and setting $z = 0$ yields

$$\vartheta^{(v)}(\chi, 0, iu/q) = W(\chi) u^{-\frac{1}{2}-v} \vartheta^{(v)}\left(\overline{\chi}, 0, -\frac{1}{qiu}\right)$$

and so

$$\sum_{n=-\infty}^{\infty} \chi(n) n^v e^{-\pi n^2 u/q} = W(\chi) u^{-\frac{1}{2}-v} \sum_{n=-\infty}^{\infty} \overline{\chi(n)} n^v e^{-\pi n^2/(qu)}.$$

Write

$$\begin{aligned}
Z(s,\chi) &= \frac{1}{2}\int_0^{\infty} \left(\sum_{n=-\infty}^{\infty} \chi(n) n^v e^{-\pi n^2 u/q}\right) u^{\frac{s+v}{2}-1}\, du \\
&= \frac{1}{2}\int_0^{1} \left(\sum_{n=-\infty}^{\infty} \chi(n) n^v e^{-\pi n^2 u/q}\right) u^{\frac{s+v}{2}-1}\, du \\
&\quad + \frac{1}{2}\int_1^{\infty} \left(\sum_{n=-\infty}^{\infty} \chi(n) n^v e^{-\pi n^2 u/q}\right) u^{\frac{s+v}{2}-1}\, du \\
&= \frac{1}{2}\int_0^{1} W(\chi) u^{-\frac{1}{2}-v} \left(\sum_{n=-\infty}^{\infty} \overline{\chi(n)} n^v e^{-\pi n^2/(qu)}\right) u^{\frac{s+v}{2}-1}\, du \\
&\quad + \frac{1}{2}\int_1^{\infty} \left(\sum_{n=-\infty}^{\infty} \chi(n) n^v e^{-\pi n^2 u/q}\right) u^{\frac{s+v}{2}-1}\, du
\end{aligned}$$

and note that the series in the improper integral over $(0, 1]$ has the bound

$$\left|\sum_{n=-\infty}^{\infty} \overline{\chi(n)} n^v e^{-\pi n^2/(qu)}\right| \leq 2\sum_{n=1}^{\infty} n e^{-n/(qu)} = 2\frac{e^{1/(qu)}}{(e^{1/(qu)} - 1)^2},$$

so the integral over $(0, 1]$ converges absolutely for every s, and similarly for the integral over $[1, \infty)$. Thus the integral representation for $Z(s, \chi)$ is valid everywhere in the complex plane. The integral

$$I_\delta(s,\chi) = \frac{1}{2}\int_\delta^{\delta^{-1}} \left(\sum_{n=-\infty}^{\infty} \chi(n) n^v e^{-\pi n^2 u/q}\right) u^{\frac{s+v}{2}-1}\, du$$

is an entire function for each $\delta > 0$, and $I_\delta(s,\chi) \to Z(s,\chi)$ uniformly on compact sets as $\delta \to 0^+$, by the bounds on the integrand over $(0,1]$ and $[1,\infty)$, so $Z(s,\chi)$ has an analytic continuation to an entire function. The calculation

$$\begin{aligned} Z(s,\chi) &= \frac{1}{2}\int_0^{\infty} \left(\sum_{n=-\infty}^{\infty} \chi(n) n^v e^{-\pi n^2 u/q}\right) u^{\frac{s+v}{2}-1}\, du \\ &= \frac{1}{2}\int_0^{\infty} u^{-\frac{1}{2}-v} \left(\sum_{n=-\infty}^{\infty} \overline{\chi(n)} n^v e^{-\pi n^2/(qu)}\right) u^{\frac{s+v}{2}-1}\, du \\ &= \frac{W(\chi)}{2}\int_\infty^{0} u^{\frac{1}{2}+v} \left(\sum_{n=-\infty}^{\infty} \overline{\chi(n)} n^v e^{-\pi n^2 u/qu}\right) u^{-\frac{s+v}{2}+1} (-u^{-2})\, du \\ &= \frac{W(\chi)}{2}\int_0^{\infty} \left(\sum_{n=-\infty}^{\infty} \overline{\chi(n)} n^v e^{-\pi n^2 u/qu}\right) u^{\frac{1-s+v}{2}-1}\, du \\ &= W(\chi) Z(1-s,\overline{\chi}) \end{aligned}$$

is valid by the absolute convergence of the integral. Note the change of variable to replace u by u^{-1}. □

The holomorphicity everywhere of $Z(s,\chi)$ for $\chi \neq \chi_0$ a primitive character implies that $L(s,\chi)$ is an entire function, since the gamma function has no zeros. But then $L(s,\chi)$ is an entire function for any character $\chi \neq \chi_0$, because $L(s,\chi)$ equals the product of $L(s,\chi^*)$ with a finite Euler product, where χ^* is the primitive character inducing χ.

The zeros of the Dirichlet L-functions have the same significance for the distribution of primes in arithmetic progressions as the zeros of the Riemann zeta function have for the distribution of primes. But several new complications arise. In particular the above calculation shows that there will in general be trivial zeros on the imaginary axis, and possibly a zero of large multiplicity at the origin, if the character is imprimitive. So it is usually assumed that χ is a primitive nonprincipal character, which is required in the form of functional equation that we have established. The fact that $Z(\sigma,\chi) \neq 0$ on the interval $\sigma \geq 1$ for all $\chi \neq \chi_0$ implies that $L(s,\chi)$ has simple zeros at $s = 0, -2, -4, \ldots$ if χ is even and at $s = -1, -3, -5, \ldots$ if χ is odd; $\gamma(s,\chi)$ has simple poles at these points. Since $\gamma(s,\chi)$ has no other poles and $L(s,\chi) = 0$ for $\sigma > 1$, all other zeros of the Dirichlet L-functions lie in the critical strip $0 \leq \sigma \leq 1$. By taking complex conjugates on each side of the functional equation, one sees that $Z(s,\chi)$ and $Z(1-\overline{s},\chi)$ have the same zeros, so the same holds for $L(s,\chi)$ and $L(1-\overline{s},\chi)$ when $0<\sigma<1$, since $\gamma(s,\chi)$ is holomorphic and free of zeros in this domain. Thus the zeros of Dirichlet L-functions are symmetric about the critical line, though not necessarily about the real axis. The L-functions with real characters are real on the real axis, and so their zeros are also symmetric about the real axis.

8.6. The Hadamard factorization theorem

Like polynomials, entire functions can be factored in terms of their zeros. Suppose that $f(z)$ is an entire function, with a zero of multiplicity $m \geq 0$ at the origin. Then $f(z)/z^m$ is also an entire function, and is nonzero at the origin. Let $(z_k)_{k=1}^{\infty}$ be the sequence of zeros of $f(z)/z^m$ counted with multiplicity, and ordered by increasing modulus, and then by increasing argument. Supposing that the product in the denominator converges uniformly on compact sets, the function

$$h(z) = \frac{f(z)}{z^m \prod_{k=1}^{\infty}\left(1-\frac{z}{z_k}\right)}$$

is entire and has no zeros. Since the complex plane is simply connected, the analytic function $\log(h(z))$ has a holomorphic branch $g(z)$ on $\mathbb{C}$ and so

$$f(z) = z^m e^{g(z)} \prod_{k=1}^{\infty}\left(1-\frac{z}{z_k}\right)$$

with $g(z)$ an entire function. This will be valid if

$$\sum_{k=1}^{\infty} \frac{1}{|z_k|} < \infty,$$

since the infinite product converges uniformly on compact sets then. If this condition is satisfied, moreover, the infinite product is absolutely convergent, so the particular ordering of the factors is actually immaterial.

When the zeros of $f(z)$ are so dense that the series of reciprocals of their moduli diverges, the above argument fails. To deal with this problem, Weierstrass introduced *primary factors* $E_J(z) = (1-z)\exp(p(J,z))$ of order J, where

$$p(J,z) = \sum_{j=1}^{J} \frac{1}{j} z^j$$

is a partial sum of the power series expansion of $-\mathrm{Log}(1-z)$ at the origin. For these an estimate $\mathrm{Log}(E_J(z)) = O(|z|^{J+1})$ holds in the disk $|z| \leq 1/2$. With the same argument as above, but using primary factors instead of linear factors, yields an absolutely and uniformly convergent product representation

$$f(z) = z^m e^{g(z)} \prod_{k=1}^{\infty} \left(1 - \frac{z}{z_k}\right) e^{p(J_k, z/z_k)} \quad \text{if} \quad \sum_{k=1}^{\infty} \frac{1}{|z_k|^{J_k+1}} < \infty.$$

For any entire function, the Weierstrass primary factors will produce convergence of the infinite product if the sequence $(J_k)_{k=1}^{\infty}$ of their orders is chosen to grow fast enough. Since it is a matter of producing convergence, we may consider only those zeros for which $|z_k| \geq 2$, and $J_k = k$ will work. Note that Weierstrass factorizations are not unique by any means.

Proposition 8.12 (Jensen's formula). *Suppose $f(z)$ is a holomorphic function in a domain that contains the closed disk $|z| \leq R$, where $f(z)$ has the zeros $z_1, \ldots, z_n$ counted with multiplicity. Then*

$$\log(|f(0)|) + \log\left(\frac{R^n}{|z_1 \cdots z_n|}\right) = \frac{1}{2\pi} \int_0^{2\pi} \log(|f(Re^{i\theta})|)\, d\theta$$

if $f(0) \neq 0$.

Proof. There is some $\varepsilon > 0$ so that $f(z)$ is holomorphic in $|z| < R + \varepsilon$ and has no more zeros in this disk than in $|z| \leq R$. The function

$$h(z) = \frac{f(z)}{\left(1 - \frac{z}{z_1}\right) \cdots \left(1 - \frac{z}{z_n}\right)}$$

is holomorphic and without zeros in $|z| < R + \varepsilon$. Then $\log(h(z))$ has a holomorphic branch $g(z)$ on the disk $|z| < R + \varepsilon$ since this is simply connected. Now

$$g(0) = \frac{1}{2\pi} \int_{-\pi}^{\pi} g(Re^{i\theta})\, d\theta$$

by the mean value version of the Cauchy integral formula. Taking the real part on both sides and noting that $\mathrm{Re}(g(0)) = \log(|h(0)|) = \log(|f(0)|)$ while

$$\begin{aligned} \mathrm{Re}\left(\frac{1}{2\pi} \int_{-\pi}^{\pi} g(Re^{i\theta})\, d\theta\right) &= \frac{1}{2\pi} \int_{-\pi}^{\pi} \log(|h(Re^{i\theta}|)\, d\theta \\ &= \frac{1}{2\pi} \int_{-\pi}^{\pi} \log(|f(Re^{i\theta})|)\, d\theta \\ &\quad - \sum_{m=1}^{n} \frac{1}{2\pi} \int_{-\pi}^{\pi} \log\left(\left|1 - \frac{Re^{i\theta}}{z_m}\right|\right) d\theta \end{aligned}$$

we see that

$$\log(|f(0)|) + \sum_{m=1}^{n} \frac{1}{2\pi} \int_{-\pi}^{\pi} \log\left(\left|1 - \frac{Re^{i\theta}}{z_m}\right|\right) d\theta = \frac{1}{2\pi} \int_{-\pi}^{\pi} \log(|f(Re^{i\theta})|)\, d\theta.$$

But

$$\int_{-\pi}^{\pi} \log\left(\left|1 - \frac{Re^{i\theta}}{z_m}\right|\right)\frac{d\theta}{2\pi} = \log\left(\frac{R}{|z_m|}\right) + \int_{-\pi}^{\pi} \log\left(\left|1 - z_m R^{-1} e^{i\theta}\right|\right)\frac{d\theta}{2\pi},$$

and the last integral is zero if $|z_m| < R$. Then $\mathrm{Log}(1 - z_m R^{-1} w)$ is holomorphic on an open disk containing $|w| \leq 1$, and the conclusion follows from the mean value formula applied on the circle $|w| = 1$, after taking the real part on both sides. When $|z_m| = R$ the same function is holomorphic on $|w| < 1$ and continuous on the closure, except for a singularity at $w_0 = R/z_m$. The mean value formula may again be applied, by uniform continuity, and again gives the value zero for the integral, except that now the path of integration along $|w| = 1$ should have a semicircular interior indentation of radius $\varepsilon > 0$ with center at w_0. However, the integral of $\mathrm{Log}(1 - z_m R^{-1} w)$ around the closed path consisting of this indentation together with the arc of $|w| = 1$ within ε of w_0 converges absolutely and is $\ll \varepsilon|\log(\varepsilon)|$, which tends to zero as $\varepsilon \to 0^+$. Thus we get the same answer, namely zero, if we integrate along $|w| = 1$ without any indentation. □

Denote by $n(r)$ the number of zeros, counted with multiplicity, of $f(z)$ in the closed disk $|z| \leq r$. Jensen's formula may be rewritten in the form

$$\log(|f(0)|) + \int_0^R \frac{n(r)}{r}\,dr = \frac{1}{2\pi}\int_0^{2\pi} \log(|f(Re^{i\theta})|)\,d\theta$$

because

$$\int_{|z_k|}^{R} \frac{dr}{r} = \log\left(\frac{R}{|z_k|}\right).$$

The most important datum about an entire function $f(z)$ is its rate of growth. The *order* ϱ of $f(z)$ is the the least $\alpha \geq 0$ for which $|f(z)| \ll \exp(|z|^{\alpha+\varepsilon})$ for every $\varepsilon > 0$. If no such α exists, the function is said to be of infinite order, and we write $\varrho = \infty$. As an example, the function $\sin(\pi z)$ has order equal to 1, as is immediately seen from its representation of the complex exponential function. From the last version of Jensen's formula it is clear that the denser the zeros of $f(z)$, the larger its order must be. We formulate a version of this principle that fits well with the Weierstrass product representation. As before, let $(z_k)_{k=1}^{\infty}$ be the sequence of zeros of $f(z)/z^m$ counted with multiplicity. The *exponent of convergence* μ of the zeros of $f(z)$ is the least $\alpha \geq 0$ for which

$$\sum_{k=1}^{\infty} \frac{1}{|z_k|^{\alpha+\varepsilon}} < \infty$$

for every $\varepsilon > 0$. If no such α exists, the exponent of convergence is said to be infinite, and we write $\mu = \infty$. As an example, the function $\sin(\pi z)$ has zeros at the integers, so its exponent of convergence is equal to 1.

Proposition 8.13. *The inequality $\mu \le \varrho$ holds for any entire function.*

Proof. We may without loss of generality assume that $f(0) = 1$. The inequality

$$n(R)\log(2) = n(R)\int_R^{2R}\frac{dr}{r} \le \int_0^{2R}\frac{n(r)}{r}dr = \frac{1}{2\pi}\int_0^{2\pi}\log(|f(2Re^{i\theta})|)\,d\theta$$

holds by Jensen's formula, since $n(r)$ is an increasing function. Then

$$\begin{aligned} n(R)\log(2) &\le \frac{1}{2\pi}\int_0^{2\pi}\log(|f(2Re^{i\theta})|)\,d\theta \\ &\le \frac{1}{2\pi}\int_0^{2\pi}\log(Ce^{(2R)^{\varrho+\varepsilon}})\,d\theta = \log(C) + (2R)^{\varrho+\varepsilon} \end{aligned}$$

for some constants $C > 0$ and $\varepsilon > 0$, and so $n(R) \ll R^{\varrho+\varepsilon}$.

Choosing $R = |z_k|$ yields $k \le n(|z_k|) \ll |z_k|^{\varrho+\varepsilon}$ and thus

$$\sum_{k=1}^{\infty}\frac{1}{|z_k|^{\alpha}} \ll \sum_{k=1}^{\infty}k^{-\alpha/(\varrho+\varepsilon)} < \infty$$

if $\alpha = \varrho + 2\varepsilon$. Since $\varepsilon > 0$ was arbitrary, the inequality $\mu \le \varrho$ follows. □

As before, let $(z_k)_{k=1}^{\infty}$ be the sequence of zeros of $f(z)/z^m$ counted with multiplicity. The *rank* λ of $f(z)$ is the least integer $\lambda \ge \mu$ if $\mu < \infty$, and otherwise $\lambda = \infty$. Clearly $\mu \le \lambda \le \mu + 1$ if $\mu < \infty$, so an entire function $f(z)$ of finite order has finite rank by Proposition 8.13, and $f(z)$ has a unique *canonical product*

$$P(z) = z^m \prod_{k=1}^{\infty}\left(1 - \frac{z}{z_k}\right)e^{p(\lambda, z/z_k)}$$

if it is of finite rank. For $f(z) \equiv 0$ we take $P(z) \equiv 0$. Moreover $f(z)$ has a *canonical factorization* $f(z) = e^{g(z)}P(z)$ in terms of its canonical product. It only remains to determine the entire function $g(z)$.

The next result is central in the theory of entire functions. Like the Weierstrass theory it provides a factorization in terms of zeros, but is far more precise. On the other hand this factorization theorem is only valid for entire functions of finite order.

Proposition 8.14 (Hadamard factorization theorem)**.** *An entire function of finite order ϱ has a canonical factorization*

$$f(z) = e^{g(z)}P(z)$$

with $g(z)$ a polynomial of degree $\deg(g) \le \varrho$.

Proof. Suppose without loss of generality that $f(0) = 1$. The function

$$f_R(z) = \frac{f(z)}{\prod_{|z_k|\le R}\left(1 - \frac{z}{z_k}\right)}$$

is entire for any $R > 0$. Since $|1 - z/z_k| \ge 1$ for $|z| = 2R$, the inequality $|f_R(z)| \le |f(z)| \ll \exp((2R)^{\varrho+\varepsilon})$ holds on $|z| = 2R$ for all sufficiently large R, no matter how small $\varepsilon > 0$ is chosen. But then the same inequality holds on $|z| \le 2R$ by the Maximum Principle.

Clearly $f_R(z)$ has no zeros on $|z| \le R$ and thus $\log(f_R(z))$ has a holomorphic branch $\phi_R(z)$ on this disk, say with $\phi_R(0) = 0$. Moreover $\mathrm{Re}(\phi_R(z)) = \log|f_R(z)| \ll R^{\varrho+\varepsilon}$ there. Then

$$|\phi_R^{\lambda+1}(z)| \ll \frac{R}{(R - |z|)^{\lambda+2}} R^{\varrho+\varepsilon}$$

for $|z| < R$ by the Borel-Carathéodory lemma. Thus $|\phi_R^{\lambda+1}(z)| \ll R^{\varrho+\varepsilon-\lambda-1}$ on the circle $|z| = R/2$.

Taking the logarithmic derivative in the canonical factorization of $f(z)$ and differentiating λ times yields

$$\begin{aligned} g^{(\lambda+1)}(z) &= \frac{d^\lambda}{dz^\lambda}\frac{f'(z)}{f(z)} + \lambda! \sum_{k=1}^{\infty} \frac{1}{(z_k - z)^{\lambda+1}} \\ &= \phi_R^{\lambda+1}(z) + \lambda! \sum_{|z_k|>R} \frac{1}{(z_k - z)^{\lambda+1}}. \end{aligned}$$

This implies that

$$|g^{(\lambda+1)}(z)| \ll R^{\varrho+\varepsilon-\lambda-1} + \sum_{|z_k|>R} |z_k|^{-\lambda-1}$$

on the circle $|z| = R/2$, and thus on $|z| \le R/2$ by the Maximum Principle. If $\varrho < \lambda + 1$ then we may choose $\varepsilon > 0$ so small that $\varrho + \varepsilon - \lambda - 1 < 0$, in which case the right-hand side of the last inequality tends to zero as $R \to +\infty$. Thus $g^{(\lambda+1)}(z) \equiv 0$ by the Liouville theorem, and so $g(z)$ must be a polynomial of degree at most $\lambda \le \varrho$.

If on the other hand $\varrho = \lambda + 1$, then choose $\varepsilon = 1/2$ and note that the above inequality yields $|g^{(\lambda+1)}(z) - g^{(\lambda+1)}(0)| \ll |z|^{1/2}$ on choosing $R = 2|z|$. Then $(g^{(\lambda+1)}(z) - g^{(\lambda+1)}(0))/z$ tends to zero as $|z| \to +\infty$, so $g^{(\lambda+1)}(z) \equiv g^{(\lambda+1)}(0)$. Thus $g(z)$ must be a polynomial of degree at most $\lambda + 1 = \varrho$. □

The function

$$\frac{1}{\Gamma(s)} = s e^{\gamma s} \prod_{n=1}^{\infty} \left(1 + \frac{s}{n}\right) e^{-s/n}$$

provides an example of the Hadamard factorization theorem. The Stirling formula in a weak form $n^{-c} \ll n!(n/e)^{-n} \ll n^c$ from Chapter 1 immediately implies that the entire function $1/\Gamma(s)$ is of order 1.

To apply the Hadamard factorization theorem to the entire function $\xi(s) = s(s-1)\pi^{-s/2}\Gamma(s/2)\zeta(s)/2$, it is necessary to determine the order of $\xi(s)$, or at least to bound its order from above. Because of the symmetry about the critical line, it is enough to estimate the latter function in the half plane $\sigma \geq 1/2$. The estimate $|s(s-1)\pi^{-s/2}\zeta(s)/2| \ll |s|^3$ holds on $\sigma \geq 1/2$ by Proposition 5.3. Moreover

$$|\Gamma(s/2)| \leq \Gamma(\sigma/2) \leq \Gamma([\sigma/2]+1) = [\sigma/2]! \leq (\sigma/2)^{\sigma/2} \ll e^{|s|^{1+\varepsilon}}$$

holds on the half plane $\sigma \geq 1/2$, for any $\varepsilon > 0$. Thus $\xi(s)$ has order 1 at most.

But $\xi(\sigma) = \sigma(\sigma-1)\pi^{-\sigma/2}\Gamma(\sigma/2)\zeta(\sigma)/2 \geq \pi^{-\sigma/2}(\sigma/4)^{\sigma/5} \geq e^{\sigma\log(\sigma)/6}$ for σ sufficiently large. Thus $\xi(s)$ and therefore also $\Xi(z) = \xi(1/2+iz)$ are entire functions of order 1.

The function $\Xi(\sqrt{z})$ is entire of order 1/2 because $\Xi(z)$ is an even function. Multiplying an entire function by a nonzero polynomial does not change its order, and so it follows from the Hadamard factorization theorem that an entire function of finite order with only finitely many zeros must have order equal to some integer. Hence $\Xi(\sqrt{z})$ and thus $\xi(s)$ have infinitely many zeros. The zeros of $\xi(s)$ coincide with the zeros of $\zeta(s)$ in the critical strip, thus $\zeta(s)$ has infinitely many nontrivial zeros.

The Hadamard factorization theorem implies that

$$\xi(s) = e^{A+Bz}\prod_{\rho}\left(1-\frac{s}{\rho}\right)e^{s/\rho},$$

where ρ ranges over the zeros of $\xi(s)$, repeated according to multiplicity. Here A and B are constants, and substituting $s=0$ yields $e^A = \xi(0) = 1/2$. The determination $B = \log(4\pi)/2 - 1 - \gamma/2$ is more complicated and is left as an exercise.

8.7. ⋆ The Phragmén-Lindelöf principle

In analytic number theory, vertical strips $\sigma_1 < \sigma < \sigma_2$ are encountered quite often, but these are not bounded domains. The function

$$f(s) = e^{e^{is}}$$

is bounded near the boundary of the strip $-\pi < \sigma < \pi$, though unbounded in the interior. So the boundary version of the Maximum Principle fails for unbounded domains. But in 1904 L. E. Phragmén discovered that it may be saved for unbounded domains by imposing a growth condition. He and E. L. Lindelöf obtained an improved version of his result in 1908. The growth condition required depends on the nature of the domain, so that one obtains various theorems of this kind, subsumed under the general title of the Phragmén-Lindelöf principle. We prove the one that is most often applied in analytic number theory.

Proposition 8.15 (Phragmén-Lindelöf theorem). *Suppose that f is holomorphic on a domain containing the strip $\sigma_1 \le \sigma \le \sigma_2$, except possibly for finitely many singularities, and assume that*

$$f(\sigma + it) \ll e^{e^{\delta|t|}}$$

uniformly in $\sigma_1 \le \sigma \le \sigma_2$, for any $\delta > 0$. If

$$f(\sigma_1 + it) \ll |t|^{k_1} \quad \text{and} \quad f(\sigma_2 + it) \ll |t|^{k_2}$$

as $|t| \to +\infty$, with constants $k_1, k_2 \ge 0$, then

$$f(\sigma + it) \ll |t|^{k(\sigma)}$$

as $|t| \to +\infty$, uniformly for $\sigma_1 \le \sigma \le \sigma_2$. Here $k(\sigma)$ is the linear function whose graph passes through (σ_1, k_1) and (σ_2, k_2).

Proof. It is enough to consider the region R determined by $\sigma_1 \le \sigma \le \sigma_2$ and $t \ge T > 0$ where T is chosen so large that f has no singularities in this region.

First consider the special case when $k_1 = k_2 = 0$. Then f is bounded on the boundary of R by assumption, say $|f| \le M$ on the boundary. Consider for arbitrary $\varepsilon > 0$ and $\eta > 0$ the function

$$g(s) = e^{-\eta e^{i\varepsilon s}} f(s),$$

with modulus

$$|g(s)| = \left| e^{-\eta e^{i\varepsilon s}} \right| |f(s)| = e^{-\eta e^{-\varepsilon t} \cos(\varepsilon\sigma)} |f(s)|.$$

Choose ε so small that $\cos(\varepsilon\sigma) > 0$ for $\sigma_1 \le \sigma \le \sigma_2$. Clearly $|g| \le M$ on the boundary of R. In the interior of R we have

$$|g(\sigma + it)| \ll e^{-\eta e^{-\varepsilon t} \cos(\varepsilon\sigma)} e^{e^{\delta t}}$$

as $t \to +\infty$, uniformly for $\sigma_1 \le \sigma \le \sigma_2$. Choosing $\delta < \varepsilon$, as we may by assumption, we see that

$$|g(\sigma + it)| \to +\infty$$

as $t \to +\infty$, uniformly for $\sigma_1 \le \sigma \le \sigma_2$. So we may choose some $U > T$ so that $|g| \le M$ for $\sigma_1 \le \sigma \le \sigma_2$ and $t \ge U$. Since $|g| \le M$ on the boundary of the rectangle given by $\sigma_1 \le \sigma \le \sigma_2$ and $T \le t \le U$, the Maximum Principle implies that $|g| \le M$ in this rectangle, and thus on the whole of R. Then

$$|f(\sigma + it)| \le |g(\sigma + it)| e^{\eta e^{-\varepsilon t} \cos(\varepsilon\sigma)} \le M e^{\eta e^{-\varepsilon t} \cos(\varepsilon\sigma)} \le M e^{\eta e^{-\varepsilon T}},$$

and so $|f| \le M$ on R by letting $\eta \to 0^+$.

For the general case, consider the auxiliary function

$$h(s) = e^{k(s)\mathrm{Log}(-is)},$$

which is holomorphic on a domain containing R. Let $k(\sigma) = a\sigma + b$. Then

$$\begin{aligned}|h(\sigma+it)| &= e^{\mathrm{Re}((k(\sigma)+iat)\mathrm{Log}(t-i\sigma))} = e^{(k(\sigma)+iat)(\frac{1}{2}\log(t^2+\sigma^2)+i\arctan(-\sigma/t))}\\ &= e^{k(\sigma)\frac{1}{2}\log(t^2+\sigma^2)+at\arctan(\sigma/t)} = e^{k(\sigma)\log(t)+O(1)} = t^{k(\sigma)}e^{O(1)},\end{aligned}$$

and so $f(s)/h(s)$ is bounded on the boundary of R and satisfies there a growth condition

$$\frac{f(\sigma+it)}{h(\sigma+it)} \ll e^{e^{\delta t}}.$$

Then f/h is bounded on R as above, and so

$$|f(\sigma+it)| \ll t^{k(\sigma)}$$

uniformly for $\sigma_1 \le \sigma \le \sigma_2$, as $t \to +\infty$. □

Supposing f satisfies the conditions of the above Phragmén-Lindelöf theorem on some vertical strip $\sigma_1 \le \sigma \le \sigma_2$, one may define the *Lindelöf characteristic* $\mu(\sigma)$ for each σ in the interval $[\sigma_1, \sigma_2]$ as the smallest nonnegative real number μ for which

$$f(\sigma+it) \ll |t|^{\mu+\delta}$$

as $|t| \to +\infty$, for any $\delta > 0$. Then the above theorem implies that $\mu(\sigma)$ is bounded on the interval, and is moreover a convex function there. In particular the Lindelöf characteristic is a continuous function.

The determination of the Lindelöf characteristic of the Riemann zeta function leads to an unsolved problem, known as the *Lindelöf Conjecture*. By the functional equation of $\zeta(s)$ the Lindelöf characteristic would be determined as

$$\mu(\sigma) = \begin{cases} 0 & \text{if } \sigma \ge 1/2,\\ 1/2-\sigma & \text{if } \sigma < 1/2,\end{cases}$$

if $\mu(1/2) = 0$, that is to say, if

$$\zeta(1/2+it) \ll |t|^{\varepsilon}$$

as $|t| \to +\infty$, for any $\varepsilon > 0$. The last statement is the Lindelöf Conjecture. The functional equation and the Phragmén-Lindelöf theorem may be used to bound from above the Lindelöf characteristic of the Riemann zeta function, and in particular to obtain the bound $\mu(1/2) \le 1/4$. The latter bound is known as the *convexity bound*, because it is obtained from information on the growth of $\zeta(s)$ in $\sigma > 1$ and in $\sigma < 0$ using only the fact that the Lindelöf characteristic is a convex function.

Proposition 8.16. *The Lindelöf characteristic $\mu(\sigma)$ of $\zeta(s)$ satisfies $\mu(\sigma) \leq 1/2 - \sigma/2$ for $0 \leq \sigma \leq 1$, while $\mu(\sigma) = 0$ for $\sigma > 1$ and $\mu(\sigma) = 1/2 - \sigma$ for $\sigma < 0$.*

Proof. It is clear from its definition that $\zeta(s)$ is bounded on every vertical line in $\sigma > 1$, so $\mu(\sigma) = 0$ there.

The functional equation and the complex version of Stirling's Formula from Section 10.1 implies that

$$\begin{aligned} |\zeta(\sigma + it)| &= \frac{(2\pi)^\sigma |\zeta(1-\sigma)|}{2\left|\cos\left(\frac{\pi\sigma + \pi it}{2}\right)\right| |\Gamma(\sigma + it)|} \\ &\ll \frac{1}{e^{\pi|t|/2}|t|^{\sigma - 1/2} e^{-\pi|t|/2}} \ll |t|^{1/2-\sigma} \end{aligned}$$

as $|t| \to +\infty$, for fixed $\sigma < 0$. This is moreover best possible, so $\mu(\sigma) = 1/2 - \sigma$ for $\sigma < 0$.

It remains to bound $\mu(\sigma)$ on $0 \leq \sigma \leq 1$, where we are less successful. Proposition 5.3 implies that $\zeta(s) \ll |t|$ in $1/2 \leq \sigma \leq 1$, and then the functional equation implies $\zeta(s) \ll |t|^{3/2}$ on $0 \leq \sigma \leq 1/2$ as above. Thus the conditions of Proposition 8.15 are satisfied. Since $\mu(1 + \epsilon) = 0$ and $\mu(-\epsilon) = 1/2 + \epsilon$ for any $\epsilon > 0$, it is clear that $\mu(\sigma) \leq 1/2 - \sigma/2$ for $0 \leq \sigma \leq 1$ by the convexity. □

8.8. Notes

Kolmogorov's example of an integrable function whose Fourier series diverges everywhere is in [**Kol26**]. For the proof of Proposition 8.1 I am indebted to a remark of Siegel in his *Lectures on Advanced Analytic Number Theory* [**Sie61**]. The conditions in Proposition 8.2 are by no means standard, and were chosen for the sake of a very simple proof based on Proposition 8.1. A more orthodox version, still with a simple proof, is due to L. J. Mordell [**Mor29**]. The proof of Proposition 8.3 is very inefficient, imposing unnecessary differentiability conditions in the statement of the result. But the proof is short and is based only on Proposition 8.2, which is a decisive consideration in view of the prerequisites assumed. A proof of the Poisson formula on $\mathbb{R}^n$ under weak conditions may be found in *Lectures on Fourier Integrals* by Salomon Bochner [**Boc59**]. On page 35 and 36 of volume I of *Harmonic Analysis on Symmetric Spaces and Applications* [**Ter85**] by A. Terras there are historical remarks and references for the Poisson summation formula.

Theta functions were introduced by C. G. J. Jacobi [**Jak29**], and some of the first of the many arithmetical applications are also due to him. Proposition 8.4 in [**Jak29**] has a similar proof based on Poisson summation. The Landsberg-Schaar formula was published by M. Schaar [**Sch49**] in 1849. He obtained it by a formal calculation based on a divergent Fourier series, amounting to an integration by parts in the "wrong" direction in the Euler-Maclaurin summation formula, thus introducing the Fourier series of the derivative of the sawtooth function $S(x)$. The following year he gave a proof based on Fourier analysis and Abel summation [**Sch50**]. The proof of the Landsberg-Schaar formula based on the Jacobi theta function is due to G. Landsberg [**Lan93**], and dates to 1893. This proof turned out to have considerable potential for generalization; Hecke used the same method to deduce the Law of Quadratic Reciprocity in algebraic number fields. There is an exposition in the last chapter of his *Lectures on the Theory of Algebraic Numbers* [**Hec81**]. Later developments in this direction are covered in *The Fourier-analytic Proof of Quadratic Reciprocity* [**Ber00**] by Michael C. Berg. That the determination of the sign of the classical Gauss sum implies the Law of Quadratic Reciprocity is due to Gauss [**Gau11**].

Euler's definition of the gamma function was by means of an infinite product

$$\Gamma(s) = \frac{1}{s}\prod_{n=1}^{\infty}\left(1+\frac{1}{n}\right)^{s}\left(1+\frac{s}{n}\right)^{-1},$$

while Gauss defined it by a limit

$$\Gamma(s) = \lim_{n\to\infty}\frac{n!n^s}{s(s+1)\cdots(s+n)}.$$

Weierstrass [**Wei56**] developed the theory of the gamma function by considering $1/\Gamma(s)$ as an entire function and expressing it by means of Euler's infinite product. Rewriting this infinite product yields the formula

$$\frac{1}{\Gamma(s)} = se^{\gamma s}\prod_{n=1}^{\infty}\left(1+\frac{s}{n}\right)e^{-s/n}$$

that is most often used as a definition today. This was obtained by F. W. Newman [**New48**] and O. Schlömilch [**Sch43**]. Proposition 8.6 occurs in *Funktionentheorie II* by K. Knopp [**Kno41**], attributed to H. Wielandt. Proposition 8.7 and Proposition 8.8 are due to Euler, and Proposition 8.9 is due to Legendre [**Leg11**].

The proof of the functional equation of $\zeta(s)$ given here is the second of the proofs that Riemann gave in his 1859 paper [**Rie59**]. There are several proofs in Chapter II of the treatise [**Tit86**], including Riemann's first proof and an interesting third proof that he never published. This was discovered in his Nachlass many years later.

In 1859 Riemann was elected a correspondent of the Royal Prussian Academy of the Sciences in Berlin. On the occasion of his election, he submitted a report on his recent researches to the journal of the Academy. This brief paper, whose title in English translation is "On the Number of Prime Numbers less than a given Magnitude", has been of supreme importance to the development of analytic number theory.

The specific objective of Riemann is an explicit formula for the counting function $\pi(x)$ of the primes. He begins by noting the Euler product formula as his point of departure, and obtains the analytic continuation and the functional equation of $\zeta(s)$ in two different ways. We used the second of Riemann's two methods to establish the functional equation in the proof of Proposition 8.10. To move the critical line to the real axis he introduces the function $\Xi(z)$ by a change of variable, and remarks that $\Xi(z)$ is an even function with a very rapidly convergent power series expansion.

The next part of the report is very sketchy, leaving room for speculation about what Riemann meant to state, and whether he had rigorous proofs of his claims. Denote by $N(T)$ the number of zeros, counted with multiplicity, of $\zeta(s)$ in the region of the critical strip determined by $0 < \sigma \leq T$. Riemann states a count of the zeros that amounts to

$$N(T) = \frac{T}{2\pi}\log\left(\frac{T}{2\pi}\right) - \frac{T}{2\pi} + O(\log(T)),$$

essentially remarking that this follows by the Argument Principle of complex analysis, but giving no details. A proof of this estimate was published in 1905 by von Mangoldt. Riemann next makes the important, but obscure, remark that *"Man findet nun in der That etwa so viel reelle Würzeln innerhalb diese Grenzen, und es ist sehr wahrscheinlich dass alle Würzeln reelle sind."* In the first part of this remark, he states that one finds in fact that $\Xi(z)$ has about the same number of real zeros for $0 < \mathrm{Re}(z) \leq T$ as the total number of zeros in this range. He does not explain in what sense this statement is to be understood, nor does he give any indication of an argument. The number of zeros $\rho = 1/2 + i\gamma$ of $\zeta(s)$ on the critical line, in the range $0 < t \leq T$, counted with multiplicity, is denoted by $N_0(T)$. Riemann's statement about the real zeros of $\Xi(z)$ pertains to the relationship between $N_0(T)$ and $N(T)$. The most straightforward guess is that he meant to claim that $N_0(T) \sim N(T)$ holds. This has never been proved, unless Riemann actually proved it. But the assumption that this is what he meant to claim is supported by a remark in a letter that he wrote to Weierstrass. Using a method originated by N. Levinson, J. B. Conrey succeeded in 1989 in showing that more than 2/5 of the nontrivial zeros lie on the critical line in the sense that $N_0(T)/N(T) \geq c > 2/5$ for all T sufficiently large. According to an observation of Heath-Brown and of Selberg, the Levinson method can be made to detect simple zeros, and the current state of knowledge about the number of zeros on the critical line is that in an asymptotic sense more than two fifths of all the nontrivial zeros lie there and are simple.

From the claim that the number of real zeros of $\Xi(z)$ is about the same as the total number of zeros, Riemann goes on to assert that it is very likely that all the zeros are real. This is equivalent to the statement that all nontrivial zeros of $\zeta(s)$ lie on the critical line $\sigma = 1/2$. That statement is known as the Riemann Hypothesis, (abbreviated RH) and is one of the most significant unsolved problems in mathematics. Riemann remarks that a rigorous proof of this would be desirable, but that after some fleeting failed attempts, he has put the search for a proof aside, as it is unnecessary for the next objective of his investigation.

Using his estimate for $N(T)$, Riemann sketches an argument to yield the canonical factorization of $\Xi(z)$ in terms of zeros. This is analogous to the canonical factorization of $\xi(s)$ that we established in Section 8.6, but Riemann prefers to work with $\Xi(z)$.

In the next part of the report, Riemann relates the distribution of the primes to the zeros of $\Xi(z)$ by means of Fourier theory. The Euler product formula and the product formula for $\Xi(z)$ are central here. Preliminary to applying Fourier analysis, he slightly redefines the counting function $\pi(x)$ of the primes. This function has jump discontinuities, and in each of these Riemann requires the value of $\pi(x)$ to be the average of the limit from the left and the limit from the right. This is customary in Fourier theory, due to the fact that the Fourier series of a nice function with a jump discontinuity converges to the average of the two one-sided limits at the discontinuity. Using the Euler product formula, Riemann establishes that

$$\frac{\log(\zeta(s))}{s} = \int_1^\infty \Pi(x) x^{-s-1}\, dx$$

with

$$\Pi(x) = \sum_{x^{1/k} \geq 2} \frac{1}{k} \pi(x^{1/k}) = \sum_{p^k \leq x} \frac{1}{k}.$$

The last function is a weighted counting function of the prime powers. Since prime powers higher than the first are scarce, $\Pi(x)$ is close to $\pi(x)$. Using the Fourier inversion theorem of analysis, Riemann finds the integral representation

$$\Pi(x) = \frac{1}{2\pi i} \int_{c-i\infty}^{c+i\infty} \log(\zeta(s)) \frac{x^s}{s}\, ds,$$

valid for $c > 1$. This representation is an instance of the Perron formula that we established in Chapter 5.

Due to the fact that $\zeta(s)$ can be expressed in terms of the product representation of $\Xi(z)$, the integral for $\Pi(x)$ may be evaluated in terms of the zeros of $\Xi(z)$ by substituting for $\zeta(s)$ in the integral. The resulting calculation is very delicate, and Riemann's treatment in this part of his report is again sketchy, especially considering the difficulty of the analysis. Landau succeeded in 1908 in giving a complete argument as sketched by Riemann, but von Mangoldt had attained the same objective by a different route in 1895. The resulting explicit formula is

$$\Pi(x) = \mathrm{Li}(x) - \sum_{\rho} \mathrm{Li}(x^{\rho}) - \log(2) + \int_x^\infty \frac{1}{u^2-1} \frac{du}{u \log(u)},$$

where the infinite series over the zeros ρ of $\xi(s)$ is only conditionally convergent and must be summed in a particular order. The logarithmic integral $\mathrm{Li}(x)$ is defined in a slightly different way from $\mathrm{li}(x)$; the interval of integration extends from 0 to x and the integral is a Cauchy principal value integral, where the integrand $1/\log(u)$ has a singularity at $u = 1$. From the above explicit formula, an explicit formula for $\pi(x)$ may be obtained by a routine application of Möbius inversion. Thus Riemann finally attains his objective of finding an explicit formula for $\pi(x)$.

Beside the notation $Z(s)$ for the completed Riemann zeta function, the notations $\Lambda(s)$, $\zeta^*(s)$, $\zeta*(s)$ and $\hat{\zeta}(s)$ are also in use. Moreover, the notation of Riemann is sometimes modified to make $\xi(s)$ the completed zeta function.

The reason that $Z(s)$ is termed *completed* is that from the viewpoint of algebraic number theory $\pi^{-s/2}\Gamma(s/2)$ is seen as a missing factor in the Euler product of $\zeta(s)$. To each prime p there corresponds a factor $(1-p^{-s})^{-1}$ in the Euler product and a non-Archimedean valuation $|a|_p = p^{-\alpha_p}$ on $\mathbb{Q}^\times$ where α_p is the exponent of p in the prime factorization of a. There is also an Archimedean valuation $|a|_\infty = |a|$ given by the ordinary absolute value. Then the simplest case

$$\prod_{p \leq \infty} |a|_p = 1$$

of the Artin-Whaples product formula holds on $\mathbb{Q}^\times$. In this product formula all valuations $|\cdots|_p$ with $p \le \infty$ stand on an equal footing, which suggests that there is a factor corresponding to $|\cdots|_\infty$ missing in the Euler product for the Riemann zeta function. This observation does not indicate what the extra factor corresponding to $|\cdots|_\infty$ should be. But in algebraic number theory there is a valuation-theoretic approach to the functional equations of L-functions, due to J. T. Tate, in which the factors $(1-p^{-s})^{-1}$ and $\pi^{-s/2}\Gamma(s/2)$ are obtained by a general and uniform procedure.

Special cases of Proposition 8.11 were established by H. Kinkelin [**Kin62**] and A. Hurwitz [**Hur82**], and the general case by R. O. S. Lipschitz [**Lip89**].

Proposition 8.12 is due to Jensen [**Jen99**]. Proposition 8.13 and Proposition 8.14 are due to Hadamard [**Had93**]. The proof of 8.14 given here is due to Landau [**Lan27**]. The first published proof of the validity of the factorization of $\Xi(z)$ is due to Hadamard [**Had93**].

An early version of the Phragmén-Lindelöf principle first appeared in [**Phr04**], and the stronger versions are in [**PL08**]. The Lindelöf Conjecture is in [**Lin08**].

Exercises

(1) Show that if

$$\int_0^1 |f(x)|^2\,dx < \infty,$$

then $\hat{f}_n \to 0$ as $|n| \to +\infty$.

(2) Write the periodization of $f(x) = e^{-|x|}$ in terms of the sawtooth function.

(3) Find the Fourier series of

$$f(x) = \int_0^x S(u)\,du$$

and show that it converges to $f(x)$ everywhere. Establish the identity

$$\sum_{n=1}^{\infty} \frac{1}{n^2} = \frac{\pi^2}{6}$$

of Euler.

(4) Let $A = [a_{ij}]$ be an n-by-n real matrix. The simultaneous inequalities

$$\left|\sum_{j=1}^{n} a_{ij}x_j\right| \le 1 \quad , \quad i = 1, 2, \ldots, n$$

always have the trivial integer solution $x_1 = x_2 = \cdots = 0$. Minkowski showed that if $0 < \det(A) < 1$, then there is at least one nontrivial

integer solution. This exercise outlines a proof due to Mordell, who applied the Poisson summation formula.
a) For

$$f(u_1, u_2, \ldots, u_n) = \prod_{j=1}^{n} \left(\frac{\sin(\pi u_j)}{\pi u_j} \right)^2,$$

calculate the Fourier transform $\hat{f}(y)$.
b) Express $\hat{f}(Ax)$ in terms of f.
c) To prove the claim, apply the Poisson formula to show that

$$\sum_{x \in \mathbb{Z}^n} \hat{f}(Ax) > 1$$

if $0 < \det(A) < 1$. A slight improvement may be obtained by taking more care.

(5) Some infinite series of the form

$$\sum_{n=-\infty}^{\infty} f(n)$$

may be summed by residue calculus.
a) Find the poles and residues of the function $\pi \cot(\pi z)$.
b) Evaluate the series

$$\sum_{n=-\infty}^{\infty} \frac{1}{a^2 + n^2}$$

for $a > 0$ by integrating

$$\frac{\pi \cot(\pi z)}{a^2 + z^2}$$

around the square with corners $\pm N \pm iN$ and letting $N \to +\infty$.

(6) An *elliptic function* is a meromorphic function $f(z)$ on the complex plane that is *doubly periodic* in the sense that $f(z)$ has two periods ω_1 and ω_2 that are linearly independent over $\mathbb{R}$. By scaling, rotation and interchange of the periods we may without loss of generality assume that $f(z)$ has periods 1 and $\tau \in \mathbb{H}$ where $\mathbb{H} = \{\tau \in \mathbb{C} | \mathrm{Im}(\tau) > 0\}$ is the upper half plane.
a) Show that an elliptic function $f(z)$ without poles must be constant.
b) Suppose that C is any parallelogram with corners $z_0, z_0 + 1, z_0 + \tau, z_0 + 1 + \tau$ on which $f(z)$ has no poles. Show that the sum of the residues of the poles of $f(z)$ inside C is zero. Also show that counted with multiplicity the number of zeros inside C is equal to the number of poles there.
c) Let C be as in part b) and $a_1, \ldots, a_k$ the zeros and $b_1, \ldots, b_k$ the poles of $f(z)$ inside C, repeated by order. Show that $a_1 + \cdots + a_k - b_1 - \cdots - b_k \in \mathbb{Z} + \mathbb{Z}\tau$. Note that by double periodicity we may arrange that

$a_1 + \cdots + a_k = b_1 + \cdots + b_k$, exchanging one of the zeros for some other zero. Assume this done in the rest of this exercise.

d) Show that

$$f(z) = c\frac{\vartheta(1/2+\tau/2+z-a_1,\tau)\cdots\vartheta(1/2+\tau/2+z-a_k,\tau)}{\vartheta(1/2+\tau/2+z-b_1,\tau)\cdots\vartheta(1/2+\tau/2+z-b_k,\tau)}$$

for some complex constant c.

e) Give an existence and uniqueness theorem for elliptic functions in terms of their zeros and poles (Jacobi).

(7) † Show that the sum of the series

$$\theta(u) = \vartheta(0, iu) = \sum_{n=-\infty}^{\infty} e^{-\pi n^2 u}$$

with $u > 0$ satisfies the functional equation

$$\theta(u) = \frac{1}{\sqrt{u}}\,\theta\left(\frac{1}{u}\right)$$

by applying the Residue Theorem to the integral of

$$f(z) = \frac{e^{-\pi z^2 u}}{e^{2\pi i z} - 1}$$

around the contour from $N+1/2+i$ to $-(N+1/2)+i$ to $-(N+1/2)-i$ to $N+1/2-i$ and back to $N+1/2+i$, and let $N \to +\infty$ (Kronecker).

(8) a) Introduce theta functions $\vartheta_{jk}(z,\tau)$ for $j,k = 0,1$ by

$$\vartheta_{jk}(z,\tau) = \sum_{n=-\infty}^{\infty} e^{\pi i (n+j/2)^2 \tau} e((n+j/2)(z+k/2))$$

and express these theta functions in terms of the Jacobi theta function defined earlier. The corresponding theta-constants are abbreviated $\vartheta_{jk} = \vartheta_{jk}(0,\tau)$.

b) Introduce the *nome* $q = e^{\pi i \tau}$ and express $\vartheta_{00}^4 - \vartheta_{01}^4$ and ϑ_{10}^4 as power series in q. Note that the coefficients count certain representations by sums of four squares or four triangular numbers respectively.

c) Define a meromorphic function of z by

$$f(z) = \frac{\vartheta_{00}^2\vartheta_{00}^2(z,\tau) - \vartheta_{01}^2\vartheta_{01}^2(z,\tau)}{\vartheta_{10}^2\vartheta_{10}^2(z,\tau)},$$

where $\vartheta_{10} \neq 0$ because $\vartheta(\tau/2,\tau) \neq 0$. Clearly f has period 1, but show that τ is also a period of $f(z)$.

d) Show that the numerator of the expression defining $f(z)$ has a zero at $z = 1/2$. Then integrate $f(z)$ around the parallelogram with corners $-\tau/2, -\tau/2+1, \tau/2+1, \tau/2$ to show that $f(z)$ has no singularities.

e) Conclude that $f(z)$ is constant, and determine this constant, for example by substituting $z = 1/2 + \tau/2$.
f) From part e) we obtain the identity $\vartheta_{00}^4 = \vartheta_{01}^4 + \vartheta_{10}^4$ by substituting $z = 0$. This is called the *Jacobi quartic identity.* Use it to show that the number of representations of an odd integer N as a sum of four squares is eight times the number of representations of $(N-1)/2$ as a sum of four triangular numbers $n(n+1)/2$ with $n \in \mathbb{Z}$ (Jacobi).
g) Deduce Lagrange's theorem that each nonnegative integer is a sum of four squares from Gauss' theorem that each nonnegative integer is a sum of three triangular numbers.

(9) Establish the Euler factorization

$$\sin(z) = z \prod_{n=1}^{\infty} \left(1 - \frac{z^2}{\pi^2 n^2}\right)$$

of the sine using the theory of the gamma function.

(10) Show that $\Gamma(\sigma + it) = O(|t|^{-k})$ uniformly as $t \to \pm\infty$ for σ in any closed bounded interval, for arbitrary k. In fact, by the complex version of Stirling's formula, much better bounds hold for $\Gamma(s)$ as $t \to \pm\infty$.

(11) Show that

$$|\Gamma(1/2 + it)| = \sqrt{\frac{2\pi}{e^{\pi t} + e^{-\pi t}}}$$

for t real.

(12) Show that

$$e^{-u} = \frac{1}{2\pi i} \int_{c-i\infty}^{c+i\infty} \Gamma(s) u^{-s}\, ds$$

for $u > 0$ and $c > 0$. This is called the *Cahen-Stieltjes integral.*

If you feel an urge to differentiate under the integral sign, keep in mind that the familiar theorem on differentiation under the integral sign is not valid for improper integrals. For a theorem that will serve, see Section 4.12 of volume II of *Introduction to Calculus and Analysis* by Richard Courant and Fritz John. Alternatively, you can pass the difference quotient under the integral sign, estimate, and take the limit at the end.

The Cahen-Stieltjes integral formula may also be established by means of residue calculus.

Under suitable conditions

$$F(s) = (\mathcal{M}f)(s) = \int_0^{\infty} f(u) u^{s-1}\, du$$

implies and is implied by

$$f(u) = (\mathcal{M}^{-1}F)(s) = \frac{1}{2\pi i}\int_{c-i\infty}^{c+i\infty} F(s)u^{-s}\,ds$$

where $\mathcal{M}^{-1}$ is the *inverse Mellin transform.* This result is the *Mellin inversion theorem,* which may be obtained from the Fourier inversion theorem by a change of variable. See *Introduction to the Theory of Fourier Integrals* by E. C. Titchmarsh. The Mellin inversion theorem implies the Cahen-Stieltjes integral formula by Proposition 8.7.

(13) Calculate $\zeta'(0)$.

(14) Show that there is some constant $c > 0$ so that $\xi(s) = O(e^{c|s|\log(|s|)})$ as $|s| \to +\infty$.

(15) Deduce the reflection formula for the gamma function from the unsymmetrical form of the functional equation of the Riemann zeta function.

(16) Check that if $0 \le \alpha < 1$ and $\mathrm{Im}(\tau) > 0$ the function

$$f(x) = \begin{cases} (x-\alpha)^{s-1}e^{2\pi i\tau(x-\alpha)} & \text{for } x > \alpha, \\ 0 & \text{for } x \le \alpha, \end{cases}$$

satisfies the conditions of Proposition 8.2 if $\sigma > 2$. Then show that

$$\sum_{n=1}^{\infty} \frac{e^{2\pi i\tau(n-\alpha)}}{(n-\alpha)^{1-s}} = \frac{\Gamma(s)}{(-2\pi i)^s}\sum_{m=-\infty}^{\infty}\frac{e^{2\pi i\alpha m}}{(\tau+m)^s}$$

if $\sigma > 1$. This is the *Lipschitz summation formula,* which may be used to prove the functional equation of the Riemann zeta function.

(17) Show that $W(\chi)W(\overline{\chi}) = 1$ and deduce what consequence this has for the Gauss sum $\tau(\chi)$.

(18) a) The series expansion

$$\Phi(\lambda,\alpha,s) = \sum_{n=0}^{\infty} e^{2\pi i\lambda n}(n+\alpha)^{-s},$$

defines the *Lerch zeta function.* Find an integral representation for the Dirichlet series $\Phi(\lambda,1,s)$.

b) Show that $\Phi(\lambda,1,s)$ is an entire function for each λ not an integer.

c) Express $L(s,\chi)$ in terms of $\Phi(\lambda,1,s)$ for rational values of λ and conclude that $L(s,\chi)$ is an entire function for any nonprincipal character χ.

(19) Show that $|L(0,\chi)| < \sqrt{q}\log(q)$ if χ is a primitive Dirichlet character modulo $q \ge 3$.

(20) Show that $L(1/2,\chi) \ll q^{1/4}\log^{1/2}(q)$ for any nonprincipal Dirichlet character χ modulo q.

(21) † Show that

$$L(1/2 + it, \chi) \ll (1 + |t|)^{1/2} q^{1/4} \log^{1/2}(q)$$

for nonprincipal characters χ modulo q. This bound can be improved both in t-aspect and q-aspect. Improvement in t-aspect means a smaller exponent than 1/2 in the first factor, and improvement in q-aspect means a smaller exponent than 1/4 in the second factor. To show that $L(s, \chi) \ll_\varepsilon (1 + |t|)^\varepsilon q^\varepsilon$ for any $\varepsilon > 0$ is an unsolved problem, known as the Lindelöf Conjecture for Dirichlet L-functions.

(22) Use the Hadamard factorization theorem to establish the Euler factorization

$$\sin(z) = z \prod_{n=1}^{\infty} \left(1 - \frac{z^2}{\pi^2 n^2}\right)$$

of the sine.

(23) It may seem that the definition of the order of an entire function involves an arbitrary choice in that growth is measured out from a particular point, namely the origin. Show that if $f(z)$ is an entire function and z_0 an arbitrary point, then $f(z)$ and $f(z + z_0)$ have the same order.

(24) a) Show that any sequence $(z_n)_{n=1}^{\infty}$ of points in the complex plane, with $|z_n| \to +\infty$, is the set of zeros of some entire function, with repetitions in the sequence corresponding to multiple zeros (Weierstrass.)
b) Prove that any meromorphic function is the quotient of two entire functions (Weierstrass.)

(25) Show that any entire function of order zero that is not a polynomial takes every complex number as a value infinitely often.

(26) a) An entire function $f(z)$ is of *exponential type* if there is some constant $c \geq 0$ so that $|f(z)| \ll \exp(c|z|)$ as $|z| \to +\infty$. Show that an entire function of order 1 with only finitely many zeros is of exponential type.
b) Show that though $\xi(s)$ is of order 1, it is not of exponential type and hence must have infinitely many zeros.

(27) Determine the order of the Jacobi theta function $\vartheta(z, \tau)$ as an entire function of z.

(28) a) Show that

$$\frac{\xi'(s)}{\xi(s)} = B + \sum_{\rho} \frac{s}{\rho(s - \rho)}$$

where B is the constant in the canonical factorization of $\xi(s)$.
b) Show that $B = -\xi'(1)/\xi(1)$.
c) Show that $B = \log(4\pi)/2 - 1 - \gamma/2$.

d) Show that

$$\frac{\zeta'(s)}{\zeta(s)} = \frac{\log(2\pi)}{2} + B + \sum_{\rho} \frac{s}{\rho(s-\rho)} - \frac{1}{s} - \frac{1}{s-1} - \frac{1}{2}\frac{\Gamma'(s/2)}{\Gamma(s/2)}$$

where B is the constant in the canonical factorization of $\xi(s)$.

(29) Show that if f is holomorphic and nonconstant on a domain, then $|f|$ cannot attain a nonzero local minimum at any point of the domain.

Chapter 9

Euler Products and Number Fields

9.1. The Dedekind zeta function

We shall obtain information about the arithmetic of the ring of integers $\mathcal{O}_K$ of a number field K by means of the Euler product formula. The latter is based on unique factorization, which does not hold generally in $\mathcal{O}_K$. But unique factorization of nonzero ideals into prime ideals does hold. For fixed $s \in \mathbb{C}$ the map $\mathfrak{a} \mapsto N(\mathfrak{a})^{-s}$ from nonzero ideals $\mathfrak{a}$ of $\mathcal{O}_K$ into $\mathbb{C}$ is totally multiplicative, so the product formula

$$\prod_{\mathfrak{p}}(1 - N(\mathfrak{p})^{-s})^{-1} = \prod_{\mathfrak{p}}\sum_{j=0}^{\infty} N(\mathfrak{p}^j)^{-s} = \sum_{\mathfrak{a}} N(\mathfrak{a})^{-s}$$

holds as a formal identity. The product is taken over nonzero prime ideals $\mathfrak{p}$ and the sum is taken over nonzero ideals $\mathfrak{a}$. Writing both expressions explicitly in terms of formal Dirichlet series yields

$$\prod_{\mathfrak{p}}(1 - N(\mathfrak{p})^{-s})^{-1} = \prod_{p}\prod_{\mathfrak{p}|p}(1 - p^{-\deg(\mathfrak{p})s})^{-1}$$

and

$$\sum_{\mathfrak{a}} N(\mathfrak{a})^{-s} = \sum_{m=1}^{\infty}\left(\sum_{N(\mathfrak{a})=m} 1\right)m^{-s} = \sum_{m=1}^{\infty} a_K(m)m^{-s}$$

where $a_K(m)$ is the *ideal counting function* that counts the number of ideals with norm m. Note that the innermost product is finite because there are at most n_K prime ideals in $\mathcal{O}_K$ with $\mathfrak{p}|p$. Comparing coefficients of the two formal Dirichlet series shows how the splitting of primes determines the ideal counting function. In particular there are only finitely many ideals with a given norm.

The *Dedekind zeta function* $\zeta_K(s)$ of a number field K will be defined by

$$\zeta_K(s) \stackrel{\text{def}}{=} \prod_{\mathfrak{p}} (1 - N(\mathfrak{p})^{-s})^{-1},$$

where the product is taken over the nonzero prime ideals of $\mathcal{O}_K$. The infinite product converges absolutely in $\sigma > 1$, for

$$\sum_{\mathfrak{p}} N(\mathfrak{p})^{-\sigma} \leq n_K \sum_{p} p^{-\sigma} \leq n_K \zeta(\sigma)$$

since a rational prime splits into at most n_K primes in K. Moreover the product defining $\zeta_K(s)$ converges uniformly on compact sets in the same half plane, so the Dedekind zeta function is holomorphic in $\sigma > 1$. Proposition 3.2 implies that the Dirichlet series expansion

$$\zeta_K(s) = \sum_{\mathfrak{a}} N(\mathfrak{a})^{-s} = \sum_{m=1}^{\infty} a_K(m) m^{-s}$$

converges absolutely for $\sigma > 1$.

Note that at this point it is definitely more convenient to define the zeta function by the product rather than the sum. We are spared having to estimate the ideal counting function before establishing convergence.

As an example, we consider the Dedekind zeta function of $\mathbb{Q}(\sqrt{-1})$. Using the information from the Summary about the splitting of rational primes in the Gaussian integers, the Euler product is seen to be

$$\begin{aligned}\zeta_{\mathbb{Q}(\sqrt{-1})}(s) &= (1 - 2^{-s})^{-1} \prod_{p \equiv 1 \bmod 4} (1 - p^{-s})^{-2} \prod_{p \equiv 3 \bmod 4} (1 - p^{-2s})^{-1} \\ &= \zeta(s) \prod_{p \equiv 1 \bmod 4} (1 - p^{-s})^{-1} \prod_{p \equiv 3 \bmod 4} (1 + p^{-s})^{-1} = \zeta(s) L(s, \chi),\end{aligned}$$

where χ is the unique nonprincipal character modulo 4. Noting that there are four units among the Gaussian integers, we obtain the Dirichlet series expansion

$$\zeta_{\mathbb{Q}(\sqrt{-1})}(s) = \frac{1}{4} \sum_{0 \neq (u,v) \in \mathbb{Z}^2} (u^2 + v^2)^{-s} = \frac{1}{4} \sum_{m=1}^{\infty} r(m) m^{-s},$$

where $r(m)$ is the number of representations of m as a sum of two squares. Comparing the Euler product and the Dirichlet series for $\zeta_{\mathbb{Q}(\sqrt{-1})}(s)$, we deduce Jacobi's formula $r = 4 * \chi$.

The Dirichlet series in this example was obtained directly using the special fact that the ring of Gaussian integers is a principal ideal domain.

Usually the ring of integers of a number field is not a PID, but the Dirichlet series could have been obtained by multiplying out the Euler product. Whenever we know how the rational primes split in a number field, we can write down an Euler product over rational primes for the Dedekind zeta function of the number field.

We now consider the Dirichlet series of the Dedekind zeta function for a number field K whose ring of integers $\mathcal{O}_K$ is not necessarily a principal ideal domain. Recall the abelian group J_K of nonzero fractional ideals of $\mathcal{O}_K$ and its subgroup P_K of principal fractional ideals. The quotient group $Cl_K = J_K/P_K$ is called the *class group* of K. This is one of the most important invariants of a number field. Clearly $\mathcal{O}_K$ is a principal ideal domain if and only if the class group is trivial. A *fractional ideal class* is a coset of P_K in J_K and an *ideal class* consists of the ideals in a fractional ideal class. We shall usually consider elements of the class group to be represented by ideals rather than fractional ideals. It is true, though by no means obvious, that the class group of the ring of algebraic integers of a number field has finite order. For more general integral domains this statement does not hold in general.

By absolute convergence the Dirichlet series of the Dedekind zeta function may be expressed as a sum

$$\zeta_K(s) = \sum_{C \in Cl_K} \zeta_K(s, C), \qquad \zeta_K(s, C) \stackrel{\text{def}}{=} \sum_{\mathfrak{a} \in C} N(\mathfrak{a})^{-s}$$

over ideal classes when $\sigma > 1$. The zeta function $\zeta_K(s, C)$ of an ideal class C is sometimes called a *partial zeta function.* We complete the Dedekind zeta function and the partial zeta function by multiplying with the *gamma factor*

$$\gamma_K(s) \stackrel{\text{def}}{=} \left(\frac{|d_K|}{2^{2r_2}\pi^n}\right)^{s/2} \Gamma(s/2)^{r_1}\Gamma(s)^{r_2},$$

where d_K is the discriminant, r_1 the number of real embeddings, $2r_2$ the number of complex embeddings, and $n = n_K$ the degree. Then

$$Z_K(s) = \gamma_K(s)\zeta_K(s) \quad \text{and} \quad Z_K(s, C) = \gamma_K(s)\zeta_K(s, C)$$

are the completions.

We are going to represent the completed partial zeta function $Z_K(s, C)$ by a Mellin transform. From this we obtain in the next section a fundamental identity that has many important consequences.

If $\mathfrak{b}$ is a fixed fractional ideal in C^{-1}, then $\mathfrak{ab} = (\beta)$ is a principal fractional ideal for every fractional ideal $\mathfrak{a}$ representing C; $\mathfrak{a}$ is an ideal if and only if $\beta \in \mathfrak{b}$. This observation enables us to replace summation over the ideals in C by summation over nonzero principal ideals contained in $\mathfrak{b}$

using $N(\mathfrak{a})^{-s}N(\mathfrak{b})^{-s} = N(\mathfrak{ab})^{-s} = N((\beta))^{-s}$, so

$$\zeta_K(s, C) = N(\mathfrak{b})^s \sum_{(\beta)\subseteq\mathfrak{b}} N((\beta))^{-s}$$

or

$$\zeta_K(s, C) = N(\mathfrak{b})^s {\sum_{\beta\in\mathfrak{b}}}' |N_K(\beta)|^{-s},$$

where the prime indicates summation over nonzero elements of $\mathfrak{b}$ up to associates.

Put $r = r_1 + r_2 - 1$,

$$c(\mathfrak{b}) = (|d_K|N(\mathfrak{b})^2)^{-1/n},$$

and $\beta^{(j)} = \sigma_j(\beta)$ for $1 \le j \le r+1$. Here $\sigma_1, \dots, \sigma_{r_1}$ are real embeddings and $\sigma_{r_1+1}, \dots, \sigma_{r_1+r_2}$ are pairwise nonconjugate complex embeddings, taken in some fixed order. Further let

$$\begin{aligned} m(\beta, v_1, \dots, v_r) &= \left|\beta^{(1)}\right|^2 e^{v_1} + \cdots + \left|\beta^{(r_1)}\right|^2 e^{v_{r_1}} \\ &\quad + 2\left|\beta^{(r_1+1)}\right|^2 e^{v_{r_1+1}} + \cdots + 2\left|\beta^{(r+1)}\right|^2 e^{v_{r+1}} \end{aligned}$$

with v_{r+1} determined by

$$v_1 + \cdots + v_{r_1} + 2v_{r_1} + \cdots + 2v_{r+1} = 0,$$

and with $v_1, \dots, v_r$ real variables.

Proposition 9.1. *The Mellin transform representation*

$$Z_K(s, C) = \frac{n}{2}\int_0^\infty x^{ns/2-1}\left(\int_{\mathbb{R}^r} {\sum_{\beta\in\mathfrak{b}}}' e^{-\pi c(\mathfrak{b})x m(\beta,\mathbf{v})}d\mathbf{v}\right)dx$$

is valid for $\sigma > 1$. If $r_2 = 0$ the right-hand side increases by a factor of 2.

Proof. We have

$$(\pi c(\mathfrak{b}))^{-s/2}\Gamma(s/2)\left|\beta^{(j)}\right|^{-s} = \int_0^\infty x_j^{s/2} e^{-\pi c(\mathfrak{b})x_j|\beta^{(j)}|^2}\frac{dx_j}{x_j}$$

for $1 \le j \le r_1$ and

$$(2\pi c(\mathfrak{b}))^{-s}\Gamma(s)\left|\beta^{(j)}\right|^{-2s} = \int_0^\infty x_j^s e^{-2\pi c(\mathfrak{b})x_j|\beta^{(j)}|^2}\frac{dx_j}{x_j}$$

for $r_1 + 1 \le j \le r+1$ by a change of variable in the integral for the gamma function. Multiplying these formulas together yields

$$\begin{aligned} &\pi^{-ns/2}2^{-r_2 s}|d_K|^{s/2}N(\mathfrak{b})^s\Gamma(s/2)^{r_1}\Gamma(s)^{r_2}|N_K(\beta)|^{-s} \\ &= \int_0^\infty \cdots \int_0^\infty (x_1 \cdots x_{r_1} x_{r_1+1}^2 \cdots x_{r+1}^2)^{s/2} e^{-\pi c(\mathfrak{b})A}\frac{dx_1}{x_1}\cdots\frac{dx_{r+1}}{x_{r+1}} \end{aligned}$$

where

$$A = x_1 \left|\beta^{(1)}\right|^2 + \cdots + x_{r_1} \left|\beta^{(r_1)}\right|^2 + 2x_{r_1+1} \left|\beta^{(r_1+1)}\right|^2 + \cdots + 2x_{r+1} \left|\beta^{(r+1)}\right|^2,$$

because $n = r_1 + 2r_2$ and the norm of β equals the product of the embeddings of β. To simplify the integrand we introduce new variables x and $v_1, \ldots, v_r$ by

$$x = (x_1 \cdots x_{r_1} x_{r_1+1}^2 \cdots x_{r+1}^2)^{1/n} \quad \text{and} \quad v_j = \log(x_j) - \log(x)$$

for $1 \le j \le r$ and determine v_{r+1} as above. Then the old variables may be expressed in terms of the new by

$$x_j = xe^{v_j}$$

for $1 \le j \le r+1$. The Jacobian of the change of variables is

$$\frac{\partial(x_1, \ldots, x_{r+1})}{\partial(v_1, \ldots, v_r, x)} = \begin{vmatrix} xe^{v_1} & 0 & 0 & \cdots & 0 & e^{v_1} \\ 0 & xe^{v_2} & 0 & \cdots & 0 & e^{v_2} \\ \vdots & \vdots & \vdots & \vdots & \vdots & \vdots \\ 0 & 0 & 0 & \cdots & xe^{v_r} & e^{v_r} \\ -\frac{x}{2}e^{v_{r+1}} & -\frac{x}{2}e^{v_{r+1}} & -\frac{x}{2}e^{v_{r+1}} & \cdots & -xe^{v_{r+1}} & e^{v_{r+1}} \end{vmatrix}.$$

This determinant may be simplified by multiplying each of the first r columns by $1/x$ and subtracting from the last column. Thus

$$\frac{\partial(x_1, \ldots, x_{r+1})}{\partial(v_1, \ldots, v_r, x)} = \begin{vmatrix} xe^{v_1} & 0 & 0 & \cdots & 0 & 0 \\ 0 & xe^{v_2} & 0 & \cdots & 0 & 0 \\ \vdots & \vdots & \vdots & \vdots & \vdots & \vdots \\ 0 & 0 & 0 & \cdots & xe^{v_r} & 0 \\ -\frac{x}{2}e^{v_{r+1}} & -\frac{x}{2}e^{v_{r+1}} & -\frac{x}{2}e^{v_{r+1}} & \cdots & -xe^{v_{r+1}} & B \end{vmatrix}$$

where

$$B = e^{v_{r+1}} + \frac{1}{2}e^{v_{r+1}} + \cdots + \frac{1}{2}e^{v_{r+1}} + e^{v_{r+1}} + \cdots + e^{v_{r+1}} = \frac{n}{2}e^{v_{r+1}}$$

because there are r_1 terms with coefficient $1/2$ and $r_2 - 1$ terms with coefficient 1 at the end. If $r_2 = 0$ this calculation must be modified, and the correct value of B is then larger by a factor of 2. Now

$$\frac{\partial(x_1, \ldots, x_{r+1})}{\partial(v_1, \ldots, v_r, x)} = xe^{v_1} \cdots xe^{v_r} B = x_1 \cdots x_r \frac{n}{2} x_{r+1} x^{-1}$$

so

$$\gamma_K(s) \frac{N(\mathfrak{b})^s}{|N_K(\beta)|^s} = \frac{n}{2} \int_0^\infty x^{ns/2-1} \left(\int_{-\infty}^\infty \cdots \int_{-\infty}^\infty e^{-\pi c(\mathfrak{b}) x m(\beta, \mathbf{v})} d\mathbf{v} \right) dx$$

by expressing A in terms of the new variables. Summing and interchanging sum and integral yields the Mellin transform representation for $Z_K(s,C)$. The bound

$$\left|\frac{n}{2}\int_a^b x^{ns/2-1}\left(\int_{\|\mathbf{v}\|\le\rho} e^{-\pi c(\mathfrak{b})xm(\beta,\mathbf{v})}d\mathbf{v}\right)dx\right|$$
$$\le \frac{n}{2}\int_0^\infty x^{n\sigma/2-1}\left(\int_{\mathbf{R}^r} e^{-\pi c(\mathfrak{b})xm(\beta,\mathbf{v})}d\mathbf{v}\right)dx = \gamma_K(\sigma)\frac{N(\mathfrak{b})^\sigma}{|N_K(\beta)|^\sigma}$$

and the convergence of

$$\sum_{\beta\in\mathfrak{b}}{}' |N_K(\beta)|^{-\sigma}$$

implies that the interchange of sum and improper multiple integral is valid. □

The series under the integral sign in Proposition 9.1 is related to a theta function. We prove a functional equation for this theta function, and in the next section use it to extract valuable information about the Dedekind zeta function and the arithmetic of algebraic number fields.

Proposition 9.2 (Hecke theta formula). *The identity*

$$\sum_{\beta\in\mathfrak{b}} e^{-\pi c(\mathfrak{b})xm(\beta,\mathbf{v})} = \frac{1}{x^{n/2}}\sum_{\beta'\in\mathfrak{b}'} e^{-\pi c(\mathfrak{b}')x^{-1}m(\beta',-\mathbf{v})}$$

holds for any fractional ideal $\mathfrak{b}$ of a number field. Here β' denotes a general element of the Dedekind complement $\mathfrak{b}'$ of $\mathfrak{b}$ with respect to $\mathbb{Z}$.

Proof. The fractional ideal $\mathfrak{b}$ has an ideal basis $\beta_1,\ldots,\beta_n$ and the Dedekind complement $\mathfrak{b}'$ has a basis $\beta'_1,\ldots,\beta'_n$ complementary to this. Then

$$m(\mathbf{j}^t\mathbf{b},\mathbf{v}) = \sum_{l=1}^n\left|\sum_{p=1}^n j_p\beta_p^{(l)}\right|^2 e^{v_l} = \sum_{l=1}^n\sum_{p=1}^n\sum_{q=1}^n j_pj_q\beta_p^{(l)}\overline{\beta_q^{(l)}}e^{v_l}$$

with $\mathbf{b}=[\beta_1,\ldots,\beta_n]^t$. Note that since the summations are extended from 1 to n rather than from 1 to $r+1$, those coefficients that were equal to 2 in the original expression for $m(\beta,v_1,\ldots,v_r)$ here correspond to pairs of repeated terms. Now

$$c(\mathfrak{b})m(\mathbf{j}^t\mathbf{b},\mathbf{v}) = \mathbf{j}^tM\mathbf{j}, \qquad M=B^*LB,$$

where L is the diagonal matrix with $c(\mathfrak{b})e^{v_1},\ldots,c(\mathfrak{b})e^{v_n}$ on the main diagonal,

$$B=\begin{bmatrix}\beta_1^{(1)} & \cdots & \beta_n^{(1)}\\ \vdots & \vdots & \vdots\\ \beta_1^{(n)} & \cdots & \beta_n^{(n)}\end{bmatrix}, \qquad B'=\begin{bmatrix}\beta_1'^{(1)} & \cdots & \beta_n'^{(1)}\\ \vdots & \vdots & \vdots\\ \beta_1'^{(n)} & \cdots & \beta_n'^{(n)}\end{bmatrix}$$

are the matrices of algebraic conjugates of the elements of the two ideal bases, and B^* the adjoint (conjugate transpose) matrix of B. From the way it is defined it is clear that M is a real matrix, and the calculation $M^\star = (B^\star LB)^\star = B^\star L^\star(B^\star)^\star = B^\star LB = M$ then implies that it is symmetric. Since $\det(B)^2 = \Delta_{K/\mathbf{Q}}(\beta_1, \ldots, \beta_n) \neq 0$, the definition of M shows that it is positive definite.

Now

$$\sum_{\beta\in\mathfrak{b}} e^{-\pi c(\mathfrak{b})xm(\beta,\mathbf{v})} = \sum_{\mathbf{j}\in\mathbb{Z}^n} e^{-\pi x\mathbf{j}^tM\mathbf{j}} = \sum_{\mathbf{h}\in\mathbb{Z}^n} \int_{\mathbb{R}^n} e^{-\pi x\mathbf{y}^tM\mathbf{y}} e(-\mathbf{h}^t\mathbf{y})\, d\mathbf{y}$$

by the Poisson summation formula. For the function being summed is the exponential of a negative definite quadratic form, and all its partial derivatives are products of this function and polynomials, so the condition of uniform convergence on compact sets in Proposition 8.3 is clearly satisfied.

The following calculation of a Fourier transform is an interesting case of a computation performed by means of the Spectral Theorem without actually having to carry out the diagonalization explicitly.

The calculation

$$\begin{aligned}
&\int_{\mathbb{R}^n} e^{-\pi x\mathbf{y}^tM\mathbf{y}} e(-\mathbf{h}^t\mathbf{y})\, d\mathbf{y} \\
&= \int_{\mathbb{R}^n} e^{-\pi x(x^{-1/2}P\mathbf{z})^tM(x^{-1/2}P\mathbf{z})} e(-\mathbf{h}^tx^{-1/2}P\mathbf{z}) \left|\det\left(x^{-1/2}P\right)\right| d\mathbf{z} \\
&= x^{-\frac{n}{2}} \int_{\mathbb{R}^n} e^{-\pi\mathbf{z}^tP^tMP\mathbf{z}} e(-\mathbf{h}^tx^{-1/2}P\mathbf{z})\, d\mathbf{z} \\
&= x^{-\frac{n}{2}} \int_{\mathbb{R}^n} e^{-\pi\mathbf{z}^t\Lambda\mathbf{z}} e(-\mathbf{h}^tx^{-1/2}P\mathbf{z})\, d\mathbf{z} \\
&= x^{-\frac{n}{2}} \int_{\mathbb{R}^n} e^{-\pi\mathbf{z}^t\Lambda\mathbf{z}} e(-x^{-1/2}\mathbf{w}^t\mathbf{z})\, d\mathbf{z} \\
&= x^{-\frac{n}{2}} \left(\int_{-\infty}^{\infty} e^{-\pi\lambda_1z_1^2-2\pi ix^{-1/2}w_1z_1}\, dz_1\right) \\
&\qquad \cdots \left(\int_{-\infty}^{\infty} e^{-\pi\lambda_nz_n^2-2\pi ix^{-1/2}w_nz_n}\, dz_n\right) \\
&= x^{-\frac{n}{2}} \left(e^{-\pi x^{-1}w_1^2/\lambda_1} \int_{-\infty}^{\infty} e^{-\pi(z_1+ix^{-1/2}w_1/\lambda_1)^2}\, dz_1\right) \\
&\qquad \cdots \left(e^{-\pi x^{-1}w_n^2/\lambda_n} \int_{-\infty}^{\infty} e^{-\pi(z_n+ix^{-1/2}w_n/\lambda_1)^2}\, dz_n\right) \\
&= x^{-\frac{n}{2}} e^{-\pi x^{-1}w_1^2/\lambda_1-\cdots-\pi x^{-1}w_n^2/\lambda_n} = x^{-\frac{n}{2}} e^{-\pi x^{-1}\mathbf{h}^tM^{-1}\mathbf{h}}
\end{aligned}$$

proceeds by a change of variable $\mathbf{y} = x^{-1/2}P\mathbf{z}$ with P a suitable matrix. This is the orthogonal matrix in a diagonalization $P^{-1}MP = P^tMP = \Lambda$

of the real and symmetric matrix M. Here $\mathbf{w} = P^t\mathbf{h}$ is a vector with components $w_1, \ldots, w_n$. All the improper integrals converge absolutely, for Λ is a positive definite diagonal matrix.

The definition of the complementary basis $\beta'_1, \ldots, \beta'_n$ implies that $B^{-1} = (B')^t$. Thus

$$\begin{aligned}
\mathbf{h}^t M^{-1}\mathbf{h} &= \mathbf{h}^t B^{-1} L^{-1} (B^\star)^{-1}\mathbf{h} = \mathbf{h}^t (B')^t L^{-1} (B^{-1})^\star \mathbf{h} \\
&= \mathbf{h}^t (B')^t L^{-1} ((B')^t)^\star \mathbf{h} = \mathbf{h}^t (B')^t L^{-1} \overline{B'}\mathbf{h} \\
&= \mathbf{h}^t (B')^\star L^{-1} B'\mathbf{h} = c(\mathfrak{b})^{-1} m(\mathbf{h}^t\mathbf{b}', -\mathbf{v}) \\
&= (|d_K| N(\mathfrak{b})^2)^{1/n} m(\beta', -\mathbf{v}) \\
&= (N(\mathfrak{d}_{K/\mathbb{Q}})^2 |d_K|^{-1} N(\mathfrak{b})^2)^{1/n} m(\beta', -\mathbf{v}) \\
&= (|d_K| N(\mathfrak{d}^{-1}\mathfrak{b}^{-1})^2)^{-1/n} m(\beta', -\mathbf{v}) \\
&= (|d_K| N(\mathfrak{b}')^2)^{-1/n} m(\beta', -\mathbf{v}) = c(\mathfrak{b}') m(\beta', -\mathbf{v})
\end{aligned}$$

with $\mathbf{b}' = [\beta'_1, \ldots, \beta'_n]^t$. Note the use of the fact that the transpose leaves one-by-one matrices (scalars) unchanged. □

9.2. The analytic class number formula

To make progress with the Mellin transform representation for $Z_K(s, C)$ from the last section, we must take units into account. Units $\varepsilon_1, \ldots, \varepsilon_l$ are *multiplicatively independent* if the relation

$$\varepsilon_1{}^{h_1} \ldots \varepsilon_l{}^{h_l} = 1$$

with $h_1, \ldots, h_l$ integers implies that $h_1 = \cdots = h_l = 0$. Clearly a unit of finite order is not multiplicatively independent, while a single unit of infinite order is. Associate to each unit ε a vector

$$\mathbf{d}_\varepsilon = \left(\log\left(\left|\varepsilon^{(1)}\right|^2\right), \ldots, \log\left(\left|\varepsilon^{(r)}\right|^2\right)\right)$$

in $\mathbb{R}^r$. The following geometric interpretation of multiplicative independence of units goes back to Dirichlet.

Proposition 9.3. *Units $\varepsilon_1, \ldots, \varepsilon_l$ are multiplicatively independent if and only if $\mathbf{d}_{\varepsilon_1}, \ldots, \mathbf{d}_{\varepsilon_l}$ are linearly independent over $\mathbb{R}$.*

Proof. The relation

$$\frac{h_1}{k}\mathbf{d}_{\varepsilon_1} + \cdots + \frac{h_l}{k}\mathbf{d}_{\varepsilon_l} = 0$$

is equivalent to the statement that all the conjugates of

$$\varepsilon = \varepsilon_1^{h_1} \ldots \varepsilon_l^{h_l}$$

lie on the unit circle. From the Summary we know that an algebraic integer has all its conjugates on the unit circle if and only if it is a root of unity. So the above relation is equivalent to

$$\varepsilon_1^{h_1 h} \dots \varepsilon_l^{h_l h} = 1$$

for some positive integer h. Thus $\varepsilon_1, \dots, \varepsilon_l$ are multiplicatively independent if and only if $\mathbf{d}_{\varepsilon_1}, \dots, \mathbf{d}_{\varepsilon_l}$ are linearly independent over $\mathbb{Q}$. Since linear independence over $\mathbb{R}$ implies linear independence over $\mathbb{Q}$, one direction of the claimed equivalence is established.

We assume $\varepsilon_1, \dots, \varepsilon_l$ to be multiplicatively independent. Let $\alpha_1, \dots, \alpha_l$ be real numbers for which

$$\alpha_1 \mathbf{d}_{\varepsilon_1} + \dots + \alpha_l \mathbf{d}_{\varepsilon_l} = 0.$$

By the Dirichlet approximation theorem, there exists integers $h_1, \dots, h_l$ with

$$\left| \alpha_j - \frac{h_j}{k} \right| \leq \frac{1}{k^{1+1/l}}, \qquad j = 1, 2, \dots, l,$$

for arbitrarily large positive integers k. Then

$$\frac{h_1}{k} \mathbf{d}_{\varepsilon_1} + \dots + \frac{h_l}{k} \mathbf{d}_{\varepsilon_l} = -\left(\alpha_1 - \frac{h_1}{k} \right) \mathbf{d}_{\varepsilon_1} - \dots - \left(\alpha_l - \frac{h_l}{k} \right) \mathbf{d}_{\varepsilon_l}$$

so

$$|h_1 \mathbf{d}_{\varepsilon_1} + \dots + h_l \mathbf{d}_{\varepsilon_l}| \leq \frac{M}{k^{1/l}}, \qquad M = |\mathbf{d}_{\varepsilon_1}| + \dots + |\mathbf{d}_{\varepsilon_l}|.$$

This yields a unit

$$\varepsilon = \varepsilon_1^{h_1} \cdots \varepsilon_l^{h_l}$$

for which the bounds

$$e^{-\frac{M}{2k^{1/l}}} \leq \left| \varepsilon^{(j)} \right| \leq e^{\frac{M}{2k^{1/l}}}, \qquad 1 \leq j \leq l$$

hold for the conjugates. Denote the minimal polynomial of ε by $m_\varepsilon(x)$. The degree of $m_\varepsilon(x)$ is bounded by n_K and the coefficients are bounded in terms of M, being symmetric functions of the algebraic conjugates. So there are only finitely many of these minimal polynomials $m_\varepsilon(x)$, hence only finitely many of the units ε. But letting $k \to +\infty$ we obtain an infinite sequence of such units all of whose conjugates tend towards the unit circle. Since this sequence contains a constant subsequence, there is such a unit

$$\varepsilon = \varepsilon_1^{h_1} \dots \varepsilon_l^{h_l}$$

all of whose conjugates lie on the unit circle. This unit is a root of unity, so there is a positive integer h so that

$$\varepsilon_1^{h_1 h} \dots \varepsilon_l^{h_l h} = 1.$$

But then $h_1 = \cdots = h_l = 0$ by assumption, so

$$|\alpha_j| \leq \frac{1}{k^{1+1/l}}, \qquad 1 \leq j \leq l,$$

for arbitrarily large k. Thus $\alpha_1 = \cdots = \alpha_l = 0$. □

Let w_K denote the number of units of finite order of $\mathcal{O}_K$. Recall that these are roots of unity and that $\mathcal{O}_K$ contains only finitely many. Then

$$w_K Z_K(s, C) = \frac{n}{2} \int_0^\infty x^{ns/2-1} \left(\int_{\mathbb{R}^r} \sum_{\beta \in \mathfrak{b}}{}'' e^{-\pi c(\mathfrak{b}) x m(\beta, \mathbf{v})} d\mathbf{v} \right) dx$$

for $\sigma > 1$, where the double prime on the sum indicates that the summation is over all nonzero elements of $\mathfrak{b}$ up to associates, such that for each β included in the sum all associates $\varepsilon\beta$ under multiplication by units ε of finite order are also included and no other associates of β are included.

Proposition 9.4 (Dirichlet unit theorem). *The unit group U_K of a number field is the internal direct product of the group of roots of unity in the field with a torsion free subgroup of rank $r = r_1 + r_2 - 1$ (called the* unit rank.)

Proof. The units of finite order are roots of unity and there are only finitely many of them. There are at most r multiplicatively independent units, because $\mathbb{R}^r$ cannot contain more than r linearly independent vectors $\mathbf{d}_{\varepsilon_1}, \ldots, \mathbf{d}_{\varepsilon_l}$. Thus the structure theorem for finitely generated abelian groups applies, and shows that U_K is the internal direct product of the group of roots of unity of K and a torsion free subgroup of rank $l \leq r$.

If ε is a unit of infinite order then

$$m(\varepsilon\beta, \mathbf{v}) = \left|\varepsilon^{(1)}\beta^{(1)}\right|^2 e^{v_1} + \cdots + 2\left|\varepsilon^{(r+1)}\beta^{(r+1)}\right|^2 e^{v_{r+1}} = m(\beta, \mathbf{v} + \mathbf{d}_\varepsilon)$$

where $\mathbf{d}_\varepsilon$ is a nonzero vector. In the integral

$$\int_{\mathbb{R}^r} \sum_{\beta \in \mathfrak{b}}{}'' e^{-\pi c(\mathfrak{b}) x m(\beta, \mathbf{v})} d\mathbf{v}$$

we may enlarge the sum by including the whole orbit under $\varepsilon^{\mathbf{Z}}$ of each element β, if we restrict $\mathbf{v}$ in the integral to the closed region S between two hyperplanes H_1 and H_2 in $\mathbb{R}^r$, where H_1 has $\mathbf{d}_\varepsilon$ as a normal and H_2 is obtained from H_1 by translation by $\mathbf{d}_\varepsilon$. The translation identifies pairs of points on the boundary of S, but this is harmless as the boundary does not contribute to the r-fold integral. Generating the torsion free subgroup by l multiplicatively independent units $\varepsilon_1, \ldots, \varepsilon_l$, we obtain

$$w_K Z_K(s, C) = \frac{n}{2} \int_0^\infty x^{ns/2-1} \left(\int_V \sum_{0 \neq \beta \in \mathfrak{b}} e^{-\pi c(\mathfrak{b}) x m(\beta, \mathbf{v})} d\mathbf{v} \right) dx$$

where V is the parallelepiped obtained by intersecting the closed regions S_j corresponding to the generating units $\varepsilon_1, \ldots, \varepsilon_l$. We note that $\mathrm{vol}(V) = \infty$ unless $l = r$. Next

$$\begin{aligned}
w_K Z_K(s, C) &= \frac{n}{2}\left(\int_0^1 + \int_1^\infty\right) x^{ns/2-1}\left(\int_V \sum_{0\neq\beta\in\mathfrak{b}} e^{-\pi c(\mathfrak{b}) x m(\beta,\mathbf{v})} d\mathbf{v}\right) dx \\
&= \frac{n}{2}\int_1^\infty x^{-ns/2+1}\left(\int_V \sum_{0\neq\beta\in\mathfrak{b}} e^{-\pi c(\mathfrak{b}) x^{-1} m(\beta,\mathbf{v})} d\mathbf{v}\right)\frac{dx}{x^2} \\
&\quad + \frac{n}{2}\int_1^\infty x^{ns/2-1}\left(\int_V \sum_{0\neq\beta\in\mathfrak{b}} e^{-\pi c(\mathfrak{b}) x m(\beta,\mathbf{v})} d\mathbf{v}\right) dx \\
&= \frac{n}{2}\int_1^\infty x^{-ns/2-1}\left(\int_V \sum_{0\neq\beta\in\mathfrak{b}} e^{-\pi c(\mathfrak{b}) x^{-1} m(\beta,\mathbf{v})} d\mathbf{v}\right) dx \\
&\quad + \frac{n}{2}\int_1^\infty x^{ns/2-1}\left(\int_V \sum_{0\neq\beta\in\mathfrak{b}} e^{-\pi c(\mathfrak{b}) x m(\beta,\mathbf{v})} d\mathbf{v}\right) dx.
\end{aligned}$$

Then

$$\begin{aligned}
&\int_1^\infty x^{-ns/2-1}\left(\int_V \sum_{0\neq\beta\in\mathfrak{b}} e^{-\pi c(\mathfrak{b}) x^{-1} m(\beta,\mathbf{v})} d\mathbf{v}\right) dx \\
&\quad = \int_1^\infty x^{-ns/2-1}\int_V \left(x^{n/2} - 1 + x^{n/2}\sum_{0\neq\beta'\in\mathfrak{b}'} e^{-\pi c(\mathfrak{b}') x m(\beta',-\mathbf{v})}\right) d\mathbf{v}\, dx \\
&\quad = \frac{2}{n}\frac{\mathrm{vol}(V)}{s(s-1)} + \int_1^\infty x^{n(1-s)/2-1}\left(\int_V \sum_{0\neq\beta'\in\mathfrak{b}'} e^{-\pi c(\mathfrak{b}') x m(\beta',-\mathbf{v})} d\mathbf{v}\right) dx
\end{aligned}$$

by the Hecke theta formula

$$\sum_{\beta\in\mathfrak{b}} e^{-\pi c(\mathfrak{b}) x m(\beta,\mathbf{v})} = \frac{1}{x^{n/2}}\sum_{\beta'\in\mathfrak{b}'} e^{-\pi c(\mathfrak{b}') x^{-1} m(\beta',-\mathbf{v})}.$$

Thus

$$\begin{aligned}
w_K Z_K(s, C) &= \frac{\mathrm{vol}(V)}{s(s-1)} + \frac{n}{2}\int_1^\infty x^{ns/2-1}\left(\int_V \sum_{0\neq\beta\in\mathfrak{b}} e^{-\pi c(\mathfrak{b}) x m(\beta,\mathbf{v})} d\mathbf{v}\right) dx \\
&\quad + \frac{n}{2}\int_1^\infty x^{n(1-s)/2-1}\left(\int_V \sum_{0\neq\beta'\in\mathfrak{b}'} e^{-\pi c(\mathfrak{b}') x m(\beta',-\mathbf{v})} d\mathbf{v}\right) dx
\end{aligned}$$

holds for $\sigma > 1$. If s is real and $s > 1$ then the left-hand side is finite while the terms on the right-hand side are nonnegative, so $\mathrm{vol}(V) < \infty$. Hence the units of infinite order cannot be multiplicatively generated by fewer than r units, thus the rank equals r. □

We call the formula at the end of the above proof the *Mellin transform identity*. It continues $Z_K(s, C)$ to the whole complex plane as a meromorphic function, with simple poles at $s = 0$ and $s = 1$. For the integrands in the two integrals decay exponentially as $x \to +\infty$, as one sees from the proof of the Hecke theta formula. A collection $\varepsilon_1, \ldots, \varepsilon_r$ of multiplicatively independent units generating the torsion free subgroup of U_K is called a *system of fundamental units.*

Proposition 9.5 (Finiteness of the class number). *The class group Cl_K has finite order. (The order $h_K = |Cl_K|$ is called the* class number *of K.)*

Proof. Assuming s real and $s > 1$, summing the Mellin transform identity over classes $C \in Cl_K$ shows that

$$w_K Z_K(s) \geq \frac{|Cl_K|\mathrm{vol}(V)}{s(s-1)}.$$

But $Z_K(s)$ is finite for $s > 1$, and the parallelepiped V has positive volume, for the vectors $\mathbf{d}_{\varepsilon_1}, \ldots, \mathbf{d}_{\varepsilon_r}$ are linearly independent over $\mathbb{R}$. □

Proposition 9.6 (Functional equation of the Dedekind zeta function). *The completed Dedekind zeta function has an analytic continuation to the whole complex plane as a meromorphic function and satisfies the functional equation*

$$Z_K(s) = Z_K(1-s)$$

there. It has two poles, at $s = 0$ and at $s = 1$, both of which are simple.

Proof. The right-hand side of the Mellin transform identity is left unchanged on replacing s by $1-s$ and $\mathfrak{b}$ by $\mathfrak{b}'$, so $Z_K(s, C) = Z_K(1-s, C)$. □

The volume $\mathrm{vol}(V)$ in the Mellin transform identity does not depend on the choice of a system of fundamental units, even though the parallelepiped V does so depend. For the computation of the residue

$$\mathrm{Res}_{s=1}(w_K Z_K(s)) = |Cl_K|\mathrm{vol}(V) \times \begin{Bmatrix} 1 & \text{if } r_2 > 0 \\ 2 & \text{if } r_2 = 0 \end{Bmatrix}$$

in accordance with Proposition 9.1 and the Mellin transform identity determines $\mathrm{vol}(V)$ solely by invariants of the number field. The same volume can also be computed in terms of any system of fundamental units, and is given by $\mathrm{vol}(V) = 2^{r_1} R_K$ where

$$R_K = \left|\det\left(\left[e_i \log\left|\varepsilon_i^{(j)}\right|\right]_{1\le i,j\le r}\right)\right|, \qquad e_i = \begin{cases} 1 & \text{for } 1 \le i \le r_1, \\ 2 & \text{for } r_1 < i \le r, \end{cases}$$

is the *regulator* of K. Note that the value of the regulator is independent both of the choice of system of fundamental units and the ordering of the embeddings. If K is $\mathbb{Q}$ or an imaginary quadratic field, the formula for the regulator is inapplicable, and $R_K = 1$ by convention. If $r_2 = 0$ the expression for R_K must be increased by a factor of 2 as above.

Proposition 9.7 (Analytic class number formula). *The identity*

$$h_K = \frac{w_K |d_K|^{1/2}}{2^{r_1(K)+r_2(K)} \pi^{r_2(K)} R_K} \lim_{s\to 1} \frac{\zeta_K(s)}{\zeta(s)}$$

holds for any number field K.

Proof. Computing the residue

$$\mathrm{Res}_{s=1}(w_K Z_K(s)) = w_K \frac{|d_K|^{1/2}}{2^{r_2}\pi^{n/2}} \pi^{r_1/2} \mathrm{Res}_{s=1}(\zeta_K(s))$$

in a second way and comparing with the first expression for the residue in terms of $\mathrm{vol}(V)$ yields the identity. □

The existence of a simple pole of the Dedekind zeta function at $s = 1$ carries implications for the distribution of prime ideals. The following result goes back to Kronecker.

Proposition 9.8. *Suppose that $K/\mathbb{Q}$ is a finite extension. Then $\mathcal{O}_K$ has infinitely many prime ideals of degree one. Let B be some set of rational primes that are divisible by at least b distinct prime ideals of degree one in $\mathcal{O}_K$. If the Dirichlet density of B in the rational primes exists, it is at most $1/b$. If $K/\mathbb{Q}$ is a normal extension, the set of rational primes that split completely in K has Dirichlet density $1/n_K$ in the rational primes.*

Proof. Taking logarithms in the Euler product one sees that

$$\log(\zeta_K(s)) = \sum_p \sum_{\mathfrak{p}|p} \sum_{k=1}^{\infty} \frac{p^{-k\deg(\mathfrak{p})\sigma}}{k} = \sum_{m=1}^{\infty} \frac{1}{m} \sum_p p^{-m\sigma} \sum_{\substack{\mathfrak{p}|p \\ \deg(\mathfrak{p})|m}} \deg(\mathfrak{p})$$

for $\sigma > 1$. Now

$$\sum_{m=2}^{\infty} \frac{1}{m} \sum_p p^{-m\sigma} \sum_{\substack{\mathfrak{p}|p \\ \deg(\mathfrak{p})|m}} \deg(\mathfrak{p}) \le \sum_{m=2}^{\infty} \frac{1}{m} \sum_p p^{-m} m n_K = n_K \sum_p \frac{p^{-2}}{1-p^{-1}}$$

for $\sigma \geq 1$. Then

$$\log(\zeta_K(\sigma)) = \sum_p p^{-\sigma} \sum_{\substack{\mathfrak{p}|p \\ \deg(\mathfrak{p})|1}} \deg(\mathfrak{p}) + O(1)$$

as $\sigma \to 1^+$. Thus

$$\frac{\log(\zeta_K(\sigma))}{\log(\zeta(\sigma))} = \frac{\sum_p p^{-\sigma}\omega_K(p) + O(1)}{\sum_p p^{-\sigma} + O(1)}, \qquad \omega_K(p) = \sum_{\substack{\mathfrak{p}|p \\ \deg(\mathfrak{p})=1}} 1,$$

and noting that $\zeta_K(s)$ and $\zeta(s)$ both have simple poles at $s = 1$, we see that

$$\delta_{B,\mathbb{P}}\, b = \lim_{\sigma\to 1^+} \frac{\sum_{p\in B} p^{-\sigma} b}{\sum_p p^{-\sigma}} \leq \lim_{\sigma\to 1^+} \frac{\sum_p p^{-\sigma}\omega_K(p)}{\sum_p p^{-\sigma}} = 1,$$

where $\mathbb{P}$ denotes the set of rational primes. The inequality $\delta_{B,\mathbb{P}} \leq 1/b$ follows.

The set of rational primes p for which $\omega_K(p) > 0$ must be infinite, otherwise the last limit above could not equal 1 since the denominator tends to $+\infty$ as $\sigma \to 1^+$. So there are infinitely many rational primes divisible by a prime ideal of $\mathcal{O}_K$ of degree one. Since a prime ideal cannot divide two distinct rational primes, there must be infinitely many prime ideals of degree one.

If $K/\mathbb{Q}$ is normal, all prime ideals $\mathfrak{p}$ in $\mathcal{O}_K$ lying above a rational prime p have the same degree. Then p is divisible by a prime ideal $\mathfrak{p}$ of degree one if and only if p splits completely into n_K prime ideals in $\mathcal{O}_K$. But in this situation $\omega_K(p)$ equals n_K or zero according as p does or does not split completely. □

We know that the rational primes $p \equiv 1 \pmod 4$ are precisely those that split completely in the extension $\mathbb{Q}(\sqrt{-1})/\mathbb{Q}$. Could there be some other normal extension of $\mathbb{Q}$ in which these are the rational primes that split completely? By the preceding result this would have to be a quadratic extension, since the Dirichlet density is $1/2$. But it is a remarkable fact that the rational primes that split completely in a finite normal extension of $\mathbb{Q}$ determine the extension.

Proposition 9.9 (Theorem of Bauer). *Let $K/\mathbb{Q}$ and $L/\mathbb{Q}$ be finite normal extensions in $\mathbb{C}$. If with the exception of a set of primes of zero Dirichlet density the same rational primes split completely in K and in L, then $K = L$.*

Proof. Let B be the set of rational primes that split completely in K. Our assumptions imply that the Dirichlet density of B exists and is equal to $1/n_K = 1/n_L$. Moreover the primes in B split completely in KL, so since the Dirichlet density of B exists, it is at most $1/n_{KL}$. Then $n_K^{-1} = n_L^{-1} \leq n_{KL}^{-1}$, and so $n_{KL} \leq n_K = n_L$. Thus $KL \subseteq K$ and $KL \subseteq L$, hence $L \subseteq K$ and $K \subseteq L$. □

9.3. ⋆ Class numbers of quadratic fields

The analytic class number formula may be made more specific whenever the limit on the right-hand side can be calculated. This may achieved in some cases by factoring $\zeta_K(s)$ in terms of L-functions. The case of quadratic number fields is the simplest. We know from the Summary that a rational prime p remains inert, splits or ramifies according as the Legendre symbol $(d_K|p)$ with fundamental discriminant d_K equals $-1, 1$ or 0. Then

$$\begin{aligned}\zeta_{\mathbb{Q}(\sqrt{d})}(s) &= \prod_{p|d_K}(1-p^{-s})^{-1} \prod_{(d_K|p)=1}(1-p^{-s})^{-2} \prod_{(d_K|p)=-1}(1-p^{-2s})^{-1} \\ &= \prod_p (1-p^{-s})^{-1} \prod_p \left(1-\chi_d(p)p^{-s}\right)^{-1} = \zeta(s)L(s,\chi_d)\end{aligned}$$

where χ_d is the Dirichlet character $\chi_d(m) = (d_K|m)$. Clearly

$$h(d) = \frac{w|d_K|^{1/2}}{2\pi} L(1,\chi_d)$$

for $d < 0$, where $h(d)$ denotes the class number of $\mathbb{Q}(\sqrt{d})$, and

$$h(d)\log(\varepsilon_d) = \frac{|d_K|^{1/2}}{2} L(1,\chi_d)$$

for $d > 0$, where ε_d denotes the smallest unit of $\mathbb{Q}(\sqrt{d})$ that is greater than 1, and is called the *fundamental unit.* We may now prove a conjecture of Gauss that was established by H. A. Heilbronn in 1934 after earlier work of M. Deuring, E. Hecke and L. J. Mordell.

Proposition 9.10 (Theorem of Heilbronn). *The class numbers $h(d)$ of imaginary quadratic fields tend to $+\infty$ as $d \to -\infty$.*

Proof. The inequality

$$h(d) \gg |d|^{1/2} L(1,\chi_d)$$

holds for $d < 0$ by the analytic class number formula. But χ_d is a quadratic character of period at most $4d$, so

$$L(1,\chi_d) \gg |d|^{-1/4}$$

say, by the theorem of Siegel. □

m	1	3	5	7
$(-2\|m)$	1	1	-1	-1

Table 2. Values of a quadratic character

Unfortunately Siegel's inequality is ineffective, so we are unable to use it to produce some numerical constant $d_0 < 0$ so that $h(d) > 1$ for $d \leq d_0$, which would enable us to reduce the class number one problem for imaginary quadratic fields to a finite amount of computation.

We may obtain an effective lower bound for $L(1, \chi_d)$ for $d < -4$ from the class number formula, using the trivial observation that $h(d) \geq 1$. The inequality is

$$L(1, \chi_d) \geq \frac{\pi}{2|d|^{1/2}},$$

and is weaker than Siegel's inequality. The proof of the latter may be interpreted as relating to the Dedekind zeta function of a biquadratic field, while the proof of the weaker inequality relates to the Dedekind zeta function of an imaginary quadratic field.

We are now going to find more explicit formulas for the class numbers of quadratic fields. For this purpose it is necessary to calculate $L(1, \chi_d)$, and to carry out this calculation we need to know that χ_d is primitive. The case when $d_K = 8D'$ with $D' \equiv 3 \pmod 4$ goes like this: The character is

$$\chi_d(m) = \left(\frac{d_K}{m}\right) = \left(\frac{m}{|D'|}\right)\left(\frac{-2}{m}\right) = \left(\frac{m}{p_1}\right)\cdots\left(\frac{m}{p_k}\right)\left(\frac{-2}{m}\right)$$

where $|D'| = p_1 p_2 \cdots p_k$ and $p_1, p_2, \ldots, p_k$ are distinct odd primes. Suppose that $q' \leq q = |d_K|$ is the smallest positive quasiperiod of χ_d. Then $q'|q$ so if $q' < q$ there must be at least one odd prime p_j with $p_j \nmid q'$ or else $8 \nmid q'$. But $(m|p_j)$ has no smaller period than p_j and Table 2 shows that $(-2|m)$ has no smaller period than 8. Since $p_1, \ldots, p_k, 8$ are pairwise coprime, $q' < q$ cannot be a quasiperiod. The other cases go through in the same way.

Proposition 9.11 (Dirichlet class number formula). *The formula*

$$h(d) = -\frac{1}{|d_K|} \sum_{1 \leq m < |d_K|} m \left(\frac{d_K}{m}\right)$$

holds for $d < -4$ and the formula

$$h(d) \log(\varepsilon_d) = - \sum_{1 \leq m < |d_K|/2} \left(\frac{d_K}{m}\right) \log\left(\sin\left(\frac{\pi m}{d_K}\right)\right)$$

holds for $d > 1$.

Proof. To avoid manipulations with conditionally convergent series, we work with $L(\sigma, \chi_d)$ for $\sigma > 1$ and take the limit as $\sigma \to 1^+$.

Since χ_d is primitive and real, it can expressed by

$$\chi_d(n) = \frac{1}{\tau(\chi_d)} \sum_{m=1}^{|d_K|} \left(\frac{d_K}{m}\right) e(mn/|d_K|)$$

and so

$$\begin{aligned} L(\sigma, \chi_d) &= \sum_{n=1}^{\infty} \chi_d(n) n^{-\sigma} = \sum_{n=1}^{\infty} \left(\frac{1}{\tau(\chi_d)} \sum_{m=1}^{|d_K|} \left(\frac{d_K}{m}\right) e(mn/|d_K|) \right) n^{-\sigma} \\ &= \frac{1}{\tau(\chi_d)} \sum_{1 \le m < |d_K|} \left(\frac{d_K}{m}\right) \sum_{n=1}^{\infty} e(mn/|d_K|) n^{-\sigma}. \end{aligned}$$

Now

$$\sum_{n=1}^{\infty} n^{-\sigma} z^n = \sigma \int_1^{\infty} \frac{1 - z^{[u]+1}}{1 - z} u^{-\sigma-1}\, du$$

by partial summation, if $z \neq 1$, $|z| \le 1$ and $\sigma > 0$. Since

$$\left| \frac{1 - z^{[u]+1}}{1 - z} \right| \le \frac{2}{|1 - z|}$$

we see that

$$\lim_{\sigma \to 1^+} \sum_{n=1}^{\infty} e(mn/|d_K|) n^{-\sigma} = \sum_{n=1}^{\infty} e(mn/|d_K|) n^{-1} = -\mathrm{Log}\left(1 - e^{2\pi i \frac{m}{|d_K|}}\right),$$

where Log is the principal branch of the logarithm. Then

$$\mathrm{Log}\left(1 - e^{2\pi i \frac{m}{|d_K|}}\right) = \frac{\pi i m}{|d_K|} + \mathrm{Log}(-2i) + \mathrm{Log}\left(\sin\left(\frac{\pi m}{|d_K|}\right)\right)$$

yields

$$\begin{aligned} &L(1, \chi_d) = \\ &\quad \frac{-1}{\tau(\chi_d)} \sum_{1 \le m < |d_K|} \left(\frac{d_K}{m}\right) \left(\frac{\pi i m}{|d_K|} + \mathrm{Log}(-2i) + \mathrm{Log}\left(\sin\left(\frac{\pi m}{|d_K|}\right)\right) \right). \end{aligned}$$

To proceed further, it is necessary to relate $(d_K||d_K|-m)$ to $(d_K|m)$. The case when $d_K = 8D'$ with $D' \equiv 3 \pmod 4$ goes like this:

$$\begin{aligned}\left(\frac{d_K}{|d_K|-m}\right) &= \left(\frac{|d_K|-m}{|D'|}\right)\left(\frac{-2}{|d_K|-m}\right) = \left(\frac{-m}{|D'|}\right)\left(\frac{-2}{|d_K|-m}\right)\\ &= \left(\frac{-1}{|D'|}\right)\left(\frac{m}{|D'|}\right)\left(\frac{-2}{|d_K|-m}\right)\\ &= \left(\frac{-1}{|D'|}\right)\left(\frac{-2}{m}\right)\left(\frac{-2}{|d_K|-m}\right)\left(\frac{d_K}{m}\right)\\ &= (-1)^{\frac{|D'|-1}{2}}(-1)^{\frac{m^2-1+(|d_K|-m)^2-1}{8}}\left(\frac{d_K}{m}\right) = (-1)^{\frac{|D'|-1}{2}}\left(\frac{d_K}{m}\right)\end{aligned}$$

with $|D'| \equiv 3 \pmod 4$ if $d > 0$ and $|D'| \equiv -3 \pmod 4$ if $d < 0$. The other cases go through in the same way, and this leads to the conclusion that $(d_K||d_K|-m) = (d_K|m)$ if $d > 0$ while $(d_K||d_K|-m) = -(d_K|m)$ if $d < 0$.

Now

$$L(1,\chi_d) = \frac{-\pi i}{|d_K|\tau(\chi_d)} \sum_{1\le m<|d_K|} m\left(\frac{d_K}{m}\right)$$

for $d < 0$ since the graph of the sine function is symmetric about the line $x = \pi/2$ and $(d_K||d_K|-m)) = -(d_K|m)$. While

$$L(1,\chi_d) = \frac{-1}{\tau(\chi_d)} \sum_{1\le m<|d_K|/2} \left(\frac{d_K}{m}\right)\operatorname{Log}\left(\sin\left(\frac{\pi m}{|d_K|}\right)\right)$$

for $d > 0$ in a similar way.

Suppose ψ_1 is a Dirichlet character modulo q_1 and that ψ_2 is a Dirichlet character modulo q_2. By the theory of linear Diophantine equations there exists for each integer x modulo q_1q_2 precisely one integer y modulo q_2 and precisely one integer z modulo q_1 for which $q_1y + q_2z = x$, and thus

$$\begin{aligned}\tau(\psi_1\psi_2) &= \sum_{x=1}^{q_1q_2}(\psi_1\psi_2)(x)e(x/(q_1q_2))\\ &= \sum_{y=1}^{q_2}\sum_{z=1}^{q_1}\psi_1(q_1y+q_2z)\psi_2(q_1y+q_2z)e((q_1y+q_2z)/(q_1q_2))\\ &= \sum_{y=1}^{q_2}\sum_{z=1}^{q_1}\psi_1(q_2z)\psi_2(q_1y)e(y/q_2)e(z/q_1)\\ &= \psi_1(q_2)\psi_2(q_1)\sum_{y=1}^{q_2}\psi_2(y)e(y/q_2)\sum_{z=1}^{q_1}\psi_1(z)e(z/q_1)\\ &= \psi_1(q_2)\psi_2(q_1)\tau(\psi_1)\tau(\psi_2).\end{aligned}$$

In the case when $d_K = 8D'$ with $D' \equiv 3 \pmod 4$ a squarefree number the calculation of the Gauss sum $\tau(\chi_d)$ goes like this: The character is

$$\chi_d(m) = \left(\frac{m}{p_1}\right)\cdots\left(\frac{m}{p_k}\right)\left(\frac{-2}{m}\right)$$

where $|d_K| = p_1 p_2 \cdots p_k$ and $p_1, p_2, \ldots, p_k$ are distinct odd primes. The last factor on the right-hand side is a Dirichlet character modulo 8, which we shall denote by ψ. Repeated use of the formula for the Gauss sum of a product of two Dirichlet characters to coprime moduli shows that

$$\tau(\chi_d) = \tau_{p_1}\tau_{p_2}\cdots\tau_{p_k}\tau(\psi)\left(\frac{|d_K|/p_1}{p_1}\right)\left(\frac{|d_K|/p_2}{p_2}\right)\cdots\left(\frac{|d_K|/p_k}{p_k}\right)\psi(|d_K|/8).$$

Denote the number of primes congruent to 3 modulo 4 among $p_1, p_2, \ldots, p_k$ by Q. Then

$$\tau(\chi_d) = i^Q\sqrt{|D'|}\tau(\psi)\left(\frac{8}{|D'|}\right)\psi(|d_K|/8)\prod_{j<\ell}\left(\frac{p_j}{p_\ell}\right)\left(\frac{p_\ell}{p_j}\right)$$

by the Landsberg-Schaar formula. Now

$$\begin{aligned}\tau(\chi_d) &= i^Q\sqrt{|D'|}i2\sqrt{2}\left(\frac{2}{|D'|}\right)^3\left(\frac{-2}{|D'|}\right)\prod_{j<\ell}(-1)^{\frac{p_j-1}{2}\frac{p_\ell-1}{2}}\\ &= i^{Q+1}\sqrt{|d_K|}\left(\frac{-1}{|D'|}\right)(-1)^{\frac{Q(Q-1)}{2}}\end{aligned}$$

by quadratic reciprocity, and thus

$$\tau(\chi_d) = \begin{cases}\sqrt{|d_K|} & \text{if } d > 0,\\ i\sqrt{|d_K|} & \text{if } d < 0.\end{cases}$$

For if $d > 0$ then $|D'| \equiv 3 \pmod 4$ so Q is odd, while if $d < 0$ then $|D'| \equiv 1 \pmod 4$ so Q is even. The other cases go through in the same way, and the Dirichlet class number formula follows. □

The class number is a positive integer so, for computational purposes, we could avoid the use of the Landsberg-Schaar formula by introducing absolute values in the Dirichlet class number formula. This would however obscure some interesting arithmetical consequences of the formula. These are apparently very difficult to deduce in full by elementary means.

The arithmetical consequences of the Dirichlet class number formula have to do with the distribution of quadratic residues and nonresidues. We shall look at a single case; that of imaginary quadratic fields with an odd discriminant, and leave other cases for the exercises.

To deduce arithmetical consequences from the Dirichlet class number formula for imaginary quadratic fields with an odd discriminant, we proceed to simplify it. The identity

$$\left(\frac{d_K}{|d_K|-m}\right) = -\left(\frac{d_K}{m}\right)$$

for d_K negative is already known to us. Then

$$\begin{aligned}
h(d) &= -\frac{1}{|d_K|}\sum_{1\le m<|d_K|} m\left(\frac{d_K}{m}\right)\\
&= -\frac{1}{|d_K|}\sum_{1\le m<|d_K|/2} m\left(\frac{d_K}{m}\right) - \frac{1}{|d_K|}\sum_{|d_K|/2\le m<|d_K|} m\left(\frac{d_K}{m}\right)\\
&= -\frac{1}{|d_K|}\sum_{1\le m<|d_K|/2} m\left(\frac{d_K}{m}\right)\\
&\quad -\frac{1}{|d_K|}\sum_{1\le m<|d_K|/2} (|d_K|-m)\left(\frac{d_K}{|d_K|-m}\right)\\
&= -\frac{1}{|d_K|}\sum_{1\le m<|d_K|/2} m\left(\frac{d_K}{m}\right) + \frac{1}{|d_K|}\sum_{1\le m<|d_K|/2} (|d_K|-m)\left(\frac{d_K}{m}\right)\\
&= \sum_{1\le m<|d_K|/2} \left(\frac{d_K}{m}\right) - \frac{2}{|d_K|}\sum_{1\le m<|d_K|/2} m\left(\frac{d_K}{m}\right),
\end{aligned}$$

while

$$\begin{aligned}
h(d) &= -\frac{1}{|d_K|}\sum_{\substack{1\le m<|d_K|\\ 2\mid m}} m\left(\frac{d_K}{m}\right)\\
&\quad -\frac{1}{|d_K|}\sum_{\substack{1\le m<|d_K|\\ 2\mid m}} (|d_K|-m)\left(\frac{d_K}{|d_K|-m}\right)\\
&= -\frac{1}{|d_K|}\sum_{\substack{1\le m<|d_K|\\ 2\mid m}} m\left(\frac{d_K}{m}\right) + \frac{1}{|d_K|}\sum_{\substack{1\le m<|d_K|\\ 2\mid m}} (|d_K|-m)\left(\frac{d_K}{m}\right)\\
&= \sum_{1\le m<|d_K|/2} \left(\frac{d_K}{2m}\right) - \frac{2}{|d_K|}\sum_{1\le m<|d_K|/2} 2m\left(\frac{d_K}{2m}\right)\\
&= \left(\frac{d_K}{2}\right)\sum_{1\le m<|d_K|/2} \left(\frac{d_K}{m}\right) - 2\left(\frac{d_K}{2}\right)\frac{2}{|d_K|}\sum_{1\le m<|d_K|/2} m\left(\frac{d_K}{m}\right)
\end{aligned}$$

if $|d_K|$ is odd. Thus

$$h(d) = -\left(\frac{d_K}{2}\right) \sum_{1 \le m < |d_K|/2} \left(\frac{d_K}{m}\right)$$
$$-2\left(\frac{d_K}{2}\right)\left(\sum_{1 \le m < |d_K|/2} \left(\frac{d_K}{m}\right) - h(d)\right),$$

and so

$$h(d) = \frac{1}{2 - \left(\frac{d_K}{2}\right)} \sum_{1 \le m < |d_K|/2} \left(\frac{d_K}{m}\right).$$

The same formula also holds if $|d_K|$ is even, but we won't need to know this.

Suppose that p is a prime with $p \equiv 3 \pmod 4$, and choose $d = -p \equiv 1 \pmod 4$. Then $d_K = d = -p$, thus $(d_K|m) = (m||d_K|) = (m|p)$ is the Legendre symbol modulo p. Furthermore $(-p|2) = -1$ if $p \equiv 3 \pmod 8$ while $(-p|2) = 1$ if $p \equiv 7 \pmod 8$. Denoting the number of quadratic residues modulo p in the interval $[1, p/2)$ by $\mathcal{R}$ and the number of nonresidues modulo p in the same interval by $\mathcal{N}$, we see that $h(-p) = (\mathcal{R} - \mathcal{N})/3$ for $p \equiv 3 \pmod 8$ while $h(-p) = \mathcal{R} - \mathcal{N}$ for $p \equiv 7 \pmod 8$.

It is certainly striking that there are such simple formulas for the class number of $\mathbb{Q}(\sqrt{-p})$ for primes $p \equiv 3 \pmod 4$. Moreover these formulas imply that there are always more quadratic residues than nonresidues in the interval $[1, p/2)$ for these primes, since the class number is positive.

Further material on the Dirichlet class number formula and other more or less explicit class number formulas may be found in *Number Theory* by Z. I. Borevich and I. R. Shafarevich, *Lectures on the Theory of Algebraic Numbers* by Erich Hecke, *Elementary and Analytic Theory of Algebraic Numbers* by W. Narkiewicz, and *Classical Theory of Algebraic Numbers* by Paulo Ribenboim. *Über die Klassenzahl abelscher Zahlkörper* by Helmut Hasse is a standard reference for the topic.

9.4. ⋆ A discriminant bound

We are going to prove a lower bound for the absolute value $|d_K|$ of the discriminant of a number field K in terms of the arithmetic data $r_1(K)$ and $r_2(K)$. First we need an integral formula for the logarithmic derivative of the gamma function, due to Gauss.

Proposition 9.12. *The integral representation*

$$\frac{\Gamma'(s)}{\Gamma(s)} = \int_0^\infty \left(\frac{e^{-u}}{u} - \frac{e^{-su}}{1 - e^{-u}}\right) du$$

holds in the half plane $\sigma > 0$.

Proof. The computations

$$\int_0^\infty \left(\frac{1}{1-e^{-u}} - \frac{1}{u}\right)e^{-u}\,du = \int_0^\infty \left(\frac{1}{e^u-1} - u^{-1}e^{-u}\right)du$$

and

$$\begin{aligned}\int_0^\infty \left(\frac{u^{s-1}}{e^u-1} - u^{s-2}e^{-u}\right)du &= \Gamma(s)\zeta(s) - \Gamma(s-1)\\ &= \Gamma(s)\zeta(s) - \frac{\Gamma(s)}{s-1} = \Gamma(s)\Big(\zeta(s) - \frac{1}{s-1}\Big)\\ &= \Gamma(s)(\gamma + O(|s-1|)) = \gamma + O(|s-1|)\end{aligned}$$

establish the identity

$$\gamma = \int_0^\infty \left(\frac{1}{1-e^{-u}} - \frac{1}{u}\right)e^{-u}\,du,$$

to be used used in the next calculation.

The series expansion

$$\log(\Gamma(s)) = -\mathrm{Log}(s) - \gamma s - \sum_{n=1}^\infty \left(\mathrm{Log}\Big(1+\frac{s}{n}\Big) - \frac{s}{n}\right)$$

may be differentiated term by term since the differentiated series

$$\frac{\Gamma'(s)}{\Gamma(s)} = -\frac{1}{s} - \gamma - \sum_{n=1}^\infty \left(\frac{1}{n+s} - \frac{1}{n}\right)$$

converges uniformly on compact sets. Then

$$\begin{aligned}\frac{\Gamma'(s)}{\Gamma(s)} &= -\int_0^\infty e^{-su}\,du - \gamma - \sum_{n=1}^\infty\left(\int_0^\infty e^{-(s+n)u}\,du - \int_0^\infty e^{-nu}\,du\right)\\ &= -\int_0^\infty e^{-su}\,du - \int_0^\infty \left(\frac{1}{1-e^{-u}} - \frac{1}{u}\right)e^{-u}\,du\\ &\quad - \sum_{n=1}^\infty \int_0^\infty (e^{-(s+n)u} - e^{-nu})\,du\\ &= \int_0^\infty \left(\frac{e^{-u}}{u} - e^{-su} - \frac{e^{-u}}{1-e^{-u}} + \sum_{n=1}^\infty (e^{-nu} - e^{-(s+n)u})\right)du\\ &= \int_0^\infty \left(\frac{e^{-u}}{u} - e^{-su} - \frac{e^{-u}}{1-e^{-u}} + (1-e^{-su})\frac{e^{-u}}{1-e^{-u}}\right)du\\ &= \int_0^\infty \left(\frac{e^{-u}}{u} - \frac{e^{-su}}{1-e^{-u}}\right)du\end{aligned}$$

for $\sigma > 0$. □

Now we are ready to establish a lower bound for the discriminant of a number field. Lower bounds similar to this one were originally proved by geometric arguments. We shall instead use an analytic method evolved by H. M. Stark and then A. M. Odlyzko. In a more elaborate form it yields the best discriminant bounds currently known.

Proposition 9.13. *The inequality*

$$\frac{1}{n}\log|d_K| > \frac{r_1(K)}{n}3.108 + \frac{2r_2(K)}{n}2.415 - 6n^{-1/2}$$

holds for $n = n_K \geq 2$.

Proof. Fix the number field K, put $f(s) = s(s-1)Z_K(s)$, and differentiate logarithmically to obtain

$$\frac{f'(s)}{f(s)} = \frac{1}{s} + \frac{1}{s-1} + \frac{\gamma_K'(s)}{\gamma_K(s)} + \frac{\zeta_K'(s)}{\zeta_K(s)},$$

where

$$\frac{\gamma_K'(s)}{\gamma_K(s)} = \frac{1}{2}\log\left(\frac{|d_K|}{2^{2r_2}\pi^n}\right) + \frac{r_1}{2}\frac{\Gamma'(s/2)}{\Gamma(s/2)} + r_2\frac{\Gamma'(s)}{\Gamma(s)}.$$

Stark discovered that $|d_K|$ can be efficiently bounded from below by means of the second formula, using the first formula and just information about the sign of the logarithmic derivatives $f'(s)/f(s)$ and $\zeta_K'(s)/\zeta_K(s)$ for $s > 1$ real.

The Mellin transform identity yields

$$\begin{aligned} w_K Z_K(s) = {} & \frac{h_K \mathrm{vol}(V)}{s(s-1)} \\ & + \frac{n}{2}\sum_{h=1}^{h_K}\int_1^\infty x^{ns/2-1}\left(\int_V \sum_{0\neq\beta\in\mathfrak{b}_h} e^{-\pi c(\mathfrak{b}_h)xm(\beta,\mathbf{v})}d\mathbf{v}\right)dx \\ & + \frac{n}{2}\sum_{h=1}^{h_K}\int_1^\infty x^{n(1-s)/2-1}\left(\int_V \sum_{0\neq\beta'\in\mathfrak{b}_h'} e^{-\pi c(\mathfrak{b}_h')xm(\beta',-\mathbf{v})}d\mathbf{v}\right)dx \end{aligned}$$

where $\mathfrak{b}_1, \ldots, \mathfrak{b}_{h_K}$ are suitable fractional ideals. Noting that V is compact, we see from the proof of the Hecke theta formula that

$$\int_V \sum_{0\neq\beta\in\mathfrak{b}_h} e^{-\pi c(\mathfrak{b}_h)xm(\beta,\mathbf{v})}d\mathbf{v} = O(e^{-\delta_h x})$$

for some constant $\delta_h > 0$, and similarly with $\mathfrak{b}_h'$. Then

$$f(s) \ll 1 + |s(s-1)|\left(\int_1^\infty x^{n\sigma/2-1}e^{-\delta x}\,dx + \int_1^\infty x^{n(1-\sigma)/2-1}e^{-\delta x}\,dx\right)$$

for some constant $\delta > 0$. Now

$$\int_1^\infty x^{n\sigma-1}e^{-\delta x}\,dx = \int_\delta^\infty \delta^{1-n\sigma}u^{n\sigma-1}e^{-u}\delta^{-1}\,du$$
$$\leq \int_0^\infty \delta^{1-n\sigma}u^{n\sigma-1}e^{-u}\delta^{-1}\,du = \delta^{-n\sigma}\Gamma(n\sigma)$$

for $\sigma > 0$, and so $f(s)$ is an entire function of order at most 1. Then

$$f(s) = e^{A+Bs}\prod_\rho\left(1-\frac{s}{\rho}\right)e^{s/\rho}$$

by the Hadamard factorization theorem. Logarithmic differentiation yields

$$\frac{f'(s)}{f(s)} = B + \sum_\rho\left(\frac{1}{s-\rho}+\frac{1}{\rho}\right),$$

but also

$$\frac{f'(s)}{f(s)} + \frac{f'(1-s)}{f(1-s)} = 0$$

by the functional equation. Hence

$$B + B + \sum_\rho\left(\frac{1}{1-\rho}+\frac{1}{\rho}\right) = 0$$

on substituting $s = 0$. Thus

$$\frac{f'(s)}{f(s)} = \sum_\rho\left(\frac{1}{s-\rho}+\frac{1}{\rho}-\frac{1}{2}\frac{1}{1-\rho}-\frac{1}{2\rho}\right) = \sum_\rho\left(\frac{1}{s-\rho}+\frac{1}{2\rho}-\frac{1}{2}\frac{1}{1-\rho}\right)$$

and so

$$2\frac{f'(s)}{f(s)} = \sum_\rho\left(\frac{1}{s-\rho}+\frac{1}{s-(1-\rho)}\right)$$

since $1-\rho$ runs through the zeros of $f(s)$ as ρ runs through the zeros, by the functional equation. But $\overline{\rho}$ also runs through the zeros of $f(s)$ as ρ does, since $f(s)$ is real on the real axis. Then

$$4\frac{f'(s)}{f(s)} = \sum_\rho\left(\frac{1}{s-\rho}+\frac{1}{s-(1-\rho)}+\frac{1}{s-\overline{\rho}}+\frac{1}{s-(1-\overline{\rho})}\right)$$

and so

$$2\frac{f'(s)}{f(s)} = \sum_\rho\left(\frac{1}{s-\rho}+\frac{1}{s-\overline{\rho}}\right) = \sum_\rho \frac{2s-2\beta}{(s-\beta)^2+\gamma^2} \geq 0$$

with $\rho = \beta + i\gamma$ and $s > 1$ real. For $\beta \leq 1$ by the Euler product formula.

Next

$$\frac{\zeta_K'(s)}{\zeta_K(s)} = -\sum_{\mathfrak{p}}\frac{\log(N(\mathfrak{p}))N(\mathfrak{p})^{-s}}{1-N(\mathfrak{p})^{-s}} < 0,$$

for $s > 1$ real, by differentiating the Euler product formula logarithmically. Hence

$$\frac{1}{s} + \frac{1}{s-1} + \frac{\gamma_K'(s)}{\gamma_K(s)} > 0,$$

thus

$$\frac{1}{s} + \frac{1}{s-1} + \frac{1}{2}\log\left(\frac{|d_K|}{2^{2r_2}\pi^n}\right) + \frac{r_1}{2}\frac{\Gamma'(s/2)}{\Gamma(s/2)} + r_2\frac{\Gamma'(s)}{\Gamma(s)} > 0,$$

and so

$$\log|d_K| > r_1\left(\log(\pi) - \frac{\Gamma'(s/2)}{\Gamma(s/2)}\right) + 2r_2\left(\log(2\pi) - \frac{\Gamma'(s)}{\Gamma(s)}\right) - 2\left(\frac{1}{s} + \frac{1}{s-1}\right)$$

for $s > 1$ real.

Proposition 9.12 yields the inequality

$$\begin{aligned}\frac{\Gamma'(b)}{\Gamma(b)} - \frac{\Gamma'(a)}{\Gamma(a)} &= \int_0^\infty \frac{e^{-au} - e^{-bu}}{1 - e^{-u}}\,du \le \int_0^\infty \frac{1}{\frac{u}{1+u}}\left(\int_{au}^{bu} e^{-v}\,dv\right)du \\ &= \int_0^\infty e^{-v}\left(\int_{v/b}^{v/a}\left(1 + \frac{1}{u}\right)du\right)dv = \log\left(\frac{b}{a}\right) + \frac{1}{a} - \frac{1}{b}\end{aligned}$$

for $a < b$. Then

$$\begin{aligned}\log|d_K| > {} & r_1\left(\log(\pi) - \frac{\Gamma'(1/2)}{\Gamma(1/2)}\right) \\ & + 2r_2\left(\log(2\pi) - \frac{\Gamma'(1)}{\Gamma(1)}\right) - 2\left(\frac{1}{s} + \frac{1}{s-1}\right) \\ & - r_1\left(\log\left(\frac{s/2}{1/2}\right) + \frac{1}{1/2} - \frac{1}{s/2}\right) - 2r_2\left(\log\left(\frac{s}{1}\right) + \frac{1}{1} - \frac{1}{s}\right) \\ \ge {} & r_1(3.108) + 2r_2(2.415) - 2\left(\frac{1}{s} + \frac{1}{s-1}\right) - n\log(s) - 2n(1-s),\end{aligned}$$

and choosing $s = 1 + n^{-1/2}$, we obtain

$$\begin{aligned}\frac{1}{n}\log|d_K| &> \frac{r_1}{n}3.108 + \frac{2r_2}{n}2.415 - 4n^{-1/2} - \frac{2n^{-1/2}}{n^{1/2}+1} - \log(1 + n^{-1/2}) \\ &> \frac{r_1}{n}3.108 + \frac{2r_2}{n}2.415 - 6n^{-1/2}\end{aligned}$$

for $n \ge 2$. □

This inequality is vacuous for small degrees n, because the right-hand side is not positive then. But it is quite satisfactory for large degrees. The first inequality of this kind is due to Minkowski, and has the constants 2 and $2 + \log(\pi/4) = 1.758$ instead of 3.108 and 2.415 respectively.

The following result plays an important role in some of the proofs of the Kronecker-Weber theorem to the effect that every finite extension of $\mathbb{Q}$ with abelian Galois group is contained in a cyclotomic extension. There are numerous counterexamples to its analogue for relative extensions K/k.

Proposition 9.14 (Theorem of Minkowski). *Each proper finite extension $K/\mathbb{Q}$ ramifies.*

Proof. We have to show that some rational prime ramifies in K, and for this it will be sufficient (and necessary), by the Dedekind discriminant theorem, to establish the inequality $|d_K| > 1$ for $n \geq 2$. The inequality in Proposition 9.13 is not good enough for small degrees for this purpose, so we must establish another inequality.

Choose $s = 3$ in

$$\begin{aligned}\log|d_K| > r_1\Big(\log(\pi) - \frac{\Gamma'(s/2)}{\Gamma(s/2)}\Big) \\ + 2r_2\Big(\log(2\pi) - \frac{\Gamma'(s)}{\Gamma(s)}\Big) - 2\Big(\frac{1}{s} + \frac{1}{s-1}\Big)\end{aligned}$$

to yield

$$\begin{aligned}\frac{1}{n}\log|d_K| &> \frac{r_1}{n}\Big(\log(\pi) - \frac{\Gamma'(3/2)}{\Gamma(3/2)}\Big) \\ &\quad + \frac{2r_2}{n}\Big(\log(2\pi) - \frac{\Gamma'(3)}{\Gamma(3)}\Big) - \frac{2}{n}\Big(\frac{1}{3} + \frac{1}{2}\Big) \\ &> \frac{r_1}{n}1.108 + \frac{2r_2}{n}0.915 - \frac{5}{3n} > 0.915 - \frac{5}{6} > 0.08\end{aligned}$$

for $n \geq 2$. □

It may be worth remarking that we cannot do without the Hadamard factorization theorem in the proof of Proposition 9.13. The function

$$F(s) = e^{-(s-1/2)^2}$$

satisfies the functional equation $F(1-s) = F(s)$. Its logarithmic derivative

$$\frac{F'(s)}{F(s)} = 1 - 2s$$

is negative for $s > 1$, but this entire function is of order 2. The use of the Mellin transform identity to show that the order of $f(s)$ is no larger than 1 is not so critical. This could also be done by means of the Phragmén-Lindelöf principle, for example.

9.5. ⋆ The Prime Ideal Theorem

The problem of counting primes $p \leq x$ leads to the Prime Number Theorem, and the problem of counting prime ideals $\mathfrak{p}$ in $\mathcal{O}_K$ for which $N(\mathfrak{p}) \leq x$ leads to a generalization of it called the Prime Ideal Theorem of Landau. In its weakest form this states that

$$\sum_{N(\mathfrak{p})\leq x} 1 \sim \frac{x}{\log(x)},$$

but the result may also be established in a more precise form with an error term. We shall establish the Prime Ideal Theorem in its weakest form, using the following theorem of analysis.

Proposition 9.15 (Theorem of Ikehara). *Suppose that $A(u)$ is a nonnegative and nondecreasing real function on $[0, \infty)$. Further assume that the Laplace transform*

$$L(s) = \int_0^\infty A(u)e^{-su}\,du$$

exists for $\sigma > 1$. If for some constant C the holomorphic function

$$L(s) - \frac{C}{s-1}$$

extends to a holomorphic function on a domain containing $\sigma \geq 1$, then

$$A(u)e^{-u} \to C$$

as $u \to +\infty$.

To appreciate the convenience of using the Ikehara theorem, we apply it with $A(u) = \psi(e^u)$. Then

$$\begin{aligned} L(s) &= \int_0^\infty \psi(e^u)e^{-su}\,du = \int_0^\infty \left(\sum_{n\leq e^u} \Lambda(n)\right) e^{-su}\,du \\ &= \sum_{n=1}^\infty \Lambda(n) \int_{\log(n)}^\infty e^{-su}\,du = \sum_{n=1}^\infty \Lambda(n)\frac{n^{-s}}{s} = -\frac{1}{s}\frac{\zeta'(s)}{\zeta(s)}, \end{aligned}$$

and so

$$L(s) - \frac{1}{s-1} = -\frac{1}{s}\frac{\zeta'(s)}{\zeta(s)} - \frac{1}{s-1}$$

has a holomorphic extension to $\sigma \geq 1$. For $-\zeta'(s)/\zeta(s)$ has a simple pole at $s = 1$ with residue 1 and no other singularities on $\sigma = 1$, since $\zeta(s)$ has no zeros on this line. Thus $\psi(e^u) \sim e^u$ by the Ikehara theorem, yielding a form of the Prime Number Theorem. It is very striking that unlike the proof in Chapter 6 this proof of the PNT does not require any bound on the growth of $\zeta(s)$ as $t \to \pm\infty$.

Proposition 9.16 (Prime Ideal Theorem of Landau). *If K is a number field then*

$$\sum_{N(\mathfrak{p})\le x} 1 \sim \frac{x}{\log(x)}$$

as $x \to +\infty$. Here $\mathfrak{p}$ ranges over the prime ideals of $\mathcal{O}_K$.

Proof. We use the Ikehara theorem, and begin by noting that

$$-\frac{\zeta_K'(s)}{\zeta_K(s)} = \sum_{k=1}^{\infty}\sum_{\mathfrak{p}} \log(N(\mathfrak{p}))N(\mathfrak{p})^{-ks} = \sum_{\mathfrak{a}} \Lambda_K(\mathfrak{a})N(\mathfrak{a})^{-s}$$

where

$$\Lambda_K(\mathfrak{a}) = \begin{cases} \log(N(\mathfrak{p})) & \text{if } \mathfrak{a} = \mathfrak{p}^k, \\ 0 & \text{otherwise,} \end{cases}$$

and

$$\psi_K(u) = \sum_{N(\mathfrak{a})\le u} \Lambda_K(\mathfrak{a}).$$

Clearly ψ_K is a nonnegative and nondecreasing function. It is piecewise constant, thus Riemann integrable, on bounded intervals. It remains to see that the improper integral giving the Laplace transform converges for $\sigma > 1$. This is easiest seen by carrying through the calculation

$$\begin{aligned} L(s) &= \int_0^{\infty} \psi_K(e^u)e^{-su}\,du = \int_0^{\infty} \left(\sum_{N(\mathfrak{a})\le e^u} \Lambda_K(\mathfrak{a}) \right) e^{-su}\,du \\ &= \sum_{\mathfrak{a}} \Lambda_K(\mathfrak{a}) \int_{\log(N(\mathfrak{a}))}^{\infty} e^{-su}\,du = \sum_{\mathfrak{a}} \Lambda_K(\mathfrak{a})\frac{N(\mathfrak{a})^{-s}}{s} = -\frac{1}{s}\frac{\zeta_K'(s)}{\zeta_K(s)}, \end{aligned}$$

first for $s > 1$ real by nonnegativity of the integrand, and then concluding that $L(s)$ exists for $\sigma > 1$ by absolute convergence.

The function $-\zeta_K'(s)/\zeta_K(s)$ has a simple pole with residue equal to 1 at $s = 1$, and is otherwise holomorphic except where $\zeta_K(s)$ is zero. To show that

$$L(s) - \frac{1}{s-1} = -\frac{1}{s}\frac{\zeta_K'(s)}{\zeta_K(s)} - \frac{1}{s-1}$$

extends to a holomorphic function on a domain containing $\sigma \ge 1$ it will be enough to show that $\zeta_K(s)$ has no zeros on the line $\sigma = 1$. Now

$$\log(|\zeta_K(\sigma + it)|) = \sum_{k=1}^{\infty}\sum_{\mathfrak{p}} N(\mathfrak{p})^{-k\sigma} \cos(kt\log(N(\mathfrak{p})))$$

by the Euler product, and thus

$$\log(|\zeta_K(\sigma)|^3|\zeta_K(\sigma+it)|^4|\zeta_K(\sigma+2it)|)$$
$$=\sum_{k=1}^{\infty}\sum_{\mathfrak{p}} N(\mathfrak{p})^{-k\sigma}(3+4\cos(kt\log(N(\mathfrak{p}))+\cos(2kt\log(N(\mathfrak{p})))) \geq 0$$

by the trigonometric inequality $3+4\cos(\theta)+\cos(2\theta)\geq 0$ from Section 6.1. If $\zeta_K(s)$ had a zero at $s=1+it$ then $\zeta_K(\sigma)^3\zeta_K(\sigma+it)^4$ would tend to zero as $\sigma\to 1^+$, which would force $\zeta_K(s)$ to have a singularity at $s=1+2it$ by the inequality

$$|\zeta_K(\sigma)|^3|\zeta_K(\sigma+it)|^4|\zeta_K(\sigma+2it)|\geq 1.$$

Since this is impossible, the Dedekind zeta function has no zeros on the line $\sigma=1$. The conditions to apply the Ikehara theorem have now been verified, so we conclude that $\psi_K(e^u)\sim e^u$. This is a form of the Prime Ideal Theorem, but not the form that we want.

Define

$$\vartheta_K(x)=\sum_{N(\mathfrak{p})\leq x}\log(N(\mathfrak{p}))$$

and note that

$$\psi_K(x)=\sum_{N(\mathfrak{p}^k)\leq x}\log(N(\mathfrak{p}))=\sum_{k=1}^{\infty}\vartheta_K(x^{1/k})$$

so that

$$\psi_K(x)-2\psi_K(x^{1/2})=\sum_{k=1}^{\infty}(-1)^{k+1}\vartheta_K(x^{1/k})\leq\vartheta_K(x)\leq\psi_K(x).$$

This yields $\vartheta_K(x)=\psi_K(x)+O(x^{1/2})$ because $\psi_K(x)=O(x)$. At last

$$\begin{aligned}\sum_{N(\mathfrak{p})\leq x}1&=\sum_{2\leq m\leq x}\sum_{N(\mathfrak{p})=m}1=\sum_{2\leq m\leq x}\frac{1}{\log(m)}\sum_{N(\mathfrak{p})=m}\log(N(\mathfrak{p}))\\&=\frac{1}{\log(x)}\sum_{m\leq x}\sum_{N(\mathfrak{p})=m}\log(N(\mathfrak{p}))\\&\quad-\int_2^x\left(\frac{1}{\log(u)}\right)'\left(\sum_{m\leq u}\sum_{N(\mathfrak{p})=m}\log(N(\mathfrak{p}))\right)du\\&=\frac{\vartheta_K(x)}{\log(x)}+\int_2^x\frac{\vartheta_K(u)}{u\log^2(u)}\,du\\&=\frac{\psi_K(x)}{\log(x)}+O\left(\frac{x^{1/2}}{\log(x)}\right)+O\left(\int_2^x\frac{du}{\log^2(u)}\right)\sim\frac{x}{\log(x)}\end{aligned}$$

by partial summation. □

Specializing the Prime Ideal Theorem to the case $K = \mathbb{Q}$ yields the version

$$\pi(x) \sim \frac{x}{\log(x)}$$

of the Prime Number Theorem. Specializing to the case $K = \mathbb{Q}(\sqrt{-1})$ yields the statement that

$$\sum_{\substack{p \le x \\ p \equiv 1 \pmod 4}} 1 \sim \frac{1}{2}\frac{x}{\log(x)}.$$

Apart from $(1+i)$ the prime ideals of $\mathcal{O}_{\mathbb{Q}(\sqrt{-1})}$ are of the form (π) with $\pi\overline{\pi} = p \equiv 1 \pmod 4$ or (p) with $p \equiv 3 \pmod 4$. The norm of (p) is $N((p)) = p^2$ so there are only $O(\sqrt{x})$ of those ideals with norm less than or equal to x. Since there are two distinct prime ideals (π) and $(\overline{\pi})$ associated to each rational prime $p \equiv 1 \pmod 4$, the statement follows, and then

$$\sum_{\substack{p \le x \\ p \equiv 3 \pmod 4}} 1 \sim \frac{1}{2}\frac{x}{\log(x)}$$

by comparison with the Prime Number Theorem.

The cubic field $K = \mathbb{Q}(\sqrt[3]{2})$ has the ring of integers $\mathbb{Z}[\sqrt[3]{2}]$ and class number $h_K = 1$. A norm form is

$$F(X,Y,Z) = N_K(X + Y\sqrt[3]{2} + Z(\sqrt[3]{2})^2) = X^3 + 2Y^3 + 4Z^3 - 6XYZ,$$

and one can use the Prime Ideal Theorem to estimate the number of primes represented by this integral cubic form. Suppose that p is a rational prime that does not ramify in K. If it remains inert, then $N((p)) = p^3$ and these prime ideals contribute only $O(x^{1/3})$ to the sum on the left-hand side in the Prime Ideal Theorem. The primes that decompose into the product of a prime ideal of degree one and one of degree two contribute $N_1(x) + O(x^{1/2})$ where $N_1(x)$ is the contribution from the prime ideals of degree one; those primes that decompose into a product of three prime ideals make a contribution of $N_2(x)$ prime ideals of degree one, where

$$N_1(x) + N_2(x) \sim \frac{x}{\log(x)}$$

by the Prime Ideal Theorem. Now $\mathbb{Z}[\sqrt[3]{2}]$ is a PID and so all the prime ideals discussed above are principal ideals (π), with $N((\pi)) = |N_K(\pi)|$. Then

$$\sum_{\substack{p \le x \\ p \in \mathcal{V}}} 1 = N_1(x) + \frac{1}{3}N_2(x) \gg \frac{x}{\log(x)}$$

where $\mathcal{V}$ is the set of integer values of the norm form $F(X,Y,Z)$. Note the crucial importance for this argument of the fact that $h_K = 1$.

This is only moderately interesting, because the values of $F(X,Y,Z)$ are quite dense, their number in $[1,x]$ growing linearly with x. Moreover every rational prime $p>3$ with $p\equiv 2 \pmod 3$ decomposes as a product of a prime ideal of the first degree and one of the second degree, so to bound the number of prime values of $F(X,Y,Z)$ from below we could instead use information about the distribution of primes congruent to 2 modulo 3. Obtaining this information from the Prime Ideal Theorem is a simple exercise.

The set of values of the incomplete norm form

$$F(X,Y,0)=X^3+2Y^3$$

is considerably sparser, the number of values in $[1,x]$ being $O(x^{2/3})$. Heath-Brown showed that this cubic form takes infinitely many prime values for positive integer values of X and Y, and thus answered a question of Hardy and Littlewood dating to the 1920s: Can infinitely many primes be written as a sum of three nonnegative cubes?

In his 1903 proof of the Prime Ideal Theorem, Landau used an analytic continuation of the Dedekind zeta function to the left of the line $\sigma=1$. At that time the functional equation and full analytic continuation of $\zeta_K(s)$ were known only in special cases. Landau instead used an estimate

$$\sum_{m\le x} a_K(m)=h_K\rho_K x+O(x^{1-1/n_K})$$

for the summatory function of the ideal counting function. This estimate was obtained by Dedekind and Weber using a geometric method. The constant ρ_K is given as

$$\rho_K=\frac{2^{r_1(K)+r_2(K)}\pi^{r_2(K)}R_K}{w_K|d_K|^{1/2}}$$

in terms of arithmetic data. Clearly the Dedekind-Weber estimate yields an analytic continuation of $\zeta_K(s)$ to $\sigma>1-1/n_K$ except for the simple pole at $s=1$.

The exponent in the error term in the Dedekind-Weber estimate may be improved to $1-2/(n_K+1)$ by the method of contour integrals. The argument is due to Landau and requires the functional equation of the Dedekind zeta function, but also a bound for a certain nonelementary integral with an oscillatory integrand. The resulting proof is at the same level of difficulty as the contour integration argument for the error term $O(x^{1/3}\log(x))$ in the Dirichlet divisor problem, and so we do not pursue it here. Instead we use the Ikehara theorem to prove a weaker version of the Dedekind-Weber result.

Proposition 9.17. *The estimate*

$$\sum_{\substack{N(\mathfrak{a})\leq x\\ \mathfrak{a}\in C}} 1 \sim \rho_K x$$

holds for each ideal class C.

Proof. The Laplace transform

$$\begin{aligned} L(s) &= \int_0^\infty \left(\sum_{\substack{N(\mathfrak{a})\leq e^u\\ \mathfrak{a}\in C}} 1 \right) e^{-su}\, du = \sum_{\mathfrak{a}\in C} \int_{\log(N(\mathfrak{a}))}^\infty e^{-su}\, du \\ &= \sum_{\mathfrak{a}\in C} \frac{N(\mathfrak{a})^{-s}}{s} = \frac{\zeta_K(s,C)}{s} \end{aligned}$$

exists for $\sigma > 1$ and the sum under the integral sign is nonnegative and nondecreasing. The partial zeta function $\zeta_K(s,C)$ is holomorphic everywhere except for a simple pole at $s = 1$ with residue ρ_K, so

$$L(s) - \frac{\rho_K}{s-1} = \frac{\zeta_K(s,C)}{s} - \frac{\rho_K}{s-1}$$

has a holomorphic extension to a domain containing $\sigma \geq 1$. Thus the conditions to apply the Ikehara theorem are satisfied, and the estimate follows. □

The main implication of this result is that in the sense of asymptotic density ideals are equally distributed among classes. Summing over all classes gives a weak version of the Dedekind-Weber estimate. This is inadequate for analytic continuation, but we have the full analytic continuation of $\zeta_K(s)$ already.

Einführung in die elementare und analytische Theorie der algebraischen Zahlen und der Ideale by Edmund Landau and *Elementary and Analytic Theory of Algebraic Numbers* by W. Narkiewicz have proofs of the Prime Ideal Theorem with good bounds for the error term.

The proof that $X^3 + 2Y^3$ represents infinitely many primes for X and Y positive integers requires sieve theory, and is more difficult than anything we do in this book. There is an exposition of this result in *Prime-Detecting Sieves* by Glyn Harman.

Some of the many textbooks of algebraic number theory that establish the finiteness of the class number and the Dirichlet unit theorem by the geometric approach also include a proof of the Dedekind-Weber estimate. This is true of *Einführung in die elementare und analytische Theorie der algebraischen Zahlen und der Ideale* by Edmund Landau, *Algebraic Number*

Theory by Serge Lang, *Algebraic Number Theory* by Robert L. Long, *Number Fields* by Daniel A. Marcus, the second edition of *Problems in Algebraic Number Theory* by M. Ram Murty and Jody Esmonde, and *Algebraic Theory of Numbers* by Hermann Weyl. Landau in addition establishes an improved error term in the Dedekind-Weber estimate by contour integration, and also shows that the exponent in the error term cannot be improved beyond $1/2 - 1/(2n_K)$. Lang presents an analytic argument leading to a very good bound for the error term conditional on the generalized Riemann hypothesis for the Dedekind zeta function. The paper *Counting integral ideals in a number field* by M. Ram Murty and Jeanine Van Order has a more extensive and precise treatment of the Dedekind-Weber estimate by the geometric method than one can find in textbooks.

9.6. ⋆ A proof of the Ikehara theorem

The Ikehara theorem was originally proved by means of Fourier transforms. We give a more recent proof due to J. Korevaar that relies on complex analysis, building on earlier work of D. J. Newman. This proof is not easier than a real analysis proof, but it is better adapted to the prerequisites assumed for this text.

Proof. We define a function

$$\rho(u) = e^{-u}A(u) - C$$

and calculate its Laplace transform

$$G(z) = \int_0^\infty \rho(u)e^{-zu}\,du = L(z+1) - \frac{C}{z}.$$

This Laplace transform exists for $\mathrm{Re}(z) > 0$ by assumption.

Define for each $\varepsilon > 0$ the function $\rho_\varepsilon(u) = e^{-\varepsilon u}\rho(u)$. Since the Laplace transform

$$\begin{aligned}\int_0^\infty A(u)e^{-(1+\varepsilon)u}\,du &\geq \int_T^\infty A(u)e^{-(1+\varepsilon)u}\,du \\ &\geq \int_T^\infty A(T)e^{-(1+\varepsilon)u}\,du = \frac{1}{1+\varepsilon}A(T)e^{-(1+\varepsilon)T}\end{aligned}$$

exists for each $\varepsilon > 0$ by assumption, the function $A(u)e^{-(1+\varepsilon)u}$ cannot be unbounded, and so $\rho_\varepsilon(u)$ is bounded.

We shall now show that $\rho(u)$ itself is bounded. For this purpose we apply an inequality for an integral of a function in terms of its Laplace transform. This inequality is the heart of the proof, where most of the work lies, and we delay its proof until later.

Let $h(u)$ be a function on $[0,\infty)$ that is integrable on closed bounded intervals with $|h(u)| \le M$ there. The Laplace transform

$$H(z) = \int_0^\infty h(u)e^{-zu}\,du$$

exists and is holomorphic on $\mathrm{Re}(z) > 0$. Suppose that for $-R \le y \le R$ we have uniform convergence

$$H(x+iy) \to H(iy)$$

as $x \to 0^+$ to a limit function $H(iy)$. Then

$$\left|\int_T^{T+\delta} h(u)\,du\right| \le \frac{4M}{R} + \frac{1}{2\pi}\left|\int_{-R}^{R} H(iy)\frac{e^{i\delta y}-1}{y}\left(1-\frac{y^2}{R^2}\right)e^{iTy}\,dy\right|$$

for any positive T and δ. We defer the proof of this inequality.

If $\rho(u)$ is never positive, it is automatically bounded, so if we define

$$M_\varepsilon = \sup_{u\ge 0}\rho_\varepsilon(u) = \sup_{u\ge 0} e^{-\varepsilon u}\rho(u),$$

we may assume that there is an ε_0 such that $M_\varepsilon > 0$ for $0 < \varepsilon \le \varepsilon_0 \le 1$. From now on we suppose that $0 < \varepsilon \le \varepsilon_0$. Our assumptions on $A(u)$ imply that the above inequality may be applied to $\rho_\varepsilon(u)$ to yield

$$\left|\int_T^{T+1} \rho_\varepsilon(u)\,du\right| \le \frac{4M_\varepsilon}{R} + \frac{1}{2\pi}\left|\int_{-R}^{R} G(\varepsilon+iy)\frac{e^{iy}-1}{y}\left(1-\frac{y^2}{R^2}\right)e^{iTy}\,dy\right|$$

on choosing $\delta = 1$. For each ε there is some $T_\varepsilon > 0$ so that $\rho_\varepsilon(T_\varepsilon) \ge M_\varepsilon/2$. The assumptions on $A(u)$ and $L(s)$ imply that C is nonnegative. Then

$$\rho_\epsilon(u) = e^{-(1+\varepsilon)u}A(u) - e^{-\varepsilon u}C \ge e^{(1+\varepsilon)(T+1)}A(T) - C = e^{-(1+\varepsilon)}\rho_\varepsilon(T) - C$$

for $T \le u \le T+1$. Choosing $T = T_\varepsilon$ yields

$$e^{-(1+\epsilon)}\frac{M_\varepsilon}{2} - C \le \frac{4M_\varepsilon}{R} + \frac{1}{2\pi}\left|\int_{-R}^{R} G(\varepsilon+iy)\frac{e^{iy}-1}{y}\left(1-\frac{y^2}{R^2}\right)e^{iTy}\,dy\right|$$

and so, recalling that $\varepsilon \le 1$,

$$\left(\frac{e^{-2}}{2} - \frac{4}{R}\right)M_\varepsilon \le C + \frac{|I_R|}{2\pi},$$

where I_R is the integral under the absolute value sign. Choosing $R = 16e^2$ we obtain an upper bound for M_ε for $0 < \varepsilon \le \varepsilon_0$ that is independent of ε, and thus $\rho(u)$ is bounded.

Supposing $|\rho(u)| \le M$ the above inequality yields

$$\left|\int_T^{T+\delta} \rho(u)\,du\right| \le \frac{4M}{R} + \frac{1}{2\pi}\left|\int_{-R}^{R} J_{\delta,R}(y)e^{iTy}\,dy\right|$$

with

$$J_{\delta,R}(y) = G(iy)\frac{e^{i\delta y}}{y}\left(1 - \frac{y^2}{R^2}\right).$$

Note that $J_{\delta,R}(y)$ is continuously differentiable on the whole real line by the assumption that $L(s)$ has an analytic continuation. For any $\varepsilon > 0$ we may choose R so large that $4M/R < \varepsilon/2$. For such R and fixed $\delta > 0$

$$\int_{-R}^{R} J_{\delta,R}(y)e^{iTy}\,dy = J_{\delta,R}(y)\frac{e^{iTy}}{iT}\bigg|_{y=-R}^{y=R} - \int_{-R}^{R} J'_{\delta,R}(y)\frac{e^{iTy}}{iT}\,dy$$

implies that

$$\int_{-R}^{R} J_{\delta,R}(y)e^{iTy}\,dy \to 0$$

as $T \to +\infty$. So there is some T_0 such that

$$\frac{1}{2\pi}\left|\int_{-R}^{R} J_{\delta,R}(y)e^{iTy}\,dy\right| < \frac{\varepsilon}{2}$$

for $t \geq T_0$. Thus

$$\int_{T}^{T+\delta} \rho(u)\,du \to 0$$

as $T \to +\infty$, for any fixed $\delta > 0$. The inequality

$$\begin{aligned}\rho(u) - \rho(T) &= e^{-u}A(u) - e^{-T}A(T) \geq e^{-u}A(T) - e^{-T}A(T)\\ &= -(1 - e^{-(u-T)})(\rho(T) + C) \geq -(u - T)(\rho(T) + C)\end{aligned}$$

holds for $u \geq T$ and implies that

$$\int_{T}^{T+\delta} \rho(u)\,du - \delta\rho(T) \geq -\int_{T}^{T+\delta} (u - T)(\rho(T) + C)\,du = -\frac{\delta^2}{2}(\rho(T) + C).$$

This yields

$$\left(\delta - \frac{\delta^2}{2}\right)\rho(T) \leq \frac{C\delta^2}{2} + \int_{T}^{T+\delta} \rho(u)\,du,$$

and so

$$\rho(T) \leq \frac{C\delta}{2 - \delta} + \frac{1}{\delta - \frac{\delta^2}{2}}\int_{T}^{T+\delta} \rho(u)\,du,$$

for $0 < \delta \leq 1$. For arbitrary $\epsilon > 0$ we may choose δ so small that $C\delta/(2-\delta) < \varepsilon/2$. Then there is some T_1 so that

$$\frac{1}{\delta - \frac{\delta^2}{2}}\int_{T}^{T+\delta} \rho(u)\,du < \frac{\varepsilon}{2}$$

for $T \geq T_1$. Thus $\rho(T) < \varepsilon$ for $T \geq T_1$, and repeating the same argument on $[T - \delta, T]$ instead of $[T, T + \delta]$ we see that there is some T_2 such that $\rho(T) > -\epsilon$ for $T \geq T_2$. Hence $\rho(T) \to 0$ as $T \to +\infty$. □

Modifying this proof by substituting the Riemann-Lebesgue Lemma for the integration by parts, the assumption on the boundary behavior may be weakened. For example to $L(s) - C/(s-1)$ extending to a continuous function on $\sigma \geq 1$.

It remains to establish the inequality that we used to prove Ikehara's theorem.

Proposition 9.18. *Let $h(u)$ be a function on $[0, \infty)$ that is integrable on closed bounded intervals with $|h(u)| \leq M$ there. The Laplace transform*

$$H(z) = \int_0^\infty h(u)e^{-zu}\,du$$

exists and is holomorphic on $\mathrm{Re}(z) > 0$. Suppose that for $-R \leq y \leq R$ we have uniform convergence

$$H(x+iy) \to H(iy)$$

as $x \to 0^+$ to a limit function $H(iy)$. Then

$$\left|\int_T^{T+\delta} h(u)\,du\right| \leq \frac{4M}{R} + \frac{1}{2\pi}\left|\int_{-R}^{R} H(iy)\frac{e^{i\delta y}-1}{y}\left(1-\frac{y^2}{R^2}\right)e^{iTy}\,dy\right|$$

for any positive T and δ.

Proof. We have

$$\int_T^{T+\delta} h(u)\,du = H_{T+\delta}(0) - H_T(0)$$

where

$$H_T(z) = \int_0^T h(y)e^{-zu}\,du.$$

By the Residue Theorem

$$2\pi i H_T(0) = \int_\Gamma H_T(z)e^{Tz}\left(\frac{1}{z}+\frac{z}{R^2}\right)dz,$$

with Γ taken as the circle $|z| = R$ traversed in the positive direction. The factor

$$e^{Tz}\left(\frac{1}{z}+\frac{z}{R^2}\right) \quad \text{rather than just} \quad \frac{1}{z}$$

was introduced by Newman and is included in the integrand to enable efficient estimates.

Our assumptions on the Laplace transform yield information about $H(0)$ rather than $H_T(0)$. But we can hope to obtain information about $H_T(0)$ by bounding the difference $H_T(0) - H(0)$. For this purpose it is convenient also to express $H(0)$ by means of a contour integral. But then we cannot integrate over Γ, since $H(z)$ may not be holomorphic to the left of $\mathrm{Re}(z) > 0$.

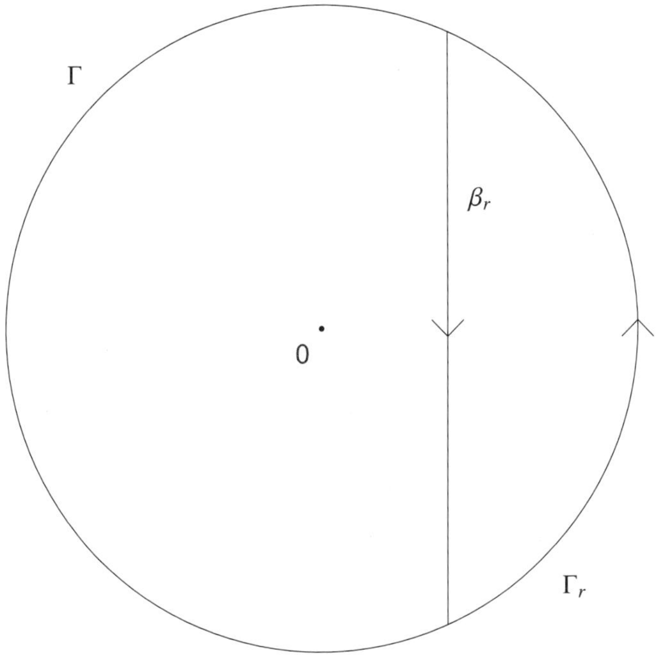

Figure 5. Contours in the proof of the Ikehara theorem

We introduce some paths of integration that will facilitate the estimates needed. Let Γ_r be the part of Γ lying in the half plane $\mathrm{Re}(z) \geq r$, and let β_r denote the downwards oriented vertical line segment through the point r on the real axis, and dividing the circular disk. Then $\Gamma_r + \beta_r$ is a closed contour traversed in the positive direction. Denote the part of Γ that lies in $\mathrm{Re}(z) \geq 0$ by Γ^+ and the rest of Γ by Γ^-. See Figure 5.

Next we use the paths of integration we have defined to split $H_T(0)$ into a sum of three terms, each of which we bound separately.

$$\begin{aligned}
2\pi i H_T(0) &= \int_\Gamma H_T(z)e^{Tz}\left(\frac{1}{z} + \frac{z}{R^2}\right)dz - \int_{\Gamma_r+\beta_r} H(z)e^{Tz}\left(\frac{1}{z} + \frac{z}{R^2}\right)dz \\
&= \int_{\Gamma_r} (H_T(z) - H(z))e^{Tz}\left(\frac{1}{z} + \frac{z}{R^2}\right)dz \\
&\quad + \int_{\Gamma-\Gamma_r} H_T(z)e^{Tz}\left(\frac{1}{z} + \frac{z}{R^2}\right)dz \\
&\quad - \int_{\beta_r} H(z)e^{Tz}\left(\frac{1}{z} + \frac{z}{R^2}\right)dz \\
&= I_1(R, r, T) + I_2(R, r, T) - I_3(R, r, T),
\end{aligned}$$

by Cauchy's theorem. This expression for $H_T(0)$ will be used to bound the difference $H_{T+\delta}(0) - H_T(0)$. First

$$|H_T(z) - H(z)| = \left|\int_T^\infty h(u)e^{-zu}\,du\right| \le \frac{M}{\mathrm{Re}(z)}e^{-T\mathrm{Re}(z)}$$

for $\mathrm{Re}(z) > 0$, and so

$$\begin{aligned}
|I_1(R,r,T)| &= \left|\int_{\Gamma_r}(H_T(z)-H(z))e^{Tz}\left(\frac{1}{z}+\frac{z}{R^2}\right)dz\right| \\
&\le \int_{\Gamma_r}|H_T(z)-H(z)||e^{Tz}|\left|\frac{1}{z}+\frac{z}{R^2}\right||dz| \\
&\le \int_{\Gamma_r}\frac{M}{\mathrm{Re}(z)}e^{-T\mathrm{Re}(z)}e^{T\mathrm{Re}(z)}\left|\frac{2\mathrm{Re}(z)}{R^2}\right||dz| \\
&\le \frac{2M}{R^2}\pi R = \frac{2\pi M}{R},
\end{aligned}$$

because the length of Γ_r is at most πR and because

$$\frac{1}{z}+\frac{z}{R^2} = \frac{2\mathrm{Re}(z)}{R^2}$$

on the circle $|z| = R$. Then

$$|I_1(R,0,T+\delta) - I_1(R,0,T)| \le \frac{4\pi M}{R}$$

in particular. Next

$$\begin{aligned}
|H_T(z)| &= \left|\int_0^T h(u)e^{-zu}\,du\right| \le \int_0^T |h(u)|e^{-\mathrm{Re}(z)u}\,du \\
&\le \int_0^T Me^{-\mathrm{Re}(z)u}\,du \le \frac{M}{|\mathrm{Re}(z)|}\max(1, e^{-T\mathrm{Re}(z)})
\end{aligned}$$

for $\mathrm{Re}(z) \ne 0$. Then

$$\begin{aligned}
|I_2(R,r,T)| &= \left|\int_{\Gamma-\Gamma_r} H_T(z)e^{Tz}\left(\frac{1}{z}+\frac{z}{R^2}\right)dz\right| \\
&= \left|\left(\int_{\Gamma^+-\Gamma_r}+\int_{\Gamma^-}\right)H_T(z)e^{Tz}\left(\frac{1}{z}+\frac{z}{R^2}\right)dz\right| \\
&\le \left(\int_{\Gamma^+-\Gamma_r}+\int_{\Gamma^-}\right)|H_T(z)||e^{Tz}|\left|\frac{1}{z}+\frac{z}{R^2}\right||dz| \\
&\le \int_{\Gamma^+-\Gamma_r}\frac{M}{\mathrm{Re}(z)}e^{T\mathrm{Re}(z)}\frac{2\mathrm{Re}(z)}{R^2}|dz| \\
&\quad + \int_{\Gamma^-}\frac{M}{|\mathrm{Re}(z)|}e^{-T\mathrm{Re}(z)}e^{T\mathrm{Re}(z)}\frac{2|\mathrm{Re}(z)|}{R^2}|dz| \\
&\le \frac{2M}{R^2}e^{Tr}2R\arcsin\left(\frac{r}{R}\right) + \frac{2M}{R^2}\pi R
\end{aligned}$$

and so

$$|I_2(R,0,T+\delta) - I_2(R,0,T)| \le \frac{4\pi M}{R}$$

after letting $r \to 0^+$. Next

$$\begin{aligned}
I_3(R, r, T+\delta) &- I_3(R,r,T) \\
&= \int_{\beta_r} H(z)(e^{(T+\delta)z} - e^{Tz})\left(\frac{1}{z} + \frac{z}{R^2}\right) dz \\
&= \int_{\beta_r} H(z)\frac{e^{\delta z}-1}{z}\left(1 + \frac{z^2}{R^2}\right) e^{Tz}\, dz \\
&= -\int_{-\sqrt{R^2-r^2}}^{\sqrt{R^2-r^2}} H(r+iy)\frac{e^{\delta r + i\delta y}-1}{r+iy}\left(1 + \frac{(r+iy)^2}{R^2}\right) e^{Tr+iTy}\, i dy \\
&\to -\int_{-R}^{R} H(iy)\frac{e^{i\delta y}-1}{iy}\left(1 + \frac{(iy)^2}{R^2}\right) e^{iTy}\, i dy \\
&= -\int_{-R}^{R} H(iy)\frac{e^{i\delta y}-1}{y}\left(1 - \frac{y^2}{R^2}\right) e^{iTy}\, dy
\end{aligned}$$

as $r \to 0^+$, because $H(r+iy) \to H(iy)$ uniformly on $-R \le y \le R$. The desired inequality follows by applying the triangle inequality to

$$\begin{aligned}
2\pi i(H_{T+\delta}(0) - H_T(0)) = (I_1 + I_2 - I_3)(R, 0, T+\delta) \\
- (I_1 + I_2 - I_3)(R,0,T)
\end{aligned}$$

and dividing by 2π. □

9.7. Induced representations

Let G be a finite group, and H a subgroup with a representation ρ and representation space $V = V_\rho$. We define a linear space W of functions $f : G \to V$ by requiring that these functions satisfy the condition $f(hg) = \rho(h)f(g)$ for all $g \in G$ and $h \in H$. The values of f on each right coset of H in G are determined by the value of f on a single element of the coset, and thus $\dim(W) = [G : H]\dim(V)$. We define the representation $\dot{\rho} = \mathrm{Ind}_H^G(\rho)$ of G *induced* from ρ to act on W by

$$(\dot{\rho}(g_0)f)(g) = f(gg_0)$$

for $g, g_0 \in G$. It is necessary to check that $\dot{\rho}(g_0)f$ is in W. We calculate

$$(\dot{\rho}(g_0)f)(hg) - f(hgg_0) - \rho(h)f(gg_0) = \rho(h)(\dot{\rho}(g_0)f)(g)$$

to see this. It is also necessary to establish that $\dot{\rho}$ is a homomorphism into $\mathrm{GL}(W)$. Because

$$(\dot{\rho}(g_1 g_2)f)(g) = f(gg_1g_2) = (\dot{\rho}(g_2)f)(gg_1) = (\dot{\rho}(g_1)\dot{\rho}(g_2)f)(g)$$

and

$$f = \dot{\rho}(e)f = \dot{\rho}(g_1 g_1^{-1})f = \dot{\rho}(g_1)\dot{\rho}(g_1^{-1})f,$$

this is clear. Note that $[G : H]\deg(\rho)$ is the degree of the induced representation. When $H = \{1\}$ and $\rho = \rho_0$, the induced representation of G is the (right) regular representation.

Proposition 9.19 (Frobenius induced character formula). *Let χ be the character of a representation ρ of a subgroup H of a group G, and let $g_1, \ldots, g_m$ be a complete collection of right coset representatives for H in G. Then*

$$\mathrm{Ind}_H^G(\chi)(g_0) = \frac{1}{|H|} \sum_{\substack{g \in G \\ g g_0 g^{-1} \in H}} \chi(g g_0 g^{-1}) = \sum_{g_j g_0 g_j^{-1} \in H} \chi(g_j g_0 g_j^{-1})$$

is the character of the induced representation $\mathrm{Ind}_H^G(\rho)$.

Proof. Extend χ to a function on G by setting it equal to zero off H. Then

$$\begin{aligned} \sum_{\substack{g \in G \\ g g_0 g^{-1} \in H}} \chi(g g_0 g^{-1}) &= \sum_{g \in G} \chi(g g_0 g^{-1}) = \sum_j \sum_{h \in H} \chi(h g_j g_0 g_j^{-1} h^{-1}) \\ &= \sum_j \sum_{h \in H} \chi(g_j g_0 g_j^{-1}) = |H| \sum_{g_j g_0 g_j^{-1} \in H} \chi(g_j g_0 g_j^{-1}) \end{aligned}$$

because χ remains invariant under conjugation after being extended to G. Now write W as a direct sum

$$W = \bigoplus_j W_j$$

of subspaces W_j the functions of which are zero off the coset Hg_j. Restricted to W_j the action of the induced representation $\dot{\rho}(g_0)$ is zero unless g_0 takes Hg_j into Hg_j by multiplication on the right. Thus the contribution to the trace is zero except for those j for which $g_j g_0 g_j^{-1} \in H$. For such j let $f \in W_j$ and $g = hg_j \in Hg_j$. Then

$$\begin{aligned} (\dot{\rho}(g_0)f)(g) = \dot{\rho}(g_0)f(hg_k) = f(hg_k g_0) &= f(hg_k g_0 g_k^{-1} h^{-1} hg_k) \\ &= \rho(hg_k g_0 g_k^{-1} h^{-1}) f(g), \end{aligned}$$

so $\dot{\rho}(g_0)$ acts on W_j like $\rho(hg_k g_0 g_k^{-1} h^{-1})$. But

$$\begin{aligned} \mathrm{tr}(\rho(hg_k g_0 g_k^{-1} h^{-1})) &= \mathrm{tr}(\rho(h)\rho(g_k g_0 g_k^{-1})\rho(h)^{-1}) \\ &= \mathrm{tr}(\rho(g_k g_0 g_k^{-1})) = \chi(g_k g_0 g_k^{-1}), \end{aligned}$$

and thus the formula is established. □

The formula for the induced character shows that equivalent representations have equivalent induced representations, since characters determine representations up to equivalence.

The *restriction* $\mathrm{Res}_H^G(\rho)$ of a representation ρ of a group G to a subgroup H is obviously a representation of H. Its character is denoted by $\mathrm{Res}_H^G(\chi)$.

Proposition 9.20 (Frobenius Reciprocity). *Let G be a finite group and H a subgroup. Then*

$$\langle \eta | \mathrm{Ind}_H^G(\chi) \rangle_{L^2(G)} = \langle \mathrm{Res}_H^G(\eta) | \chi \rangle_{L^2(H)}$$

holds for any character χ of H and η of G.

Proof. We have

$$\begin{aligned}
\langle \eta | \mathrm{Ind}_H^G(\chi) \rangle_{L^2(G)} &= \frac{1}{|G|} \sum_{g \in G} \overline{\eta(g)} \mathrm{Ind}_H^G(\chi)(g) \\
&= \frac{1}{|G|} \sum_{g \in G} \overline{\eta(g)} \frac{1}{|H|} \sum_{\substack{t \in G \\ tgt^{-1} \in H}} \chi(tgt^{-1}) \\
&= \frac{1}{|H|} \sum_{h \in H} \chi(h) \frac{1}{|G|} \sum_{\substack{t \in G \\ tgt^{-1} = h}} \overline{\eta(t^{-1}ht)} \\
&= \frac{1}{|H|} \sum_{h \in H} \chi(h) \frac{1}{|G|} \sum_{\substack{t \in G \\ tgt^{-1} = h}} \overline{\eta(h)} \\
&= \frac{1}{|H|} \sum_{h \in H} \chi(h) \overline{\eta(h)} \frac{1}{|G|} \sum_{\substack{t \in G \\ t^{-1}ht \in G}} 1 \\
&= \frac{1}{|H|} \sum_{h \in H} \chi(h) \overline{\mathrm{Res}_H^G(\eta)(h)} = \langle \mathrm{Res}_H^G(\eta) | \chi \rangle_{L^2(H)}
\end{aligned}$$

by changing the order of summation. □

Note how Frobenius Reciprocity allows us to determine the multiplicities of the irreducibles in an induced representation without having to carry out the induction.

By Proposition 3.13 the characters span the class functions, and so Frobenius Reciprocity may be extended to class functions. If $v : H \to \mathbb{C}$ is a class function and the *induced class function* on G is defined by

$$\mathrm{Ind}_H^G(v)(g_0) = \frac{1}{|H|} \sum_{\substack{g \in G \\ gg_0g^{-1} \in H}} v(gg_0g^{-1}),$$

then

$$\langle u|\mathrm{Ind}_H^G(v)\rangle_{L^2(G)} = \langle \mathrm{Res}_H^G(u)|v\rangle_{L^2(H)}$$

for any class function $u : G \to \mathbb{C}$. The restriction of a class function is defined in the obvious way.

Suppose that G is a finite group and N a normal subgroup. A representation ρ of the factor group G/N yields a representation $\widetilde{\rho}$ of G by $\widetilde{\rho}(g) = \rho(gN)$. The process is called *inflation.* In the same way characters of G/N yield characters of G by inflation.

9.8. Artin L-functions

Recall the factorization

$$\zeta_{\mathbb{Q}(\sqrt{d})}(s) = \zeta(s)L(s, \chi_d), \qquad \chi_d(m) = \left(\frac{d_K}{m}\right)$$

of the Dedekind zeta function of a quadratic number field. It was discovered in the nineteenth century that the Dedekind zeta function of any abelian extension of $\mathbb{Q}$ can be written as a product of Dirichlet L-functions. The problem of factoring the Dedekind zeta function of a normal extension with a nonabelian Galois group was studied by E. Artin in the 1920s. It turns out that Dirichlet L-functions are inadequate for this task, and Artin was led to introduce new L-functions defined in terms of representations of Galois groups. Since Galois groups may be nonabelian, these new L-functions are sometimes called *nonabelian L-functions,* though nowadays they are more commonly termed *Artin L-functions.*

Assume that K/k is a normal extension of number fields and ρ a finite-dimensional complex representation of $G = \mathrm{Gal}(K/k)$. Recall that the character $\psi = \psi_\rho$ determines ρ up to equivalence. We are going to define an Artin L-function $L(s, \psi, K/k)$ by means of an Euler product. Unlike Dirichlet L-functions and the Dedekind zeta function, Artin L-functions do not have tractable Dirichlet series expansions. So for the first time, we encounter a situation where a definition by means of an Euler product is not merely convenient but essential.

Let $\mathfrak{p}$ be a nonzero prime ideal of $\mathcal{O}_k$ and let $\mathfrak{P}$ be any prime ideal lying above $\mathfrak{p}$ in the extension K/k. Associated to $\mathfrak{P}$ is a Frobenius element $\mathrm{Frob}_{\mathfrak{P}}$ in the quotient $Z_{\mathfrak{P}}/T_{\mathfrak{P}}$ of the decomposition group and inertia group of $\mathfrak{P}$ in G. The situation is simpler when $\mathfrak{p}$ does not ramify in the extension, in which case $T_{\mathfrak{P}}$ is trivial for any $\mathfrak{P}$ lying above $\mathfrak{p}$. We define a *local factor*

$$\frac{1}{\det\left(\mathrm{id} - \rho(\mathrm{Frob}_{\mathfrak{P}})N(\mathfrak{p})^{-s}\right)}$$

at $\mathfrak{p}$, and argue that this is well defined under the assumption of no ramification. The operator $\mathrm{id} - \rho(\mathrm{Frob}_{\mathfrak{P}})N(\mathfrak{p})^{-s}$ on the representation space

V_ρ is specified by the prime $\mathfrak{p}$ but depends on the choice of prime ideal $\mathfrak{P}$ lying above it. However, choosing a different prime ideal $\mathfrak{P}$ above $\mathfrak{p}$ replaces $\mathrm{Frob}_{\mathfrak{P}}$ by a conjugate in G, and the operator by a conjugate under the action of $\mathrm{GL}(V_\rho)$ on V_ρ since ρ is a representation. Thus the determinant of the operator does not depend on which $\mathfrak{P}$ above $\mathfrak{p}$ is chosen. The same kind of reasoning shows that the choice of ρ among the equivalent representations with character ψ does not matter either.

For prime ideals $\mathfrak{p}$ that ramify in the extension, the situation is more complicated. The Frobenius element of $\mathfrak{P}$ is a whole coset of $T_{\mathfrak{P}}$ in $Z_{\mathfrak{P}}$. We may resolve this ambiguity by passing to the subspace

$$V^{\mathfrak{P}} = V_\rho^{T_{\mathfrak{P}}} = \{v \in V_\rho \mid \rho(\sigma)v = v \text{ for all } \sigma \in T_{\mathfrak{P}}\}$$

of points fixed by the image of $T_{\mathfrak{P}}$ in $\mathrm{GL}(V_\rho)$ under ρ. If $\sigma \in T_{\mathfrak{P}}$ and $v \in V^{\mathfrak{P}}$, then

$$\begin{aligned}\rho(\mathrm{Frob}_{\mathfrak{P}})v = \rho(\mathrm{Frob}_{\mathfrak{P}})\rho(\sigma)v &= \rho(\mathrm{Frob}_{\mathfrak{P}})\rho(\sigma)\rho(\mathrm{Frob}_{\mathfrak{P}})^{-1}\rho(\mathrm{Frob}_{\mathfrak{P}})v \\ &= \rho(\mathrm{Frob}_{\mathfrak{P}}\sigma\mathrm{Frob}_{\mathfrak{P}}^{-1})\rho(\mathrm{Frob}_{\mathfrak{P}})v,\end{aligned}$$

and so $\rho(\mathrm{Frob}_{\mathfrak{P}})v \in V^{\mathfrak{P}}$ because $T_{\mathfrak{P}}$ is normal in G. Thus $V^{\mathfrak{P}}$ is a ρ-invariant subspace, and ρ restricts to a subrepresentation on $V^{\mathfrak{P}}$. Then

$$\frac{1}{\det\left(\mathrm{id} - \rho(\mathrm{Frob}_{\mathfrak{P}})N(\mathfrak{p})^{-s}\Big|_{V_\rho^{T_{\mathfrak{P}}}}\right)}$$

is a well defined local factor obtained by restricting the operator to $V^{\mathfrak{P}}$. By an abuse of language, we will speak of ramified and unramified local factors.

The Artin L-function is

$$L(s,\psi,K/k) = \prod_{\mathfrak{p}} \frac{1}{\det\left(\mathrm{id} - \rho(\mathrm{Frob}_{\mathfrak{P}})N(\mathfrak{p})^{-s}\Big|_{V_\rho^{T_{\mathfrak{P}}}}\right)},$$

where the product is over the nonzero prime ideals of $\mathcal{O}_k$. It is a theorem of Dedekind that there are only finitely many ramified prime ideals for any extension K/k of number fields. Thus for some questions, such as the distribution of prime ideals, the ramified local factors may be ignored. For other questions, such as the functional equations of Artin L-functions, the ramified local factors are important.

The Euler product of $L(s,\psi,K/k)$ converges absolutely on $\sigma > 1$ and uniformly on any compact set in this half plane. This can be seen by a comparison with the Euler product of the Dedekind zeta function. Restricting an operator to an invariant subspace may remove eigenvalues, but does not create new ones. This observation enables us to apply the same kind of reasoning to the ramified local factors as we shall carry out for the unramified local factors. Alternatively, we may rely on the theorem of Dedekind to see that the ramified local factors have no significance for the convergence, since there are only finitely many.

The eigenvalues $\lambda_j(\mathfrak{p})$ of $\rho(\mathrm{Frob}_{\mathfrak{P}})$ for $1 \le j \le d = \deg(\rho)$ are roots of unity, for G is a finite group. The determinant $\det(\mathrm{id} - \rho(\mathrm{Frob}_{\mathfrak{P}})N(\mathfrak{p})^{-s})$ is the product of the eigenvalues of the operator $\mathrm{id} - \rho(\mathrm{Frob}_{\mathfrak{P}})N(\mathfrak{p})^{-s}$, whose eigenvalues are of the form $1 - \lambda_j(\mathfrak{p})N(\mathfrak{p})^{-s}$. But

$$\prod_j \frac{1}{1 - |\lambda_j(\mathfrak{p})N(\mathfrak{p})^{-s}|} = \left(\frac{1}{1 - N(\mathfrak{p})^{-\sigma}}\right)^d,$$

so the Euler product that defines the Artin L-function converges absolutely, and uniformly on compact sets, in $\sigma > 1$ by comparison with the Euler product of the Dedekind zeta function. The local factors have no singularities nor any zeros in $\sigma > 1$, so the Artin L-function is holomorphic and nonzero there.

The Artin L-function identities below are correct, but will be proved here only up to (finitely many) ramified local factors.

The calculation

$$\begin{aligned}
\log\left(\frac{1}{\det(\mathrm{id} - \rho(\mathrm{Frob}_{\mathfrak{P}})N(\mathfrak{p})^{-s})}\right) &= \sum_j \log\left(\frac{1}{1 - \lambda_j(\mathfrak{p})N(\mathfrak{p})^{-s}}\right)\\
&= \sum_{m=1}^{\infty} \frac{N(\mathfrak{p})^{-ms}}{m} \sum_j \lambda_j(\mathfrak{p})^m\\
&= \sum_{m=1}^{\infty} \frac{N(\mathfrak{p})^{-ms}}{m} \mathrm{tr}\left(\rho\left(\mathrm{Frob}_{\mathfrak{P}}^m\right)\right)\\
&= \sum_{m=1}^{\infty} \psi\left(\mathrm{Frob}_{\mathfrak{P}}^m\right) \frac{N(\mathfrak{p})^{-ms}}{m}
\end{aligned}$$

expresses the logarithm of an unramified local factor in terms of the character of the representation. This leads to the formula

$$\log(L(s,\psi,K/k)) = \sum_{\mathfrak{p}} \sum_{m=1}^{\infty} \psi\left(\mathrm{Frob}_{\mathfrak{P}}^m\right) \frac{N(\mathfrak{p})^{-ms}}{m}$$

after modifying the character ψ to be the trace of the restriction of ρ to $V^{\mathfrak{P}}$ when $\mathfrak{P}$ lies over a $\mathfrak{p}$ that ramifies. As a consequence

$$L(s,\psi_1+\psi_2,K/k) = L(s,\psi_1,K/k)L(s,\psi_2,K/k),$$

by exponentiation.

Denoting the trivial character on $G = \mathrm{Gal}(K/k)$ by ψ_0, one sees that

$$L(s,\psi_0,K/k) = \zeta_k(s)$$

because

$$\frac{1}{\det(\mathrm{id} - \rho_0(\mathrm{Frob}_{\mathfrak{P}})N(\mathfrak{p})^{-s})} = \frac{1}{1 - N(\mathfrak{p})^{-s}},$$

since $\rho_0 \equiv \mathrm{id}$ on $\mathbb{C}$.

Suppose that $L/K/k$ is a tower of normal extensions of number fields. The subgroup $\mathrm{Gal}(L/K)$ is normal in $\mathrm{Gal}(L/k)$ and $\mathrm{Gal}(L/k)/\mathrm{Gal}(L/K) \cong \mathrm{Gal}(K/k)$ by restriction. So any character ψ of $\mathrm{Gal}(K/k)$ yields a character of $\mathrm{Gal}(L/k)$ by inflation, which we denote by $\widetilde{\psi}$. Then $\widetilde{\psi}$ is the character of the inflation $\widetilde{\rho}$ of the representation ρ of ψ. Suppose that $\mathfrak{Q}$ is a prime ideal of $\mathcal{O}_L$ lying above a prime ideal $\mathfrak{P}$ of $\mathcal{O}_K$ lying above a prime ideal $\mathfrak{p}$ of $\mathcal{O}_k$. Then $\widetilde{\rho}(\mathrm{Frob}_{\mathfrak{Q}}) = \rho(\mathrm{Frob}_{\mathfrak{P}})$ because $\mathrm{Frob}_{\mathfrak{Q}}$ restricts to $\mathrm{Frob}_{\mathfrak{P}}$ as K/k was assumed normal. This implies the property

$$L(s, \widetilde{\psi}, L/k) = L(s, \psi, K/k)$$

of Artin L-functions under inflation of characters.

Let $L/K/k$ be a tower of extensions of number fields with L/k normal, $G = \mathrm{Gal}(L/k)$ and $H = \mathrm{Gal}(L/K)$. Then

$$L(s, \mathrm{Ind}_H^G(\psi), L/k) = L(s, \psi, L/K)$$

for any character ψ of H.

We will prove this identity up to ramified local factors by considering

$$\log\left(\frac{1}{\det(\mathrm{id} - \mathrm{Ind}_H^G(\rho)(\mathrm{Frob}_{\mathfrak{Q}})N(\mathfrak{p})^{-s})}\right) = \sum_{m=1}^{\infty} \mathrm{Ind}_H^G(\psi)\,(\mathrm{Frob}_{\mathfrak{Q}}^m)\,\frac{N(\mathfrak{p})^{-ms}}{m}$$

for an unramified local factor. To simplify notation, put $\sigma = \mathrm{Frob}_{\mathfrak{Q}}$. Let

$$\mathfrak{p} = \prod_{j=1}^{g} \mathfrak{P}_j$$

be the prime factorization of $\mathfrak{p}$ in K. As usual f_j denotes the residue class degree of $\mathfrak{P}_j$ over $\mathfrak{p}$. For each j with $1 \le j \le g$ let τ_j be an element of G such that $\tau_j\mathfrak{Q}$ lies over $\mathfrak{P}_j$. Then $\tau_j\sigma^i$ for $0 \le i < f_j$ and $1 \le j \le g$ is a complete collection of right coset representatives of H in G. For if $\tau_{j_1}\sigma^{i_1}H = \tau_{j_2}\sigma^{i_2}H$ then $\tau_{j_2}\sigma^{i_2-i_1}\tau_{j_1}^{-1} \in H$. Thus

$$\tau_{j_2}\sigma^{i_2-i_1}\tau_{j_1}^{-1}\tau_{j_1}\mathfrak{Q} = \tau_{j_2}\sigma^{i_2-i_1}\mathfrak{Q} = \tau_{j_2}\mathfrak{Q}.$$

But $j_1\mathfrak{Q}$ lies over $\mathfrak{P}_1$ and $j_2\mathfrak{Q}$ lies over $\mathfrak{P}_2$, while each element of H fixes $\mathfrak{P}_1$ and $\mathfrak{P}_2$, so $j_1 = j_2$. Then $\sigma^{i_2-i_1} \in H$, which implies that $i_2 = i_1$. Thus the cosets are distinct, hence

$$\sum_{j=1}^{g} f_j = n_{K/k},$$

so the collection is complete.

Next

$$\begin{aligned}\operatorname{Ind}_H^G(\psi)\,(\sigma^m) &= \sum_{\tau_j\sigma^i\sigma^m\sigma^{-i}\tau_j^{-1}\in H} \psi(\tau_j\sigma^i\sigma^m\sigma^{-i}\tau_j^{-1})\\ &= \sum_{\tau_j\sigma^m\tau_j^{-1}\in H} f_j\psi(\tau_j\sigma^m\tau_j^{-1})\\ &= \sum_{\tau_j\sigma^m\tau_j^{-1}\in H} f_j\psi(\tau_j\sigma^m\tau_j^{-1})\end{aligned}$$

by the induced character formula. For $\tau_j\sigma^m\tau_j^{-1} = (\tau_j\sigma\tau_j)^m$, and $\tau_j\sigma\tau_j^{-1}$ generates the decomposition group of $\mathfrak{P}_j$ over $\mathfrak{p}$, which is of order f_j since $\mathfrak{p}$ does not ramify in K. Thus $\tau_j\sigma^m\tau_j^{-1} \in H$ if and only if $f_j|m$. Now

$$\begin{aligned}\sum_{m=1}^{\infty} \operatorname{Ind}_H^G(\psi)(\operatorname{Frob}_{\mathfrak{Q}}^m)\,\frac{N(\mathfrak{p})^{-ms}}{m} &= \sum_{m=1}^{\infty}\frac{N(\mathfrak{p})^{-ms}}{m}\sum_{\substack{1\le j\le g\\ f_j|m}} f_j\psi(\tau_j\sigma^m\tau_j^{-1})\\ &= \sum_{j=1}^{g}\sum_{f_j|m} f_j\psi(\tau_j\sigma^m\tau_j^{-1})\frac{N(\mathfrak{p})^{-ms}}{m}\\ &= \sum_{j=1}^{g}\sum_{\ell=1}^{\infty} f_j\psi(\tau_j\sigma^{\ell f_j}\tau_j^{-1})\frac{N(\mathfrak{p})^{-\ell f_j s}}{\ell f_j}\\ &= \sum_{j=1}^{g}\sum_{\ell=1}^{\infty}\psi\left((\tau_j\sigma^{f_j}\tau_j^{-1})^{\ell}\right)\frac{(N(\mathfrak{p})^{f_j})^{-\ell s}}{\ell}\\ &= \sum_{j=1}^{g}\sum_{\ell=1}^{\infty}\psi\left((\tau_j\sigma^{f_j}\tau_j^{-1})^{\ell}\right)\frac{N(\mathfrak{P}_j)^{-\ell s}}{\ell}\end{aligned}$$

where

$$\tau_j\sigma^{f_j}\tau_j^{-1} = (\tau_j\operatorname{Frob}_{\mathfrak{Q}}\tau_j^{-1})^{f_j} = (\operatorname{Frob}_{\mathfrak{Q}_j})^{f_j} = \operatorname{Frob}_{\mathfrak{P}_j}.$$

Thus

$$\begin{aligned}&\log\left(\frac{1}{\det(\mathrm{id} - \operatorname{Ind}_H^G(\rho)(\operatorname{Frob}_{\mathfrak{Q}})N(\mathfrak{p})^{-s})}\right)\\ &= \sum_{j=1}^{g}\log\left(\frac{1}{\det(\mathrm{id} - \rho(\operatorname{Frob}_{\mathfrak{P}_j})N(\mathfrak{P}_j)^{-s})}\right),\end{aligned}$$

and the desired formula follows by summing over unramified $\mathfrak{p}$ and exponentiating.

Now recall the regular representation ρ_{reg} of G and its character which is a linear combination

$$\psi_{\mathrm{reg}} = \psi_1(1)\psi_1 + \cdots + \psi_m(1)\psi_m$$

of the irreducible characters $\psi_1, \ldots, \psi_m$ with multiplicities $\psi_1(1), \ldots, \psi_m(1)$ that are positive integers. Then

$$L(s, \psi_{\text{reg}}, K/k) = L(s, \psi_1, K/k)^{\psi_1(1)} \cdots L(s, \psi_m, K/k)^{\psi_m(1)}$$

by the formula for the Artin L-function of a sum of two characters. But on the other hand the above result on the behavior of the Artin L-function under induction implies that

$$L(s, \psi_{\text{reg}}, K/k) = \zeta_K(s),$$

since the regular character is obtained by induction from the trivial subgroup; just take $K = L$, $H = \{1\}$ and $\psi = \psi_0$ in the formula for the Artin L-function of an induced character. Finally we obtain the *Artin factorization*

$$\zeta_K(s) = L(s, \psi_1, K/k)^{\psi_1(1)} \cdots L(s, \psi_m, K/k)^{\psi_m(1)}.$$

Suppose that $K/\mathbb{Q}$ is a finite abelian extension, i.e., an extension with an abelian Galois group. The Kronecker-Weber theorem states that there is a root of unity ζ such that $\mathbb{Q}(\zeta)/K/\mathbb{Q}$ is a tower of extensions of number fields. The Artin factorization yields

$$\zeta_K(s) = L(s, \psi_1, K/\mathbb{Q})^{\psi_1(1)} \cdots L(s, \psi_m, K/\mathbb{Q})^{\psi_m(1)}$$

with $\psi_1, \ldots, \psi_m$ irreducible characters of $\text{Gal}(K/\mathbb{Q})$. Furthermore

$$L(s, \psi_j, K/\mathbb{Q}) = L(s, \widetilde{\psi}_j, \mathbb{Q}(\zeta)/\mathbb{Q})$$

by inflation of characters. It is an easy exercise to see that the inflation of an irreducible character is itself irreducible. The Galois group of a cyclotomic extension $\mathbb{Q}(\zeta)/\mathbb{Q}$ is isomorphic to the multiplicative group of reduced residue classes to some modulus. Thus the irreducible characters of this Galois group are Dirichlet characters, so $\widetilde{\psi}_i = \chi_j$ for some Dirichlet character χ_j. Then there is a factorization

$$\zeta_K(s) = L(s, \chi_1)^{\psi_1(1)} \cdots L(s, \chi_m)^{\psi_m(1)}$$

of the the Dedekind zeta function of an abelian extension of the rationals as a product of Dirichlet L-functions. But the situation is abelian and Artin L-functions are not needed to establish this result. It can be achieved directly in a much more precise form.

We obtained the analytic continuations of the Dirichlet L-functions and Dedekind zeta functions from their Dirichlet series expansions. It would seem hopeless to analytically continue even $\zeta(s)$ solely from its Euler product, without knowledge of the associated Dirichlet series. As one might expect from this observation, Artin L-functions have not thus far been analytically continued directly from their definition. By arguments of an indirect nature, which we cannot carry out here, it is known that they have continuations to the whole complex plane as meromorphic functions.

It is believed that the Artin L-functions of nontrivial irreducible characters are entire. This important *Artin Conjecture* is still open. The Artin Conjecture would imply the older, but also still unproved, Dedekind Conjecture to the effect that $\zeta_K(s)/\zeta_k(s)$ is always entire for any extension K/k of number fields. The Artin and Dedekind conjectures are known in many special cases. In particular there is a theorem of H. Aramata and R. Brauer stating that $\zeta_K(s)/\zeta_k(s)$ is entire if K/k is a normal extension of number fields.

There is an exposition of the functional equations of Artin L-functions in *Algebraic Number Theory* by Jürgen Neukirch. In combination with a proof of the Brauer induction theorem as in *Linear Representations of Finite Groups* by Jean-Pierre Serre, the meromorphic continuation of Artin L-functions is also established. In *Nonvanishing of L-functions and Applications* by M. Ram Murty and V. Kumar Murty there are results in the direction of the Artin and Dedekind conjectures, and much other material on L-functions. The use of Artin L-functions to establish relations between Dedekind zeta functions of different number fields is explained in *Algebraic Number Fields* by A. Fröhlich and M. J. Taylor.

9.9. Notes

The Dedekind zeta function was introduced in [**Ded79**]. Proposition 9.1 and 9.2 are due to Hecke [**Hec17**].

Proposition 9.3 and Proposition 9.4 are due to Dirichlet [**Dir46**]. The proof of Proposition 9.4 given here combines elements of the proof outlined by Stark in [**Sta75**] and the classical proof going back to Dirichlet. For binary quadratic forms the finiteness of the class number is due to Gauss in articles 174 and 185 of the *Disquisitiones.* The finiteness of the class number for all number fields is due to Dedekind [**Ded79**]. The proof of Proposition 9.5 given here is taken from [**Sta75**]. The functional equation of the Dedekind zeta function in the general case is due to Hecke [**Hec17**], and the proof of Proposition 9.6 is his. Jacobi [**Jak32**] conjectured the class number formula for binary quadratic forms on the basis of theoretical and numerical evidence, and Dirichlet established it in [**Dir40**]. The analytic class number formula for number fields is due to Dedekind [**Ded79**]. Proposition 9.8 is due to Kronecker [**Kro80**] and Proposition 9.9 to M. Bauer [**Bau03a, Bau03b**].

The explicit class number formulas for binary quadratic forms are due to Dirichlet [**Dir40**].

Proposition 9.12 is due to Gauss, see volume III page 159 of his collected works [**Gau33**]. The method of proof, and the specific bound, of Proposition 9.13 is due to Stark [**Sta74**]. Proposition 9.14 is due to Minkowski [**Min91b, Min91a**], and a weaker bound analogous to Proposition 9.13 may be found there.

The theorem of Ikehara [**Ike31**] was proved within an approach to Tauberian theory based on Fourier transforms, due to N. Wiener. Ikehara was a student of Wiener, and the theorem is sometimes called the Wiener-Ikehara theorem. The

Prime Ideal Theorem was established by Landau in [**Lan03a**]. The Dedekind-Weber estimate is in [**Web08**]; but see [**MO07**] for a more extensive treatment along the same lines.

The proof of the Ikehara theorem given here is due to J. Korevaar [**Kor06**].

Induced representations and characters are due to Frobenius [**Fro98**], and the results of this section are also his.

The definition of Artin L-functions up to ramified local factors, and their basic properties, are in [**Art23**]. Artin introduced the ramified local factors, and found the functional equations of Artin L-functions, in [**Art30**].

Exercises

(1) Show that

$$\frac{a_K(m)}{m^\sigma} \to 0$$

as $m \to +\infty$ for every $\sigma > 1$.

(2) Show that the ideal counting function is multiplicative.

(3) Calculate the Dedekind zeta function $\zeta_{\mathbb{Q}(\omega)}(s)$ for $\omega = e^{2\pi i/3}$ in terms of L-series. Give a Dirichlet series expansion for it in terms of the number of representations of integers by a particular binary quadratic form. The ring of algebraic integers in $\mathbb{Q}(\omega)$ is known as the *Eisenstein integers,* and like the Gaussian integers it is well studied.

(4) Show that every fractional ideal class contains a nonzero integral ideal, and that if two fractional ideal classes contain the same integral ideals, then they are equal.

(5) Show that two ideals $\mathfrak{a}$ and $\mathfrak{b}$ represent the same ideal class in Cl_K if and only if there exists nonzero $\alpha, \beta \in \mathcal{O}_K$ with $(\alpha)\mathfrak{a} = (\beta)\mathfrak{b}$.

(6) Which number fields have unit rank equal to one?

(7) a) Let K be a number field and $\omega_1, \ldots, \omega_n$ an integral basis of $\mathcal{O}_K$. Show that there is some constant A such that

$$|N(\alpha)| \leq A \max(|a_1|, \ldots, |a_n|)^n$$

for any algebraic integer $\alpha = a_1\omega_1 + \cdots + a_n\omega_n$ of $\mathcal{O}_K$.
b) Let C be an ideal class of $\mathcal{O}_K$ and $\mathfrak{b}$ an ideal in C^{-1}. Show that there is an element β of $\mathfrak{b}$ with $|N(\beta)| < AN(\mathfrak{b})$.
c) Show that there is an ideal $\mathfrak{a}$ of C such that $N(\mathfrak{a}) < A$.
d) Show that there are only finitely many ideals of $\mathcal{O}_K$ with a given norm. Conclude that the class number of a number field is finite (Dedekind).

(8) Show that in a sense analogous to Dirichlet density the average number of prime ideals of degree one in $\mathcal{O}_K/\mathbb{Q}$ that divide a given rational prime equals one.

(9) Calculate the class number of $\mathbb{Q}(\sqrt{d})$ for $d = \pm 2, \pm 3, \pm 5$.

(10) Show that $\mathbb{Q}(\sqrt{-p})$ has odd class number for each prime $p \equiv 3 \pmod 4$.

(11) For $K = \mathbb{Q}(\sqrt{d})$ a real quadratic field show that

$$\eta_d = \frac{\prod_{\substack{1 \le m < |d_K| \\ (d_K|m) = -1}} \sin\left(\frac{\pi m}{d_K}\right)}{\prod_{\substack{1 \le m < |d_K| \\ (d_K|m) = 1}} \sin\left(\frac{\pi m}{d_K}\right)}$$

is a unit, which coincides with the fundamental unit if $h_K = 1$.

(12) Show that

$$h(d) = \frac{1}{2 - \left(\frac{d_K}{2}\right)} \sum_{1 \le m < |d_K|/2} \left(\frac{d_K}{m}\right)$$

holds also if $|d_K|$ is even.

(13) Establish the bound

$$n < 7 + \frac{2}{3} \log |d_K|$$

for the degree of K.

(14) Show that $\mathbb{Q}(\sqrt{-1}, \sqrt{-5})$ is an unramified extension of $\mathbb{Q}(\sqrt{-5})$.

(15) The other asymptotic estimates that we have found relating to number fields K include a constant depending on arithmetic data for K, but the Prime Ideal Theorem does not. The estimate is the same for all K. What is the explanation?

(16) Apply the Prime Ideal Theorem to show that

$$\sum_{\substack{p \le x \\ p \equiv 1 \pmod 3}} 1 \sim \sum_{\substack{p \le x \\ p \equiv 2 \pmod 3}} 1 \sim \frac{1}{2} \frac{x}{\log(x)}$$

as $x \to +\infty$.

(17) a) Apply the Ikehara theorem to calculate the asymptotic density of the sequence of integers that are not divisible by the cube of any prime.
b) Apply the Ikehara theorem to find the average number of solutions n of the equation $\phi(n) = m$ as $m \to +\infty$.

(18) Apply the Ikehara theorem to show that

$$\sum_{\substack{n \le x \\ n \equiv a \bmod q}} \Lambda(n) \sim \frac{x}{\phi(q)}$$

as $x \to +\infty$, when q and a are coprime positive integers. This is a very modest version of the Prime Number Theorem for arithmetic progressions, strengthening Dirichlet's theorem by replacing Dirichlet density with asymptotic density.

(19) The group S_3 of permutations on three letters has three subgroups of order two and one subgroup of order three. Find the characters of S_3 induced from the characters of the subgroup of order three and from one of the subgroups of order two.

(20) Show that an inflation of an irreducible character is irreducible.

(21) a) Show that $\mathrm{Ind}_H^G(\chi_{\rho\oplus\rho'}) = \mathrm{Ind}_H^G(\chi_\rho) + \mathrm{Ind}_H^G(\chi_{\rho'})$.
b) Show that $\mathrm{Ind}_H^G(\rho \oplus \rho') \cong \mathrm{Ind}_H^G(\rho) \oplus \mathrm{Ind}_H^G(\rho')$.
c) Show that a representation that induces an irreducible is itself irreducible.

(22) Show that

$$\mathrm{Ind}_H^G(\chi \cdot \mathrm{Res}_H^G(\eta)) = \mathrm{Ind}_H^G(\chi) \cdot \eta$$

if χ is a character of H and η a character of η. This is called the *projection formula* for induced characters.

(23) Show that

$$\mathrm{Ind}_K^G(\chi) = \mathrm{Ind}_H^G(\mathrm{Ind}_K^H(\chi))$$

if K is a subgroup of H, which is a subgroup of G, and χ is a character of K. This is called the *theorem on induction in stages.*

(24) Factorize $\zeta_K(s)$ as a product of Artin L-functions, where K is the splitting field of $x^3 - 2$ over $\mathbb{Q}$.

(25) The Chebotarev Density Theorem is a major result in algebraic number theory, generalizing Dirichlet's theorem on primes in arithmetic progressions. It is too difficult for us to prove here. But in this exercise we will see that it is an easy consequence of the Artin Conjecture.
a) Assuming the Artin Conjecture, show that the analytic continuation of $L(s, \psi, K/k)$ is nonzero at $s = 1$.
b) Assuming part a), determine the behavior of

$$\sum_{\mathfrak{p}} \psi(\mathrm{Frob}_{\mathfrak{P}}) N(\mathfrak{p})^{-s}$$

in the limit as $s \to 1^+$. Here, and in the remainder of this exercise, the sum is taken over $\mathfrak{p}$ that do not ramify. Recall that only finitely many $\mathfrak{p}$ ramify.
c) Let $\mathfrak{p}$ be an unramified prime ideal of k and C any conjugacy class of $G = \mathrm{Gal}(K/k)$. Explain why the truth or falsity of the statement $\mathrm{Frob}_{\mathfrak{P}} \in C$ for a prime ideal $\mathfrak{P}$ lying above $\mathfrak{p}$ does not depend on the choice of $\mathfrak{P}$.

d) Show that

$$\sum_{\psi} \overline{\psi(C)} \sum_{\mathfrak{p}} \psi(\mathrm{Frob}_{\mathfrak{P}}) N(\mathfrak{p})^{-s} = \frac{|G|}{|C|} \sum_{\substack{\mathfrak{p} \\ \mathrm{Frob}_{\mathfrak{P}} \in C}} N(\mathfrak{p})^{-s}$$

for $\sigma > 1$.

Suppose that P is a set of prime ideals of a number field. The limit

$$\delta = \delta_P = \lim_{s \to 1^+} \frac{\sum_{\mathfrak{p} \in P} N(\mathfrak{p})^{-s}}{\sum_{\mathfrak{p}} N(\mathfrak{p})^{-s}}$$

is called the *Dirichlet density* of P in the prime ideals, if it exists.

e) Let K/k be a normal extension of number fields and $G = \mathrm{Gal}(K/k)$. Assuming part b), show that the set of unramified prime ideals $\mathfrak{p}$ of k for which the Frobenius elements $\mathrm{Frob}_{\mathfrak{P}}$ lie in a conjugacy class C of G has Dirichlet density $|C|/|G|$. This is the Chebotarev Density Theorem, proved in 1922 in difficult circumstances: "I devised my best result while carrying water from the lower part of town (Peresypi in Odessa) to the higher part, or buckets of cabbages, which my mother sold to feed the entire family" (Nikolai Grigorievich Chebotarev).

(26) In this exercise we collect some applications of the Chebotarev Density Theorem.

a) Obtain Dirichlet's theorem on primes in arithmetic progressions as a special case of the Chebotarev Density Theorem.

b) Show that if K/k is a normal extension of number fields and

$$\sum_{\mathfrak{p}} N(\mathfrak{p})^{-1} < \infty,$$

with the sum taken over all prime ideals $\mathfrak{p}$ of $\mathcal{O}_k$ that split completely in the extension, then $K = k$.

c) For any abelian extension K/k of number fields there is a homomorphism

$$\left(\frac{K/k}{\quad}\right) : J_k^S \to \mathrm{Gal}(K/k)$$

from the group J_k^S of fractional ideals of k coprime with an exceptional set (i.e. a set of prime ideals of k that contains all those that ramify.) It is constructed by

$$\left(\frac{K/k}{\mathfrak{p}}\right) = \mathrm{Frob}_{\mathfrak{P}},$$

noting that the conjugacy classes of $\mathrm{Gal}(K/k)$ are singletons, and extending the map multiplicatively. This is called the Artin map.

Determine the image of the Artin map. The determination of the kernel of the Artin map is harder. The theorem that describes the kernel and the image of the Artin map is called the Artin Reciprocity Theorem.

Chapter 10

Explicit Formulas

10.1. The von Mangoldt formula

Using the method of contour integrals together with the information about the Riemann zeta function already obtained, we find an explicit formula for $\psi(x)$ in terms of the nontrivial zeros ρ of $\zeta(s)$. The formula that we shall prove is a truncated version of the von Mangoldt explicit formula

$$\psi(x) = x - \sum_{\xi(\rho)=0} \frac{x^\rho}{\rho} - \log(2\pi) - \frac{1}{2}\log(1 - x^{-2}).$$

The von Mangoldt formula exhibits the distribution of the primes as determined by the nontrivial zeros of the Riemann zeta function. But it is difficult to use, because the infinite series over the zeros ρ is only conditionally convergent. Actually, the series should be summed in order of increasing $|\mathrm{Im}(\rho)|$, and then the right-hand side of the von Mangoldt formula converges pointwise to $\psi(x)$ for all $x > 1$ except at the prime powers. Figure 6 on page 308 contains a graph of the approximation to $\psi(x)$ obtained from the von Mangoldt formula by including the terms corresponding to the 200 nontrivial zeros of $\zeta(s)$ that lie closest to the real axis.

As an alternative, we could prove an explicit formula for the smoothed Chebyshev function $\psi_1(x)$ from Chapter 6. This leads to an absolutely convergent series in the explicit formula, which is easier to deal with, and would enable us to establish slightly finer results for $\psi_1(x)$ than for $\psi(x)$.

To prove the explicit formula, we require a bound for the logarithmic derivative of the gamma function and a local bound for the number of zeros of the Riemann zeta function in the critical strip.

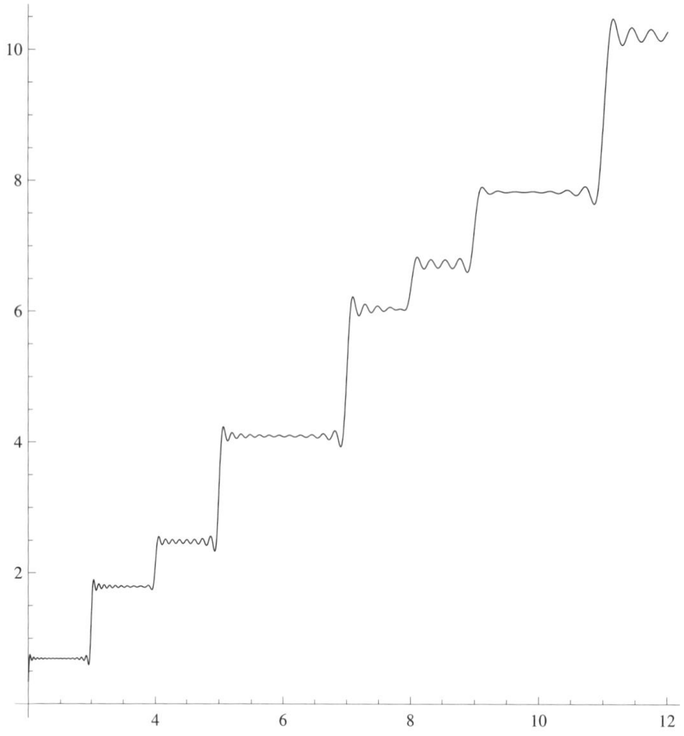

Figure 6. The truncated von Mangoldt formula

Proposition 10.1. *The estimate*

$$\frac{\Gamma'(s)}{\Gamma(s)} = \mathrm{Log}(s) - \frac{1}{2s} + O_\delta\left(\frac{1}{|s|^2}\right)$$

holds on $|\mathrm{Arg}(s)| < \pi/2 - \delta$ *for any fixed* $\delta > 0$.

Proof. The calculation

$$\begin{aligned}\frac{1}{s} &= \int_1^s \frac{dw}{w} = \int_1^s \left(\int_0^\infty e^{-wu}\,du\right)dw \\ &= \int_0^\infty \left(\int_1^s e^{-wu}\,dw\right)du = \int_0^\infty \left(\frac{e^{-u}}{u} - \frac{e^{-su}}{u}\right)du\end{aligned}$$

is valid for $\sigma > 0$. Then

$$\begin{aligned}\frac{\Gamma'(s)}{\Gamma(s)} &= \int_0^\infty \left(\frac{e^{-u}}{u} - \frac{e^{-su}}{1-e^{-u}}\right)du \\ &= \int_0^\infty \left(\frac{e^{-u}}{u} - \frac{e^{-su}}{u} - \frac{e^{-su}}{2} + \frac{e^{-su}}{2} + \frac{e^{-su}}{u} - \frac{e^{-su}}{1-e^{-u}}\right)du \\ &= \mathrm{Log}(s) - \frac{1}{2s} + \int_0^\infty \left(\frac{1}{2} + \frac{1}{u} - \frac{1}{1-e^{-u}}\right)e^{-su}\,du\end{aligned}$$

where

$$\begin{aligned}\int_0^\infty \left(\frac{1}{2}+\frac{1}{u}-\frac{1}{1-e^{-u}}\right)e^{-su}\,du \\ &= \left(\frac{1}{2}+\frac{1}{u}-\frac{1}{1-e^{-u}}\right)\frac{e^{-su}}{-s}\Big|_0^\infty - \int_0^\infty \left(\frac{e^{-u}}{(1-e^{-u})^2}-\frac{1}{u^2}\right)\frac{e^{-su}}{-s}du \\ &= \frac{1}{s}\int_0^\infty \left(\frac{e^{-u}}{(1-e^{-u})^2}-\frac{1}{u^2}\right)\frac{e^{-su}}{-s}du\end{aligned}$$

and

$$\left|\frac{e^{-u}}{(1-e^{-u})^2}-\frac{1}{u^2}\right| \le \frac{1}{12}$$

so that

$$\left|\frac{\Gamma'(s)}{\Gamma(s)} - \mathrm{Log}(s) + \frac{1}{2s}\right| \le \frac{1}{12\sigma|s|}$$

for $\sigma > 0$. The desired estimate follows because $|\mathrm{Arg}(s)| \le \pi/2-\delta$ implies $\sigma \ge |s|\cos(\delta)$. □

Rewriting this estimate and integrating from $+\infty$ on the real axis to s, one obtains

$$\log(\Gamma(s)) = \left(s-\frac{1}{2}\right)\mathrm{Log}(s) - s + \frac{1}{2}\log(2\pi) + O_\delta(|s|^{-1})$$

as $s \to \infty$ on $|\mathrm{Arg}(s)| < \pi/2-\delta$ for any fixed $\delta > 0$. This is a complex version of Stirling's formula.

Denote by $N(T)$ the number of zeros of $\zeta(s)$, counted with multiplicity, in the rectangle given by $0 \le \sigma \le 1$ and $0 < t \le T$.

Proposition 10.2. *$N(T+1) - N(T) \ll \log(T)$ as $T \to +\infty$.*

Proof. The square defined by the inequalities $0 < \sigma < 1$ and $T < t \le T+1$ is contained in the closed disk around $s = 2+iT$ with radius $\sqrt{5}$. We are going to apply Jensen's formula to $\zeta(s)$ on a slightly larger disk, and for $T \ge 3$ obtain

$$\log(|\zeta(2+iT)|) + \log\left(\frac{(5/2)^n}{|\rho_1\cdots\rho_n|}\right) = \frac{1}{2\pi}\int_{-\pi}^{\pi}\log\left(\left|\zeta(2+iT+\frac{5}{2}e^{i\theta})\right|\right)\,d\theta,$$

where $\rho_1,\ldots,\rho_n$ are the zeros, counted with multiplicity, of $\zeta(s)$ in the disk $|s-(2+iT)| \le 5/2$. An argument similar to the proof of Proposition 5.3 shows that $\zeta(s) \ll |t|^2$ holds for $\sigma \ge -1/2$ except in a neighborhood of $s=1$, and thus $\log((5/2)^n/|\rho_1\cdots\rho_n|) \ll \log(T)$. Suppose that the zeros $\rho_1,\ldots,\rho_m$ lie in the disk $|s-(2+iT)|\sqrt{5}$ and the zeros $\rho_{m+1},\ldots,\rho_n$ in the annulus $\sqrt{5} < |s-(2+iT)| \le 5/2$. Then $\log((5/2)^m/|\rho_1\cdots\rho_m|) \le \log((5/2)^n/|\rho_1\cdots\rho_n|) \ll \log(T)$ and so $m\log((5/2)/\sqrt{5}) \ll \log(T)$, thus $N(T+1)-N(T) \le m \ll \log(T)$. □

When proving the Prime Number Theorem in Chapter 6, we moved the contour of integration in the Perron formula just slightly into the half plane $\sigma < 1$, and were careful to stay to the right of the zeros of $\zeta(s)$, since these are singularities of the integrand. This time we shall move the contour far to the left, picking up contributions from the residues at the poles of the integrand.

The original von Mangoldt explicit formula

$$\psi(x) = x - \lim_{T\to+\infty} \sum_{|\mathrm{Im}(\rho)|<T} \frac{x^\rho}{\rho} - \log(2\pi) - \frac{1}{2}\log(1-x^{-2})$$

is obtained from the truncated version below by letting $T \to +\infty$.

Proposition 10.3 (Truncated von Mangoldt formula)**.** *Let $x > e$ and not a prime power. Then*

$$\psi(x) = x - \sum_{|\mathrm{Im}(\rho)|<T} \frac{x^\rho}{\rho} - \log(2\pi) - \frac{1}{2}\log(1-x^{-2}) + R(x,T)$$

holds for all $T \geq 3$ with the bound

$$R(x,T) \ll \frac{x\log^2(T)}{T\log(x)} + \log(x)\min\left(1, \frac{x}{T\langle x\rangle}\right) + \frac{x\log^2(x)}{T}$$

for the remainder term. Here $\langle x\rangle$ denotes the distance from x to the nearest prime power.

Proof. Let $\sigma_x = 1 + 1/\log(x)$ and choose $T \geq 3$ and $U \leq -1$ so that no zeros of $\zeta(s)$ lie on the rectangular contour C_U from $\sigma_x - iT$ to $\sigma_x + iT$ to $U + iT$ to $U - iT$ and back to $\sigma_x - iT$. Then

$$\frac{1}{2\pi i}\oint_{C_U}\left(-\frac{\zeta'(s)}{\zeta(s)}\right)\frac{x^s}{s}\,ds = x - \sum_{|\mathrm{Im}(\rho)|<T}\frac{x^\rho}{\rho} - \frac{\zeta'(0)}{\zeta(0)} + \sum_{2m<-U}\frac{x^{-2m}}{2m}$$

by the Residue Theorem, since the integrand has residues

$$\begin{aligned}
\mathrm{Res}_{s=0}\left(-\frac{\zeta'(s)}{\zeta(s)}\right)\frac{x^s}{s} &= -\frac{\zeta'(0)}{\zeta(0)}\\
\mathrm{Res}_{s=1}\left(-\frac{\zeta'(s)}{\zeta(s)}\right)\frac{x^s}{s} &= x\\
\mathrm{Res}_{s=-2m}\left(-\frac{\zeta'(s)}{\zeta(s)}\right)\frac{x^s}{s} &= \frac{x^{-2m}}{2m}\\
\mathrm{Res}_{s=\rho}\left(-\frac{\zeta'(s)}{\zeta(s)}\right)\frac{x^s}{s} &= -\frac{x^\rho}{\rho}.
\end{aligned}$$

at the poles $s = 0$ and $s = 1$, and at the trivial zeros $s = -2, -4, \ldots$ and nontrivial zeros $s = \rho$ of $\zeta(s)$.

Logarithmically differentiating the functional equation of the Riemann zeta function yields

$$\frac{\zeta'(s)}{\zeta(s)} = \log(2\pi) + \frac{\pi}{2}\cot\left(\frac{\pi s}{2}\right) - \frac{\Gamma'(1-s)}{\Gamma(1-s)} - \frac{\zeta'(1-s)}{\zeta(1-s)}.$$

Except for the logarithmic derivative of the gamma function, the terms on the right-hand side are uniformly bounded in the half plane $\sigma \le -1$ after removing neighborhoods $|s-2m| < 1/2$ of the even integers $2m$, these being poles of $\cot(\pi s/2)$. Choosing $s = U + it$ with U an odd negative integer one sees that

$$\begin{aligned}\left|\int_{U+iT}^{U-iT}\left(-\frac{\zeta'(s)}{\zeta(s)}\right)\frac{x^s}{s}\,ds\right| &\ll x^U\int_{U-iT}^{U+iT} |\log(2-s)|\frac{|ds|}{|s|}\\ &\ll x^U\log(2-U+T)\frac{T}{|U|} \to 0\end{aligned}$$

as $U \to -\infty$ through odd negative integers, by Proposition 10.1. Thus letting $U \to -\infty$ through odd negative integers yields the formula

$$\frac{1}{2\pi i}\oint_C\left(-\frac{\zeta'(s)}{\zeta(s)}\right)\frac{x^s}{s}\,ds = x - \sum_{|\mathrm{Im}(\rho)|<T}\frac{x^\rho}{\rho} - \frac{\zeta'(0)}{\zeta(0)} - \frac{1}{2}\log(1-x^{-2})$$

where the contour C starts at $\sigma_x - iT$ and runs vertically to $\sigma_x + iT$ and horizontally to $-\infty + iT$, then horizontally from $-\infty - iT$ back to $\sigma_x - iT$.

Choosing $s = \sigma + iT$ with $-\infty < \sigma \le -1$ one sees that

$$\begin{aligned}\left|\int_{-1+iT}^{-\infty+iT}\left(-\frac{\zeta'(s)}{\zeta(s)}\right)\frac{x^s}{s}\,ds\right| &\ll \int_{-1+iT}^{-\infty+iT} |\log(2-s)|\frac{x^\sigma}{|s|}\,|ds|\\ &\ll \int_{-\infty}^{-1}\log(2-\sigma+T)\frac{x^\sigma}{T}\,d\sigma \ll \frac{\log(T)}{Tx\log(x)},\end{aligned}$$

again by Proposition 10.1. Logarithmically differentiating the expression for $\zeta(s)$ in terms of $\xi(s)$ and using Proposition 10.1 one more time yields

$$\frac{\zeta'(s)}{\zeta(s)} = \frac{1}{2}\log(\pi) + \frac{\xi'(s)}{\xi(s)} - \frac{1}{s} - \frac{1}{s-1} - \frac{1}{2}\frac{\Gamma'(s/2)}{\Gamma(s/2)} = \frac{\xi'(s)}{\xi(s)} + O(\log(T))$$

for $s = \sigma + iT$ with $-1 \le \sigma \le 2$. Now

$$\left|\frac{\xi'(s)}{\xi(s)}\right| = \left|b + \sum_\rho\left(\frac{1}{s-\rho} + \frac{1}{\rho}\right)\right| \ll \sum_\rho \frac{|s|}{|\rho||s-\rho|}$$

by logarithmic differentiation of the canonical factorization of $\xi(s)$. If T is restricted to suitable values, then

$$\begin{aligned}\sum_\rho \frac{|s|}{|\rho||s-\rho|} &= \sum_{|\gamma-T|\le 1} \frac{|s|}{|\rho||s-\rho|} + \sum_{|\gamma-T|>1} \frac{|s|}{|\rho||s-\rho|} \\ &\ll \log(T) \sum_{|\gamma-T|\le 1} \frac{|s|}{|\rho|} + \sum_\rho \frac{|s|}{(1+|\rho|)(1+|s-\rho|)} \\ &\ll \log^2(T) + \sum_{\gamma\ge 0} \frac{T}{(1+\gamma)(1+|T-\gamma|)} \\ &\ll \log^2(T) + \sum_{n=1}^{\infty} \frac{T}{n(1+|T-n|)} \sum_{n-1\le\gamma<n} 1 \\ &\ll \log^2(T) + \int_1^T \frac{T\log(u)}{u(1+T-u)}du + \int_T^\infty \frac{T\log(u)}{u(1+u-T)}du \\ &\ll \log^2(T) + \int_1^T \frac{T\log(u)}{\left(\frac{T+1}{2}\right)^2}du + \int_T^{2T} \frac{T\log(2T)}{u(1+u-T)}du \\ &\quad + \int_{2T}^\infty \frac{T\log(u)}{\frac{u^2}{2}}du \ll \log^2(T).\end{aligned}$$

It is a consequence of Proposition 10.2 that every interval of length 1 contains some value of T so that $|\gamma - T| \gg 1/\log(T)$ for every zero $\rho = \beta + i\gamma$ of $\xi(s)$. Thus $\zeta'(s)/\zeta(s) = O\left(\log^2(T)\right)$ for $s = \sigma + iT$ with $-1 \le \sigma \le 2$ and T restricted as above. Then the bound

$$\begin{aligned}\left|\int_{\sigma_x+iT}^{-1+iT} \left(-\frac{\zeta'(s)}{\zeta(s)}\right)\frac{x^s}{s}\,ds\right| &\le \int_{\sigma_x+iT}^{-1+iT} O\left(\log^2(T)\right)\frac{x^\sigma}{|s|}\,|ds| \\ &\ll \frac{\log^2(T)}{T}\int_{-1}^{\sigma_x} x^\sigma\,d\sigma \ll \frac{x\log^2(T)}{T\log(x)}\end{aligned}$$

holds for that part of the contour that crosses the critical strip.

The inequality

$$\left|\sum_{n<x}\Lambda(n) - \frac{1}{2\pi i}\int_{\sigma_x-iT}^{\sigma_x+iT}\left(-\frac{\zeta'(s)}{\zeta(s)}\right)\frac{x^s}{s}\,ds\right| \le 2x^{\sigma_x}\sum_{n=1}^\infty \frac{\Lambda(n)n^{-\sigma_x}}{\max(1, T|\log(x/n)|)}$$

follows from Proposition 5.2. Here

$$\begin{aligned}x^{\sigma_x}\sum_{|\log(x/n)|\ge\frac12} \frac{\Lambda(n)n^{-\sigma_x}}{\max(1,T|\log(x/n)|)} &\ll x\sum_{n=1}^\infty \frac{\Lambda(n)n^{-\sigma_x}}{T} \ll \frac{x}{T}\left(-\frac{\zeta'(\sigma_x)}{\zeta(\sigma_x)}\right) \\ &\ll \frac{x}{T}\frac{1}{\sigma_x - 1} \ll \frac{x\log(x)}{T}\end{aligned}$$

since $-\zeta'(s)/\zeta(s)$ has a simple pole at $s = 1$. The estimate

$$\left|\log\left(\frac{x}{n}\right)\right| = \left|\log\left(1 + \frac{n-x}{x}\right)\right| \gg \frac{|x-n|}{x}$$

holds for $|\log(x/n)| < 1/2$ and implies that

$$x^{\sigma_x} \sum_{|\log(x/n)|<\frac{1}{2}} \frac{\Lambda(n)n^{-\sigma_x}}{\max(1, T|\log(x/n)|)} \ll \sum_{|\log(x/n)|<\frac{1}{2}} \frac{\Lambda(n)}{\max(1, T|x-n|/x)}$$
$$\ll \sum_{|\log(x/n)|<\frac{1}{2}} \Lambda(n) \min\left(1, \frac{x}{T|x-n|}\right).$$

Here

$$\sum_{|n-x|\leq\frac{1}{2}} \Lambda(n) \min\left(1, \frac{x}{T|x-n|}\right) \ll \log(x) \min\left(1, \frac{x}{T\langle x\rangle}\right)$$

since $\Lambda(n) \leq \log(n)$ for all n, while $\Lambda(n) = 0$ if n is not a prime power. Similarly

$$\sum_{\substack{|\log(x/n)|<\frac{1}{2} \\ |n-x|>\frac{1}{2}}} \Lambda(n) \min\left(1, \frac{x}{T|x-n|}\right) \ll \frac{x}{T} \sum_{\substack{|\log(x/n)|<\frac{1}{2} \\ |n-x|\geq\frac{1}{2}}} \frac{\log(n)}{|n-x|}$$
$$\ll \frac{x}{T} \sum_{x+\frac{1}{2}<n<e^{1/2}x} \frac{\log(n)}{n-x} \ll \frac{x\log^2(x)}{T}.$$

Collecting all the information obtained yields a bound

$$\left|\psi(x) - \left(x - \sum_{|\mathrm{Im}(\rho)|<T} \frac{x^\rho}{\rho} - \frac{\zeta'(0)}{\zeta(0)} - \frac{1}{2}\log(1-x^{-2})\right)\right|$$
$$\ll \frac{\log(T)}{Tx\log(x)} + \frac{x\log^2(T)}{T\log(x)} + \frac{x\log(x)}{T}$$
$$+ \log(x)\min\left(1, \frac{x}{T\langle x\rangle}\right) + \frac{x\log^2(x)}{T}$$
$$\ll \frac{x\log^2(T)}{T\log(x)} + \log(x)\min\left(1, \frac{x}{T\langle x\rangle}\right) + \frac{x\log^2(x)}{T}$$

valid for all $x > e$ that are not prime powers and values of $T \geq 3$ restricted as above. Since

$$\sum_{|\mathrm{Im}(\rho)-T|\leq 1} \frac{|x^\rho|}{|\rho|} \ll \frac{x\log(T)}{T}$$

by Proposition 10.2, we may remove the restriction on T at the cost of adding a term $x\log(T)/T$ to the right-hand side. But

$$\frac{x\log(T)}{T} \leq \sqrt{\frac{x\log^2(T)}{T\log(x)}\frac{x\log^2(x)}{T}} \ll \frac{x\log^2(T)}{T\log(x)} + \frac{x\log(x)}{T}$$

by the inequality between the geometric and the arithmetic mean. So the term $x\log(T)/T$ is dominated by the sum of two other terms and may be omitted. The determination of the constant $\zeta'(0)/\zeta(0)$ is left as an exercise. □

10.2. The primes and RH

There is a precise relationship between the size of the error term in the Prime Number Theorem and the real parts of the zeros of the Riemann zeta function.

Proposition 10.4 (Theorem of von Koch). *If $\zeta(s) \neq 0$ in the half plane $\sigma > \alpha$, then $\psi(x) = x + O(x^\alpha \log^2(x))$. While if $\psi(x) = x + O(x^{\alpha+\varepsilon})$ for any $\varepsilon > 0$, then $\zeta(s) \neq 0$ in the half plane $\sigma > \alpha$.*

Proof. Assume that $\zeta(s) \neq 0$ for $\sigma > \alpha$ and choose $T = x = [x] + 1/2 \geq 3$ in Proposition 10.3. Then

$$\psi(x) = x - \sum_{|\mathrm{Im}(\rho)|<T} \frac{x^\rho}{\rho} + O(\log^2(x))$$

where

$$\begin{aligned}\left|\sum_{|\mathrm{Im}(\rho)|<T} \frac{x^\rho}{\rho}\right| &\leq \sum_{|\mathrm{Im}(\rho)|<T} \frac{|x^\rho|}{|\rho|} \ll x^\alpha \sum_{n\leq T} \frac{N(n)-N(n-1)}{n} \\ &\ll x^\alpha \sum_{n\leq T} \frac{\log(n)}{n} \ll x^\alpha \log^2(T) = x^\alpha \log^2(x)\end{aligned}$$

by Proposition 10.3.

If $\psi(x) = x + O_\varepsilon(x^{\alpha+\varepsilon})$ for any $\varepsilon > 0$, then

$$\sum_{n\leq x}(1-\Lambda(n)) = [x] - \psi(x) = O_\varepsilon(x^{\alpha+\varepsilon}),$$

and so the Dirichlet series

$$\sum_{n=1}^{\infty}(1-\Lambda(n))n^{-s} = \zeta(s) + \frac{\zeta'(s)}{\zeta(s)}$$

converges in $\sigma > \alpha$. Then the sum is holomorphic in this half plane, and thus $\zeta(s) \neq 0$ there, since $\zeta(s)$ is holomorphic everywhere except for a pole at $s = 1$. □

There are infinitely many zeros of $\zeta(s)$ in the critical strip $0 \leq \sigma \leq 1$, and these lie symmetrically about the critical line $\sigma = 1/2$ by the functional equation. The theorem of von Koch implies that the horizontal distribution of nontrivial zeros of $\zeta(s)$ that would be the most favorable for the error term in the Prime Number Theorem is that they all lie on the critical line $\sigma = 1/2$.

That all the zeros of $\zeta(s)$ in $0 \leq \sigma \leq 1$ lie on $\sigma = 1/2$ was conjectured by Riemann in the 1859 paper where he initiated the study of $\zeta(s)$, introduced the method of contour integrals and gave the first explicit formula in prime number theory. This problem is known as the *Riemann Hypothesis* and is still unresolved after 150 years.

The Riemann Hypothesis is reckoned to be one of the most important open problems in mathematics.

10.3. The Guinand-Weil formula

Around 1942 A. P. Guinand realized that the von Mangoldt and similar explicit formulas are instances of a more general principle, relating a sum of a function taken over prime powers to a sum of another function taken over zeros of the zeta function. In 1952 A. Weil gave a very general explicit formula in this vein in terms of the Fourier transform. We are going to prove a version of the Guinand-Weil explicit formula in terms of the Mellin transform. The Fourier and Mellin transforms are related by a change of variable, so the difference between the version that we prove and Weil's original version is not essential.

For a function $f : (0, \infty) \to \mathbb{C}$ on the positive reals, we may write down its *Mellin transform*

$$F(s) = (\mathcal{M}f)(s) = \int_0^\infty f(u)u^s \frac{du}{u},$$

which may however fail to exist. We impose the condition that f shall be infinitely many times continuously differentiable on $(0, \infty)$ and that

$$\lim_{u \to 0+} f^{(m)}(u)u^{-n} = 0 \quad \text{and} \quad \lim_{u \to +\infty} f^{(m)}(u)u^{n} = 0$$

for any nonnegative integers m and n. The latter condition states that $f^{(m)}$ decays faster than polynomially at both endpoints of the interval. It clearly implies that the integral converges uniformly on any compact set in the complex plane, and thus that the Mellin transform of f is an entire function. We call functions f that satisfy the above conditions *test functions.*

Integrating by parts k times yields

$$(\mathcal{M}f)(s) = \frac{(-1)^k}{s(s+1)\cdots(s+k-1)} \int_0^\infty f^{(k)}(u) u^{s+k} \frac{du}{u}$$

since the boundary terms vanish, due to the faster than polynomial decay of f and its derivatives at the endpoints. Clearly the last integral is bounded on any vertical strip $a \leq \sigma \leq b$, so the Mellin transform of a test function decays faster than polynomially uniformly as $s \to \infty$ on any fixed vertical strip.

The involution

$$f(u) \mapsto f^*(u) = \frac{1}{u} f\left(\frac{1}{u}\right)$$

takes test functions to test functions. Then

$$\begin{aligned}(\mathcal{M}f^*)(s) &= \int_0^\infty f^*(u) u^s \frac{du}{u} = \int_0^\infty u^{-1} f(u^{-1}) u^s \frac{du}{u} \\ &= \int_\infty^0 v f(v) v^{-s} \frac{-dv}{v^2} = (\mathcal{M}f)(1-s)\end{aligned}$$

by a change of variable. Note that $s \mapsto 1-s$ is the involution from the functional equation of the zeta function.

Proposition 10.5 (Guinand-Weil formula). *The formula*

$$\begin{aligned}(\mathcal{M}f)(1) - \sum_\rho (\mathcal{M}f)(\rho) + (\mathcal{M}f)(0) &= \sum_{p^k} \log(p)(f(p^k) + f^*(p^k)) \\ &+ \frac{\log(4\pi) + \gamma}{2}(f(1) + f^*(1)) \\ &+ \int_1^\infty \left(f(u) + f^*(u) - \frac{f(1)}{u} - \frac{f^*(1)}{u}\right) \frac{u\,du}{u^2 - 1}\end{aligned}$$

holds for all test functions f. The sum over ρ is taken over all nontrivial zeros of $\zeta(s)$ counted with multiplicity.

Proof. We consider the integral

$$\frac{1}{2\pi i} \oint_{R_T} F(s) \frac{Z'(s)}{Z(s)}\, ds$$

where F is the Mellin transform of f, Z is the completed Riemann zeta function, $T > 0$ is not the ordinate of any zero of Z, and R_T is the rectangular path from $3/2 + iT$ through $-1/2 + iT$ and $-1/2 - iT$ and $3/2 - iT$ and back to $3/2 + iT$. It may be computed in two different ways; by means of the Euler product formula, or by means of the Hadamard factorization of $\xi(s)$.

Logarithmically differentiating $Z(s)$ and using the Hadamard factorization of $\xi(s)$ yields

$$\frac{Z'(s)}{Z(s)} = -\frac{1}{s} - \frac{1}{s-1} + (\log(2\pi)/2 - 1 - \gamma/2) + \sum_\rho \left(\frac{1}{s-\rho} + \frac{1}{\rho}\right).$$

The series over ρ converges uniformly on R_T by Proposition 10.2. Integrating $Z'(s)/Z(s)$ against $F(s)$ around R_T, interchanging sum and integral by uniform convergence, and applying Cauchy's formula gives

$$\frac{1}{2\pi i}\oint_{R_T} F(s)\frac{Z'(s)}{Z(s)}\,ds = -F(0) - F(1) + \sum_\rho F(\rho).$$

By an argument from the proof of Proposition 10.2 there exists a sequence $\mathcal{T}$ of values of T tending to infinity such that

$$\frac{Z'(s)}{Z(s)} = O(\log^2(T))$$

on $\mathcal{T}$. In fact there is at least one such value of T in each interval of length 1. Denoting the horizontal seqments of R_T by H_T we conclude that

$$\frac{1}{2\pi i}\int_{H_T} F(s)\frac{Z'(s)}{Z(s)}\,ds \to 0$$

as $T \to +\infty$ on $\mathcal{T}$. For $F(s)$ decays faster than polynomially as $s \to \infty$ in the strip $-1/2 \le \sigma \le 3/2$.

Next

$$\begin{aligned}\frac{1}{2\pi i}\int_{-1/2+iT}^{-1/2-iT} F(s)\frac{Z'(s)}{Z(s)}\,ds &= \frac{1}{2\pi i}\int_{3/2-iT}^{3/2+iT} F(1-w)\frac{Z'(1-w)}{Z(1-w)}(-1)\,dw \\ &= \frac{1}{2\pi i}\int_{3/2-iT}^{3/2+iT} F(1-w)\frac{Z'(w)}{Z(w)}\,dw\end{aligned}$$

by the functional equation of $Z(s)$.

Logarithmically differentiate

$$Z(s) = \pi^{-s/2}\Gamma\Big(\frac{s}{2}\Big)\prod_p (1-p^{-s})^{-1}$$

to obtain

$$\frac{Z'(s)}{Z(s)} = -\frac{\log(\pi)}{2} + \frac{1}{2}\frac{\Gamma'(s/2)}{\Gamma(s/2)} - \sum_{p^k} \log(p)p^{-ks}$$

for $\sigma > 1$.

To proceed, it is necessary to calculate the integral along the vertical line from $3/2 - i\infty$ to $3/2 + i\infty$ of each term on the right-hand side against

$F(s)$. We begin with integrals of the form

$$\int_{3/2-i\infty}^{3/2+i\infty} F(s)e^{\alpha s}\,ds$$

with α real. Now

$$\begin{aligned}\int_{3/2-i\infty}^{3/2+i\infty} F(s)e^{\alpha s}\,ds &= \int_{3/2-i\infty}^{3/2+i\infty} \left(\frac{1}{s(s+1)}\int_0^\infty f''(u)u^{s+2}\frac{du}{u}\right)e^{\alpha s}\,ds \\ &= \int_0^\infty uf''(u)\left(\int_{3/2-i\infty}^{3/2+i\infty}\frac{e^{(\alpha+\log(u))s}}{s(s+1)}ds\right)du\end{aligned}$$

using integration by parts, and an interchange of the order of integration. To justify the interchange, note that the improper double integral

$$\int_{3/2-i\infty}^{3/2+i\infty}\int_0^\infty \frac{f''(u)u^{s+1}e^{\alpha s}}{s(s+1)}\,du\,ds$$

is absolutely convergent. For any $\epsilon > 0$ it may be approximated with an error of less than $\epsilon/3$ by the corresponding double integral over a rectangle given by $U^{-1} \le u \le U$ and $-T \le t \le T$ with $s = 3/2 + it$, for all T and U sufficiently large. Since the integrand is continuous, the latter proper double integral equals the two corresponding proper iterated integrals in either order, by a result from multivariable calculus. But the two improper iterated integrals are also absolutely convergent, and each may likewise be approximated by proper iterated integrals with an error less than $\epsilon/3$ for all T and U sufficiently large. Alternatively, though less elementarily, we may apply Fubini's theorem.

The innermost integral above may be evaluated by residue calculus. When $\alpha + \log(u) < 0$ consider the integral

$$\int_{B_T}\frac{e^{(\alpha+\log(u))s}}{s(s+1)}\,ds = 0$$

where B_T is the rectangular contour with corners at $3/2 - iT, 3/2 + iT, T + iT, T - iT$. The integral is zero by Cauchy's theorem. The integrand is $O(T^{-2})$ on the three rightmost sides of the rectangle, and the total length of these three sides is $O(T)$. Thus the contribution from this part of B_T disappears as $T \to +\infty$. Hence

$$\int_{3/2-i\infty}^{3/2+i\infty}\frac{e^{(\alpha+\log(u))s}}{s(s+1)}\,ds = 0$$

if $\alpha + \log(u) < 0$. When $\alpha + \log(u) > 0$ consider the integral

$$\int_{C_T}\frac{e^{(\alpha+\log(u))s}}{s(s+1)}\,ds = 2\pi i\left(1 - e^{-\alpha-\log(u)}\right)$$

where C_T is the rectangular contour with corners at $3/2-iT, 3/2+iT, -T+iT, -T-iT$ for $T>1$. The value of the integral is determined by the residues of the integrand at the singularities $s=0,1$ inside C_T. The integrand is $O(T^{-2})$ on the three leftmost sides of the rectangle, and the total length of these three sides is $O(T)$. Thus the contribution from this part of C_T disappears as $T\to+\infty$. Hence

$$\int_{3/2-i\infty}^{3/2+i\infty}\frac{e^{(\alpha+\log(u))s}}{s(s+1)}\,ds = 2\pi i\Big(1-e^{-\alpha-\log(u)}\Big)$$

if $\alpha+\log(u)>0$. Then

$$\begin{aligned}\int_{3/2-i\infty}^{3/2+i\infty}F(s)e^{\alpha s}\,ds &= \int_{e^{-\alpha}}^{\infty}uf''(u)2\pi i\Big(1-e^{-\alpha-\log(u)}\Big)du\\ &= \int_{e^{-\alpha}}^{\infty}f''(u)2\pi i\big(u-e^{-\alpha}\big)du = 2\pi i f(e^{-\alpha})\end{aligned}$$

using integration by parts. Now

$$\frac{1}{2\pi i}\int_{3/2-i\infty}^{3/2+i\infty}F(s)\frac{\log(\pi)}{2}\,ds = \frac{\log(\pi)}{2}f(1)$$

and

$$\frac{1}{2\pi i}\int_{3/2-i\infty}^{3/2+i\infty}F(1-s)\frac{\log(\pi)}{2}\,ds = \frac{\log(\pi)}{2}f^*(1).$$

Furthermore

$$\frac{1}{2\pi i}\int_{3/2-i\infty}^{3/2+i\infty}F(s)\log(p)p^{-ks}\,ds = \log(p)f(p^k)$$

and

$$\frac{1}{2\pi i}\int_{3/2-i\infty}^{3/2+i\infty}F(1-s)\log(p)p^{-ks}\,ds = \log(p)f^*(p^k).$$

We wish to conclude that

$$\begin{aligned}\frac{1}{2\pi i}\int_{3/2-i\infty}^{3/2+i\infty}F(s)\sum_{p^k}\log(p)p^{-ks}\,ds &= \sum_{p^k}\frac{1}{2\pi i}\int_{3/2-i\infty}^{3/2+i\infty}F(s)\log(p)p^{-ks}\,ds\\ &= \sum_{p^k}\log(p)f(p^k)\end{aligned}$$

and similarly for the integral against $F(1-s)$. The series is uniformly convergent, but this is not enough to immediately permit an interchange of series and integral, for the integral is improper. But the integral is absolutely convergent, so the interchange is valid by the discussion in Section 8.4.

It remains to calculate the integral

$$\frac{1}{2\pi i}\int_{3/2-i\infty}^{3/2+i\infty} F(s)\frac{1}{2}\frac{\Gamma'(s/2)}{\Gamma(s/2)}\,ds,$$

which we note is absolutely convergent by Proposition 10.1. Now

$$\begin{aligned}\int_{3/2-i\infty}^{3/2+i\infty} F(s)\frac{\Gamma'(s/2)}{\Gamma(s/2)}\,ds \\ &= \int_{3/2-i\infty}^{3/2+i\infty}\left(\frac{1}{s(s+1)}\int_0^\infty f''(u)u^{s+2}\frac{du}{u}\right)\frac{\Gamma'(s/2)}{\Gamma(s/2)}\,ds \\ &= \int_0^\infty uf''(u)\left(\int_{3/2-i\infty}^{3/2+i\infty}\frac{u^s}{s(s+1)}\frac{\Gamma'(s/2)}{\Gamma(s/2)}\,ds\right)du\end{aligned}$$

using integration by parts, and an interchange of the order of integration as before. The innermost integral may be evaluated by residue calculus. When $u<1$ consider the integral

$$\int_{B_T}\frac{u^s}{s(s+1)}\frac{\Gamma'(s/2)}{\Gamma(s/2)}\,ds = 0$$

where B_T is the contour used earlier. The integral is zero by Cauchy's theorem. By Proposition 10.1 the integrand is $O(T^{-2}\log(T))$ on the three rightmost sides of the rectangle, and the total length of these three sides is $O(T)$. Thus the contribution from this part of B_T disappears as $T\to+\infty$. Hence

$$\int_{3/2-i\infty}^{3/2+i\infty}\frac{u^s}{s(s+1)}\frac{\Gamma'(s/2)}{\Gamma(s/2)}\,ds = 0$$

if $u<1$.

One sees directly from the definition of the gamma function that

$$\frac{\Gamma'(s)}{\Gamma(s)} = -\frac{1}{s} - \gamma + O(|s|)$$

as $s\to 0$. This permits us to calculate the residue

$$\operatorname{Res}_{s=0}\left(\frac{u^s}{s(s+1)}\frac{\Gamma'(s/2)}{\Gamma(s/2)}\right) = 2-\gamma-2\log(u)$$

at the double pole $s=0$ of the integrand. The other residues are

$$\operatorname{Res}_{s=-1}\left(\frac{u^s}{s(s+1)}\frac{\Gamma'(s/2)}{\Gamma(s/2)}\right) = -\frac{1}{u}\frac{\Gamma'(-1/2)}{\Gamma(-1/2)}$$

and

$$\operatorname{Res}_{s=-2n}\left(\frac{u^s}{s(s+1)}\frac{\Gamma'(s)}{\Gamma(s)}\right) = -\frac{2u^{-2n}}{2n(2n-1)}$$

at the simple poles $s=-1$ and $s=-2n$ of the integrand, as one sees from the definition of the gamma function.

Choosing $T = 2N+1$ with $N \geq 1$ an integer and integrating around the contour C_T used before, we obtain

$$\begin{aligned}\frac{1}{2\pi i}\int_{C_T} \frac{u^s}{s(s+1)}\frac{\Gamma'(s/2)}{\Gamma(s/2)}\,ds \\ = 2-\gamma-2\log(u) - \frac{1}{u}\frac{\Gamma'(-1/2)}{\Gamma(-1/2)} - \sum_{n=1}^{N}\frac{u^{-2n}}{n(2n-1)}\end{aligned}$$

by the Residue Theorem. The integrand is $O(T^{-2}\log(T))$ on the two horizontal segments of C_T, by Proposition 10.1, while their total length is $O(T)$, so their contribution to the integral around C_T tends to zero as $N \to +\infty$. We need to bound the contribution from the leftmost vertical segment of C_T, but Proposition 10.1 will not serve here. Logarithmically differentiating the definition of $\Gamma(s)$ yields

$$\frac{\Gamma'(s/2)}{\Gamma(s/2)} = -\frac{2}{s} - \gamma + s\sum_{n=1}^{\infty}\frac{1}{n(s+2n)}.$$

Due to the special choice of T we have $|s+n| \geq 1$ on the leftmost vertical segment, and we may bound the logaritmic derivative of the gamma function by

$$\begin{aligned}\left|\frac{\Gamma'(s/2)}{\Gamma(s/2)}\right| &\leq \frac{2}{|s|} + \gamma + |s|\sum_{n=1}^{N}\frac{1}{n} + |s|\sum_{n=N+1}^{\infty}\frac{1}{n(2n-2N-1)} \\ &= O(T\log(T))\end{aligned}$$

there. Then the integrand is

$$O(u^{-T}T^{-1}\log(T))$$

on that segment, while its length is $O(T)$, so the integral over the leftmost vertical segment tends to zero as $N \to +\infty$. Hence

$$\begin{aligned}\frac{1}{2\pi i}\int_{3/2-i\infty}^{3/2+i\infty} \frac{u^s}{s(s+1)}\frac{\Gamma'(s/2)}{\Gamma(s/2)}\,ds \\ = 2-\gamma-2\log(u) - \frac{1}{u}\frac{\Gamma'(-1/2)}{\Gamma(-1/2)} - \sum_{n=1}^{\infty}\frac{u^{-2n}}{n(2n-1)} \\ = 2-\gamma-2\log(u) - \frac{2-\gamma}{u} + \frac{1}{u}\sum_{n=1}^{\infty}\frac{1}{n(2n-1)} - \sum_{n=1}^{\infty}\frac{u^{-2n}}{n(2n-1)}\end{aligned}$$

if $u > 1$. Then

$$\begin{aligned}
\frac{1}{2\pi i}\int_{3/2-i\infty}^{3/2+i\infty} & F(s)\frac{1}{2}\frac{\Gamma'(s/2)}{\Gamma(s/2)}\,ds \\
&= \int_1^\infty f''(u)\left(\left(1-\frac{\gamma}{2}\right)(u-1) - u\log(u) + \sum_{n=1}^{\infty}\frac{1-u^{-2n+1}}{2n(2n-1)}\right)du \\
&= \int_1^\infty f'(u)\frac{\gamma+\log(u^2-1)}{2}\,du
\end{aligned}$$

by integration by parts. To prepare for integration by parts a second time rewrite as

$$\begin{aligned}
\int_1^\infty f'(u)\frac{\gamma+\log(u^2-1)}{2}\,du = & \int_1^\infty \left(f'(u)+\frac{f(1)}{u^2}\right)\frac{\gamma+\log(u^2-1)}{2}\,du \\
& - \int_1^\infty \frac{f(1)}{u^2}\frac{\gamma+\log(u^2-1)}{2}\,du
\end{aligned}$$

and note that

$$\int_1^\infty \frac{f(1)}{u^2}\frac{\gamma+\log(u^2-1)}{2}\,du = \frac{\gamma}{2}f(1) + \log(2)f(1).$$

Integrating by parts a second time yields

$$\begin{aligned}
\int_1^\infty & \left(f'(u)+\frac{f(1)}{u^2}\right)\frac{\gamma+\log(u^2-1)}{2}\,du \\
&= \left(f(u)-\frac{f(1)}{u}\right)\frac{\gamma+\log(u^2-1)}{2}\Bigg|_1^\infty \\
&\quad - \int_0^\infty \left(f(u)-\frac{f(1)}{u}\right)\frac{u\,du}{u^2-1} \\
&= -\int_0^\infty \left(f(u)-\frac{f(1)}{u}\right)\frac{u\,du}{u^2-1}
\end{aligned}$$

by the differentiability of f at $u = 1$.

Assembling the results of the various calculations gives the Guinand-Weil explicit formula. □

10.4. Notes

Proposition 10.1 is due to T. J. Stieltjes [**Sti99**]. Proposition 10.2 is a corollary of a more precise result established by von Mangoldt [**vM05**] in 1905. The von Mangoldt explicit formula is in [**vM95**]. Truncated forms like Proposition 10.3 were established by N. H. von Koch [**vK10**] and Landau [**Lan12**].

Proposition 10.4 is due to von Koch [**vK01**].

The earliest work bearing on the truth or falsity of the Riemann Hypothesis was found in Riemann's papers many years after his death. In particular it was discovered that he had verified that the first zero above the real axis lies on the critical line, and had calculated this zero $\rho = 1/2 + 14.13\ldots i$ to several decimal places. Today it is known, due to work of X. Gourdon and P. Demichel, that the first ten trillion zeros lie on the critical line.

There *are* theoretical results that offer a measure of evidence in favor of the Riemann hypothesis, and perhaps a faint clue as to why it may hold true. To formulate one such result, we introduce yet another counting function $N(\sigma, T)$ for the zeros of $\zeta(s)$. This is the number, counted with multiplicity, of such zeros $\rho = \beta + i\gamma$ that satisfy the inequalities $\beta > \sigma$ and $0 < \gamma \leq T$. In 1914 H. A. Bohr and Landau proved that $N(\sigma, T) = O_\sigma(T)$ for any $\sigma > 1/2$. This carries the striking implication that any ε-neighborhood, however small, of the critical line contains all but an infinitesimal fraction of the nontrivial zeros of $\zeta(s)$ in the asymptotic sense. Though the Bohr-Landau theorem may seem persuasive evidence in favor of the Riemann Hypothesis, there is an example that tends to weaken it. This is a function $F(s)$, discovered by H. Davenport and H. A. Heilbronn in 1936, that resembles $\zeta(s)$ closely in many respects, being a linear combination of two Dirichlet L-functions. The Davenport-Heilbronn function has a Dirichlet series expansion and a functional equation, but no Euler product, and has infinitely many zeros in the half plane $\sigma > 1$. From later work of S. M. Voronin it is known that the number of zeros of $F(s)$ in any rectangle of the form $1/2 < \sigma_1 < \sigma < \sigma_2 < 1$ and $0 < t \leq T$ is both $\gg T$ and $\ll T$. But Voronin also showed that the number of zeros of the Davenport-Heilbronn function on $\sigma = 1/2$ up to the ordinate T grows faster than any constant multiple of T. Then the functional equation of $F(s)$ implies that any ε-neighborhood, however small, of the critical line contains all but an infinitesimal fraction of the zeros of $F(s)$ in $0 < \sigma < 1$ in the asymptotic sense, and yet the Davenport-Heilbronn example badly fails its analogue of the Riemann Hypothesis.

Rather more important for the Riemann Hypothesis are the stronger assertions about $N(\sigma, T)$ known as zero density theorems. Many such results have been proved, but we will only quote A. E. Ingham's zero density theorem

$$N(\sigma, T) = O_\sigma\left(T^{3(1-\sigma)/(2-\sigma)} \log^5(T)\right), \qquad 1/2 < \sigma < 1.$$

The Davenport-Heilbronn example violates its analogue of the Ingham zero density theorem. It seems very significant that the proofs of the zero-density estimates depend indirectly (via the Möbius function) on the Euler product for $\zeta(s)$, while the proof of the Bohr-Landau theorem does not.

Proposition 10.5 is due in a slightly different form to Weil [**Wei52**], as a special case of a much more general explicit formula for Hecke L-functions. K. Barner [**Bar81**] obtained a simplification significant for applications of the formula. The explicit formula of Guinand is in [**Gui42**].

Exercises

(1) Show that $\zeta'(0)/\zeta(0) = \log(2\pi)$.

(2) Show that the multiplicity of any nontrivial zero ρ of $\zeta(s)$ is $O(\log|\rho|)$.

(3) † a) Find an explicit formula for

$$\psi_1(x) = \sum_{n\le x}(x-n)\Lambda(n)$$

in terms of the nontrivial zeros of the Riemann zeta function. Explain why we cannot obtain it easily just by integrating the von Mangoldt explicit formula.
b) Deduce that $\psi_1(x) \sim x^2/2$ and then that $\psi(x) \sim x$. Explain why this mode of reasoning cannot be applied to obtain $\psi(x) \sim x$ directly from the von Mangoldt explicit formula.
c) Show that

$$\psi_1(x) - 2\psi_1(x/2) + 3\psi_1(x/3) - \cdots = \frac{\log(2)}{2}x^2 + O(x)$$

as $x \to +\infty$.

(4) Find an explicit formula for $M(x)$ by a formal calculation on the assumption that $\zeta(s)$ has only simple zeros. The structure of the formula indicates that $M(x)$ may be harder to bound than $\psi(x) - x$, and so it turns out in practice.

(5) Show that the estimate

$$\log(\Gamma(s)) = \left(s - \frac{1}{2}\right)\log(s) - s + \frac{1}{2}\log(2\pi) + O_\delta\left(\frac{1}{|s|}\right)$$

holds on $|\mathrm{Arg}(s)| < \pi/2 - \delta$ for any fixed $\delta > 0$. (Hint: Proceed with care and integrate from $+\infty$ on the real axis in the estimate for $\Gamma'(s)/\Gamma(s)$.)

(6) a) Show that

$$-\frac{\zeta'(s)}{\zeta(s)} = -\sum_\rho\left(\frac{1}{\rho} + \frac{1}{s-\rho}\right) + O(\log|t|)$$

for $s = \sigma + it$ with $-1 \le \sigma \le 2$ and $|t| \ge 3$.
b) † Show that $\zeta(s) \ne 0$ in the region $\sigma \ge 1 - c'/\log(4+|t|)$ with some $c' > 0$.
c) Show that

$$\psi(x) = x + O\left(x\, e^{-c\sqrt{\log(x)}}\right)$$

with some $c > 0$ (de la Vallée Poussin).

(7) Show that if $\psi(x) = x + O_\varepsilon(x^{1/2+\varepsilon})$ for any $\varepsilon > 0$, then $\psi(x) = x + O(x^{1/2}\log^2(x))$.

(8) Show that no estimate $\psi(x) = x + O(x^\theta)$ can hold with $\theta < 1/2$.

(9) Show that if $M(x) \ll x^{\alpha+\varepsilon}$ for all $\varepsilon > 0$, then $\zeta(s) \neq 0$ for $\sigma > \alpha$. The opposite implication also holds, but this is harder to prove.

(10) Show that no estimate $M(x) \ll x^\theta$ can hold with $\theta < 1/2$. It is believed that the bound is also false for $\theta = 1/2$, but this is as yet unproved.

(11) Show that
$$\pi(x) = \mathrm{li}(x) + O(\sqrt{x}\log(x))$$
if the Riemann Hypothesis holds.

(12) Establish Riemann's integral representation
$$\Pi(x) = \frac{1}{2\pi i}\int_{c-i\infty}^{c+i\infty} \log(\zeta(s))\frac{x^s}{s}\,ds,$$
for any $x > 0$ not an integer and any $c > 1$ by means of the Perron formula.

Chapter 11

Supplementary Exercises

Exercises

(1) Show that the k-th prime p_k satisfies $0.71{\cdot}k\log(k) \leq p_k \leq 1.45{\cdot}k\log(k)$ for all k sufficiently large.

(2) Show that the bound

$$\sum_{n\leq x}\Lambda(n)n^{-\sigma} = O\Big(\frac{x^{1-\sigma}}{1-\sigma}\Big)$$

holds uniformly in σ for $0 \leq \sigma < 1$.

(3) Let $f(k,n)$ be the count of the number of positive integers m with $k|m|n$. Find an asymptotic estimate for the sum of $f(k,n)$ over $k, n \leq x$.

(4) Show that there are infinitely many consecutive pairs of squarefrees.

(5) Let f and w be arithmetic functions. Suppose that the values of f are positive integers, and that $f(n) \to +\infty$ as $n \to +\infty$. Let

$$c_m = \sum_{f(n)=m} w(n)$$

be the weighted number of solutions of the equation $f(n) = m$. Express the formal Dirichlet series

$$\sum_{m=1}^{\infty} c_m m^{-s}$$

in terms of f and w and find a formal Euler product formula when f and w are assumed totally multiplicative.

(6) Expand $A(s) = \zeta'(s) + \zeta^2(s) - 2\gamma\zeta(s)$ into a Dirichlet series and show that this series converges at least in the half plane $\sigma > 1/2$. What is the abscissa of absolute convergence? (T. J. Stieltjes)

(7) Calculate the Dirichlet density of the squarefrees contained in the arithmetic progression $n \equiv 1 \pmod{q}$, relative to the set of all squarefrees.

(8) a) Find a set of positive integers that contains arbitrarily long finite arithmetic progressions, but does not contain any infinite arithmetic progression. Show that the primes do not contain any infinite arithmetic progression. The recent Green-Tao theorem states that the primes contain arbitrarily long finite arithmetic progressions.
b) The unproved Hypothesis H of A. B. M. Schinzel states that if

$$P_1(x), P_2(x), \ldots, P_r(x)$$

are irreducible polynomials in $\mathbb{Z}[x]$ and their product has no fixed prime divisor, then the values $P_1(n), P_2(n), \ldots, P_r(n)$ should be simultaneously prime for infinitely many integers n. It is clear that Dirichlet's theorem on primes in arithmetic progressions is a consequence. Show that the Green-Tao theorem would also follow from Hypothesis H.

(9) Establish the estimate

$$\int_0^1 |f(\alpha)|^4 \, d\alpha \ll N^{2+\varepsilon}$$

for

$$f(\alpha) = \sum_{m \le N} e(\alpha m^k)$$

with $k \ge 2$ without using Hua's lemma.

(10) Show that a necessary, though not sufficient, condition for the Weyl inequality to be nontrivial is that the degree k of P be very small in terms of N; make this conclusion quantitative.

(11) Let F be defined by

$$\log(x) = \sum_{n \le x} F(x/n)$$

for $x \ge 1$. Find an absolutely convergent contour integral for $F(x)$ when x is not an integer.

(12) a) Assume the PNT in the form that for every $\varepsilon > 0$ there exists some $x_0 = x_0(\varepsilon)$ so that $|\psi(x) - x| \le \varepsilon x$ for all $x \ge x_0$. Show that for every $\varepsilon > 0$

$$\left| \zeta(s) + \frac{\zeta'(s)}{\zeta(s)} \right| \le C_\varepsilon \frac{|s|}{\sigma} + \varepsilon \frac{|s|}{\sigma - 1}$$

on $\sigma > 1$ for some $C_\epsilon > 0$.
b) Prove by means of part a) that $\zeta(s)$ has no zero on the line $\sigma = 1$.

(13) Show that there is some constant c such that

$$\sum_{n\le x}\frac{\Lambda(n)}{n} = \log(x) + c + O_m\left(\frac{1}{\log^m(x)}\right)$$

for each positive integer m.

(14) a) Use

$$\frac{1}{\phi(q)}\sum_{\chi \bmod q}\left(-\frac{L'(s,\chi)}{L(s,\chi)}\right) = \sum_{n\equiv 1 \bmod q}\Lambda(n)n^{-s}$$

to show that the analytic continuation of the series on the right-hand side has a simple pole in any point where one of the Dirichlet L-functions modulo q has a zero. Explain why it is unnecessary here to assume that distinct Dirichlet L-functions modulo q have distinct zeros.
b) Show that if for some α an estimate

$$\psi(x;q,1) = \frac{x}{\phi(q)} + O(x^{\alpha+\varepsilon})$$

holds for each $\varepsilon > 0$, then $L(s,\chi) \neq 0$ on $\sigma > \alpha$ for all χ modulo q. With some work this implies that the same estimate holds for $\psi(x;q,a)$ for each a coprime with q.

(15) Use the Siegel-Walfisz theorem to find an upper bound for the least prime in any arithmetic progression $n \equiv a \pmod q$ with a and q coprime. Show that a much better bound may be achieved if there is no exceptional zero modulo q.

(16) Show that for any $\varepsilon > 0$ there exists a positive constant $c(\varepsilon)$ such that $|L(s,\chi)| \geq c(\varepsilon) > 0$ holds for all Dirichlet characters χ when $\sigma > 1+\varepsilon$.

(17) Determine for which of $s = -1, -2, \ldots$ the Dedekind zeta function $\zeta_K(s)$ is zero, and determine the multiplicities of the zeros. The answers depend on the field K.

(18) Suppose that $K/\mathbb{Q}$ is a normal extension of prime degree. Find the Dirichlet density of the set of rational primes that remain inert in the extension.

(19) Use the Ikehara theorem to show that $M(x) = o(x)$.

(20) Show that the class number of an imaginary quadratic field satisfies

$$h(d) \leq \sqrt{-4d}\log(-4d)$$

if $d < -4$ and d is squarefree.

(21) Assuming that the Riemann Hypothesis holds, how small can we take T in terms of x in the truncated von Mangoldt formula and still obtain the expected error term?

Solutions

(1) Note that $\pi(p_k) = k$ and that

$$0.69\frac{x}{\log(x)} \le \pi(x) \le 1.39\frac{x}{\log(x)}$$

for all x sufficiently large. Then

$$k = \pi(p_k) \le 1.39\frac{p_k}{\log(p_k)}$$

so

$$p_k \ge \frac{1}{1.39}k\log(p_k) \ge 0.71k\log(k)$$

for all k sufficiently large. Also

$$k = \pi(p_k) \ge 0.69\frac{p_k}{\log(p_k)}$$

so

$$p_k \le \frac{1}{0.69}k\log(p_k) \le 1.4493k\log(p_k)$$

for all k sufficiently large. Now

$$\begin{aligned} p_k &\le k\log(1.4493k\log(p_k)) \\ &\le 1.4493k\log(1.4993) + 1.4993k\log(k) + 1.4493k\log\log(p_k) \\ &\le 1.45k\log(k) \end{aligned}$$

for all k sufficiently large, since $\sqrt{p_k} \le \pi(p_k) = k$ then.

(2) Since $x \to \infty$, we may assume that $x \ge 1$. There exists some constant $C \ge 1$ so that $\psi(x) \le Cx$ for all $x \ge 1$. Then

$$\begin{aligned} \sum_{n\le x}\Lambda(n)n^{-\sigma} &= \psi(x)x^{-\sigma} - \int_1^x \psi(u)(-\sigma)u^{-\sigma-1}\,du \\ &\le Cx{\cdot}x^{-\sigma} + \sigma\int_1^x Cu{\cdot}u^{-\sigma-1}\,du \\ &= Cx^{1-\sigma} + \sigma C\left(\frac{x^{1-\sigma}}{1-\sigma} - \frac{1}{1-\sigma}\right) \le 2C\frac{x^{1-\sigma}}{1-\sigma}, \end{aligned}$$

by partial summation. The constant $2C$ does not depend on σ, and neither does the interval $x \ge 1$ on which the bound is valid.

(3) Only an asymptotic estimate is asked for, so we need not strive for a small error term. Now

$$\sum_{k\le x}\sum_{n\le x} f(k,n) = \sum_{k\le x}\sum_{n\le x}\sum_{k|m|n} 1 = \sum_{m\le x}\sum_{\substack{n\le x\\ m|n}}\sum_{k|m} 1 = \sum_{m\le x} d(m)\sum_{\substack{n\le x\\ m|n}} 1$$

$$= \sum_{m\le x} d(m)\left(\frac{x}{m} + O(1)\right) = x\sum_{m\le x}\frac{d(m)}{m} + O(x\log(x)),$$

by $D(x) = x\log(x) + O(x)$. Furthermore

$$\sum_{m\le x}\frac{d(m)}{m} = \frac{D(x)}{x} - \int_1^x D(u)\left(-\frac{1}{u^2}\right)du$$

$$= O(\log(x)) + \int_3^x \left(\frac{\log(u)}{u} + O\left(\frac{1}{u}\right)\right)du$$

$$= \frac{1}{2}\log^2(x) + O(\log(x)),$$

by partial summation and the same estimate for $D(x)$. This gives the asymptotic estimate

$$\sum_{k\le x}\sum_{n\le x} f(k,n) \sim \frac{x}{2}\log^2(x)$$

as $x \to \infty$.

(4) Suppose that A is any sequence that does not contain infinitely many pairs of consecutive terms. Then for all sufficiently large $a \in A$ we have $a+1 \notin A$. So if A has asymptotic density, then its asymptotic density is at most $1/2$. But the sequence of squarefrees has asymptotic density $6/\pi^2 > 1/2$.

(5) Have

$$\sum_{m=1}^{\infty} c_m m^{-s} = \sum_{m=1}^{\infty} m^{-s}\sum_{f(n)=m} w(n) = \sum_{n=1}^{\infty} w(n)f(n)^{-s}$$

since $m = f(n)$. Now

$$\sum_{m=1}^{\infty} c_m m^{-s} = \sum_{n=1}^{\infty} w(n)f(n)^{-s} = \prod_p\sum_{\alpha=0}^{\infty} w(p^\alpha)f(p^\alpha)^{-s}$$

$$= \prod_p\sum_{\alpha=0}^{\infty} w(p)^\alpha f(p)^{-\alpha s} = \prod_p \frac{1}{1-\omega(p)f(p)^{-s}}$$

by total multiplicativity. Note that $f(p) = 1$ cannot hold for any prime, because then $f(p^\alpha) = 1$ for all $\alpha \ge 1$, and this violates the requirement that $f(n) \to +\infty$ as $n \to +\infty$.

(6) Have

$$\begin{aligned}A(s) &= \zeta'(s) + \zeta^2(s) - 2\gamma\zeta(s)\\ &= \sum_{n=1}^{\infty}(-\log(n))n^{-s} + \sum_{n=1}^{\infty} d(n)n^{-s} - 2\gamma\sum_{n=1}^{\infty} n^{-s}\\ &= \sum_{n=1}^{\infty}(d(n) - \log(n) - 2\gamma)n^{-s},\end{aligned}$$

so the summatory function of the coefficients of this Dirichlet series equals $D(x) - T(x) - 2\gamma[x] = \Delta(x) + O(\log(x))$. Since $\Delta(x) = O(x^{1/2})$, the abscissa of convergence is at most $1/2$.

Next we consider the series

$$\begin{aligned}\sum_{n=1}^{\infty} |d(n) - \log(n) - 2\gamma|n^{-\sigma} &\geq \sum_{p}^{\infty} |d(p) - \log(p) - 2\gamma|p^{-\sigma}\\ &= \sum_{p}^{\infty} |2 - \log(p) - 2\gamma|p^{-\sigma},\end{aligned}$$

which diverges for $\sigma \leq 1$ because

$$\sum_{p} \frac{1}{p}$$

diverges, and converges for $\sigma > 1$ since

$$\begin{aligned}&\sum_{n=1}^{\infty} |d(n) - \log(n) - 2\gamma|n^{-\sigma}\\ &\qquad \leq \sum_{n=1}^{\infty} d(n)n^{-\sigma} + \sum_{n=1}^{\infty} \log(n)n^{-\sigma} + 2\gamma n^{-\sigma}.\end{aligned}$$

Thus the abscissa of absolute convergence equals 1.

(7) The desired density is

$$\delta = \lim_{\sigma\to 1^+} \left(\sum_{n=1}^{\infty} I_{q,1}(n)\mu^2(n)n^{-\sigma}\right) \Big/ \left(\sum_{n=1}^{\infty} \mu^2(n)n^{-\sigma}\right).$$

Here

$$\sum_{n=1}^{\infty} I_{q,1}(n)\mu^2(n)n^{-\sigma} = \frac{1}{\phi(q)}\sum_{\chi}\sum_{n=1}^{\infty} \chi(n)\mu^2(n)n^{-\sigma},$$

where

$$\sum_{n=1}^{\infty} \chi(n)\mu^2(n)n^{-\sigma} = \prod_{p}(1 + \chi(p)p^{-\sigma}) = \frac{L(\sigma,\chi)}{L(2\sigma,\chi^2)}.$$

Thus

$$\delta = \frac{1}{\phi(q)} \lim_{\sigma\to 1^+} \sum_{\chi} \frac{L(\sigma,\chi)\zeta(2\sigma)}{L(2\sigma,\chi^2)\zeta(\sigma)} = \frac{1}{\phi(q)} \lim_{\sigma\to 1^+} \frac{L(\sigma,\chi_0)\zeta(2\sigma)}{L(2\sigma,\chi_0^2)\zeta(\sigma)}$$

$$= \frac{1}{\phi(q)} \lim_{\sigma\to 1^+} \frac{\prod_{p|q}(1-p^{-\sigma})\zeta(\sigma)\zeta(2\sigma)}{\prod_{p|q}(1-p^{-2\sigma})\zeta(2\sigma)\zeta(\sigma)} = \frac{1}{q\prod_{p|q}(1-p^{-2})}$$

is the density.

(8) a) Let

$$A = \{1^3+0, 2^3+0, 2^2+1, 3^3+0, 3^3+1, 3^3+2, \dots\}.$$

Clearly A contains arbitrarily long arithmetic progressions. On the other hand

$$\sum_{a\in A} \frac{1}{a} \le \sum_{k=1}^{\infty} \frac{k}{k^3} < \infty,$$

while if A contained an infinite arithmetic progression, then the series of reciprocals would diverge, by a limit comparison convergence test from calculus, and the divergence of the harmonic series.

The sequence of primes cannot contain an infinite arithmetic progression, for then there would be some small positive constant c such that $\pi(x) \ge cx$ as $x \to +\infty$.

b) If p is a fixed prime divisor of $(n+d)(n+2d)\cdots(n+rd)$, then $-d, -2d, \dots, -rd$ contains all residues modulo p, so $p \le r$. Denoting the primes less than or equal to r by $p_1, p_2, \dots, p_s$, it is clear that

$$(n+d)(n+2d)\cdots(n+rd)$$

has no fixed prime divisor for the choice $d = p_1p_2\cdots p_s$.

(9) We have

$$\int_0^1 |f(\alpha)|^4\,d\alpha = \int_0^1 f(\alpha)^2\,\overline{f(\alpha)}^2\,d\alpha$$

$$= \sum_{x_1}\cdots\sum_{x_4}\int_0^1 e(\alpha(x_1^k + x_2^k - x_3^k - x_4^k))\,d\alpha$$

$$= \sum_{x_1^k+x_2^k-x_3^k-x_4^k=0} 1,$$

where the integers x_1, x_2, x_3, x_4 range independently through the interval $[1, N]$. Thus we need to count solutions of

$$x_1^k - x_3^k = x_4^k - x_2^k$$

with x_1, x_2, x_3, x_4 integers in $[1, N]$. If $x_4 = x_2$, then $x_1 = x_3$, and clearly there are only $O(N^2)$ solutions of this kind. So let us assume that $x_4 \neq x_2$, and rewrite the equation as

$$(x_1 - x_3)(x_1^{k-1} + \cdots + x_3^{k-1}) = x_4^k - x_2^k,$$

and denote the right-hand side by $n(x_4, x_2)$. There are $O(N^2)$ admissible pairs x_4, x_2 with $x_4 \neq x_2$. For each divisor d of $n(x_4, x_2)$ (positive or negative) the possible values of x_1 and x_3 are determined by

$$x_1 - x_3 = d$$
$$x_1^{k-1} + \cdots + x_3^k = \frac{n(x_4, x_2)}{d},$$

and this system of equations is equivalent to a polynomial equation of degree $k-1$, and thus has at most $k-1$ distinct solutions. Since the number of divisors d is $O(N^{k\delta})$ for arbitrary $\delta > 0$, we are finished.

(10) It is certainly necessary that

$$N^\epsilon N^{-1/K} \ll 1$$

should hold. Then $(1/K - \epsilon)\log(N)$ should be large, which cannot be true unless k is very small, since $K = 2^{k-1}$. Working this out, we see that k cannot even be as large as $\log\log(N)/\log(2)$.

(11) We have

$$F(x) = \sum_{n \leq x} \mu(n) \log(x/n)$$

by the second Möbius inversion formula. Then

$$F(x) = \frac{1}{2\pi i} \int_{c-i\infty}^{c+i\infty} \frac{1}{\zeta(s)} \frac{x^s}{s^2}\, ds$$

for $c > 1$, and $x > 1$ not an integer, by a corollary of the Perron formula, and the fact that

$$\frac{1}{\zeta(s)} = \sum_{n=1}^{\infty} \mu(n) n^{-s}.$$

It remains to verify that the contour integral is absolutely convergent. We have

$$\left|\frac{1}{\zeta(s)}\right| = \prod_p |1 - p^{-s}| \leq \prod_p (1 + p^{-\sigma}) = \frac{\zeta(\sigma)}{\zeta(2\sigma)}$$

for $\sigma > 1$, so

$$\left|\int_{c-i\infty}^{c+i\infty} \frac{1}{\zeta(s)} \frac{x^s}{s^2}\, ds\right| \le \int_{c-i\infty}^{c+i\infty} \frac{1}{|\zeta(s)|} \frac{|x^s|}{|s|^2}\, |ds|$$
$$\le \frac{\zeta(c)}{\zeta(2c)} x^c \int_{c-i\infty}^{c+i\infty} \frac{|ds|}{|s|^2}$$
$$= \frac{\zeta(c)}{\zeta(2c)} x^c \int_{-\infty}^{\infty} \frac{dt}{c^2+t^2} < \infty,$$

since $s = \sigma + it = c + it$ on the line from $c - i\infty$ to $c + i\infty$. Note that $|ds| = |idt| = dt$ since the real part of s is constant.

(12) a) First

$$\int_1^\infty (\psi(x) - [x])x^{-s-1}\, dx = \int_1^\infty \left(\sum_{n\le x} (\Lambda(n) - 1)\right) x^{-s-1}\, dx$$
$$= \sum_{n=1}^{\infty} (\Lambda(n) - 1) \int_n^\infty x^{-s-1}\, dx$$
$$= \sum_{n=1}^{\infty} (\Lambda(n) - 1) \frac{n^{-s}}{-s} = \frac{1}{s}\left(\zeta(s) + \frac{\zeta'(s)}{\zeta(s)}\right),$$

for $\sigma > 1$. Given $\varepsilon > 0$ there is some $x_0 = x_0(\varepsilon) \ge 1$ such that $|\psi(x) - [x]| \le \varepsilon x$ for $x \ge x_0$, and some C_ε such that $|\psi(x) - [x]| \le C_\varepsilon$ for $x \le x_0$. Then

$$\left|\int_1^\infty (\psi(x) - [x])x^{-s-1}\, dx\right| \le \left|\int_1^{x_0} (\psi(x) - [x])x^{-s-1}\, dx\right|$$
$$+ \left|\int_{x_0}^\infty (\psi(x) - [x])x^{-s-1}\, dx\right|$$
$$\le \int_1^{x_0} C_\varepsilon x^{-\sigma-1}\, dx + \int_{x_0}^\infty \varepsilon x x^{-\sigma+1}\, dx$$
$$\le \frac{C_\varepsilon}{\sigma} + \frac{\varepsilon}{\sigma - 1},$$

which proves the claim, by the previous calculation.

b) We know from part a) that

$$\left|\zeta(s) + \frac{\zeta'(s)}{\zeta(s)}\right| \le C_\varepsilon \frac{|s|}{\sigma} + \frac{\varepsilon |s|}{\sigma - 1}.$$

If $\zeta(s)$ has a zero at $s_0 = 1 + it_0$, then $\zeta(s) + \zeta'(s)/\zeta(s)$ has a simple pole at s_0 with residue a positive integer (equal to the multiplicity of the zero.) But this leads to a contradiction when we choose ε so small that $\varepsilon|s_0| < 1$, and let $s \to s_0$ from the right.

(13) Have

$$\begin{aligned}\sum_{n\le x}\frac{\Lambda(n)}{n} &= \frac{\psi(x)}{x} - \int_1^x \psi(u)\Big(-\frac{1}{u^2}\Big)du\\ &= \frac{\psi(x)}{x} + \int_1^x (u+\psi(u)-u)\frac{1}{u^2}\,du\\ &= \frac{\psi(x)}{x} + \log(x) + \int_1^\infty \frac{\psi(u)-u}{u^2}\,du - \int_x^\infty \frac{\psi(u)-u}{u^2}\,du\end{aligned}$$

by partial summation. The improper integrals converge absolutely because

$$\psi(x) = x + O(xe^{-c\log^{1/10}(x)}),$$

and the same estimate implies that

$$\frac{\psi(x)}{x} = 1 + O\Big(e^{-c\log^{1/10}(x)}\Big) = 1 + O\Big(\frac{1}{\log^m(x)}\Big),$$

and that

$$\begin{aligned}\int_x^\infty \frac{\psi(u)-u}{u^2}\,du &= \int_x^\infty \frac{O\Big(ue^{-c\log^{1/10}(u)}\Big)}{u^2}\,du\\ &= O\Big(\int_x^\infty \frac{du}{u\log^{m+1}(u)}\Big)\\ &= O\Big(\int_{\log(x)}^\infty \frac{e^t dt}{e^t\log^{m+1}(e^t)}\Big)\\ &= O\Big(\frac{1}{m\log^m(x)}\Big) = O\Big(\frac{1}{\log^m(x)}\Big).\end{aligned}$$

This yields the claim.

(14) a) The logarithmic derivative $L'(s,\chi)/L(s,\chi)$ has a simple pole with residue m at every zero of $L(s,\chi)$. Thus the sum on the left-hand side of the identity has a simple pole with residue a negative integer at any point which is a zero of at least one $L(s,\chi)$. This integer is the sum of the contributions from the various $L(s,\chi)$ that have a zero at that point, and all these contributions are negative integers. The reason that it is unnecessary to assume that distinct Dirichlet L-functions modulo q have distinct nontrivial zeros is that the arithmetic progression $n \equiv 1 \pmod{q}$ has a special property; the Fourier coefficients with respect to the Dirichlet characters modulo q of the indicator function of this arithmetic progression are all equal to 1. So the residues cannot cancel, and must accumulate. To make the same argument work for $n \equiv a \pmod{q}$ with $a \neq 1$, we would assume distinct zeros, so that cancellation cannot take place.

b) The summation by parts

$$\begin{aligned}\sum_{n\equiv 1 \bmod q} \Lambda(n)n^{-s} &= \lim_{x\to+\infty} \frac{\psi(x;q,a)}{x^s} - \int_1^\infty \psi(u;q,a)(-s)u^{-s-1}\,du \\ &= s\int_1^\infty \left(\frac{u}{\phi(q)} + \psi(u;q,a) - \frac{u}{\phi(q)}\right)u^{-s-1}\,du \\ &= \frac{1}{\phi(q)}\frac{s}{s-1} + \int_1^\infty O(u^{\alpha+\varepsilon})u^{-s-1}\,du\end{aligned}$$

is valid for $\sigma > 1$, because $\psi(u;q,a) = O(u)$. But the last integral converges uniformly on $\sigma > \alpha + \varepsilon$ for each $\varepsilon > 0$. Thus the Dirichlet series has an analytic continuation to $\sigma > \alpha$, with a simple pole at $s = 1$. From a) we then conclude that $L(s,\chi)$ is free of zeros in $\sigma > \alpha$ for each Dirichlet character modulo q.

If we had such a wide zero-free region for all the Dirichlet L-functions, the same kind of argument as in Chapter 7 would enable us to establish the stated estimate for $\psi(x;q,a)$, and this would be easier than our work there, as exceptional zeros would be absent. Alternatively, we could proceed as in Chapter 10 and establish an explicit formula for $\psi(x;q,a)$.

(15) It will be sufficient to make $\psi(x;q,a) \geq \delta x$ for some constant $\delta > 0$, and all x sufficiently large. For the contribution to $\psi(x;q,a)$ from prime powers higher than the first is only $O(x^{1/2}\log(x))$. This we may obtain by requiring that

$$e^{-C_A\sqrt{\log(x)}} \leq \frac{k}{q}$$

for a sufficiently small constant $k > 0$. Working this out, we see that

$$x \geq e^{\log^2(q/k)/C_A^2}$$

is required. But this is not sufficient, because $q \ll \log^A(x)$ in the Siegel-Walfisz theorem. The latter condition leads to the requirement that

$$x \geq e^{m_A q^{1/A}}$$

for any $A > 0$, with some constant $m_A > 0$. Clearly this condition is far more stringent than the previous one, and we conclude that each arithmetic progression $n \equiv a \pmod q$ with a and q coprime contains a prime

$$p \leq e^{m_A q^{1/A}}.$$

This is a weak result, and we note that we are unable to do better because we are not allowed to choose q as large as we would like in terms of x in the Siegel-Walfisz theorem.

Now assume that there is no exceptional zero belonging to the modulus q. Then only

$$x \geq e^{\log^2(q/k)/C_0^2}$$

is required, with suitable positive constants k and C_0, by Proposition 7.11. That would yield a prime

$$p \leq e^{\log^2(q/k)/C_0^2}$$

in the arithmetic progression. Much better than the previous bound, but not as good as the Linnik bound, which we cannot establish, lacking the necessary tools.

(16) Have

$$|L(s,\chi)| = \prod_p |1-\chi(p)p^{-s}|^{-1} \geq \prod_p (1+p^{-\sigma})^{-1} = \frac{\zeta(2\sigma)}{\zeta(\sigma)},$$

so $L(s,\chi)| \geq \varepsilon/(1+\varepsilon)$ on $\sigma > 1+\varepsilon$. Here $\zeta(2\sigma) \geq 1$ and $\zeta(\sigma)$ has been bounded from above by the Euler-Maclaurin summation formula

$$\sum_{n=1}^{\infty} n^{-\sigma} = \int_1^{\infty} x^{-\sigma}\,dx + \frac{1}{2} + \int_1^{\infty} S(x)(-\sigma)x^{-\sigma-1}\,dx,$$

for $\sigma > 1$.

(17) Have

$$\left(\frac{|d_K|}{2^{2r_2}\pi^n}\right)^{s/2} \Gamma(s/2)^{r_1}\Gamma(s)^{r_2}\zeta_K(s)$$
$$= \left(\frac{|d_K|}{2^{2r_2}\pi^n}\right)^{(1-s)/2} \Gamma((1-s)/2)^{r_1}\Gamma(1-s)^{r_2}\zeta_K(1-s)$$

by the functional equation. When $s = -1, -2, \ldots$, the point $1-s$ lies in the half plane $\sigma > 1$. Every factor on the right-hand side is then finite and nonzero. The factor containing $|d_K|$ on the left-hand side is also finite and nonzero. Thus the zeros of $\zeta_K(s)$ at $s = -1, -2, \ldots$ are determined by the poles of $\Gamma(s/2)^{r_1}\Gamma(s)^{r_2}$. The gamma function has no zeros, only simple poles at the negative integers, and no other singularities. If s is an odd negative integer, then $\Gamma(s/2)$ is finite, so $\zeta_K(s)$ has a zero of multiplicity r_2 at s if $r_2 > 0$, otherwise it has no zero there. If s is an even negative integer, then $\zeta_K(s)$ has a zero of multiplicity $r_1 + r_2$ there, in the same way.

(18) Since the extension is assumed normal, the formula $n_{K/\mathbb{Q}} = efg$ holds. Since the degree $n_{K/\mathbb{Q}}$ is assumed prime, a rational prime p that does not ramify in the extension must either split completely or stay inert. The Dirichlet density of the primes that split completely is $1/n_K$ by Proposition 9.8. The Dirichlet density of the rational primes that ramify

in the extension is zero, since there are only finitely many. Thus the Dirichlet density of the primes that stay inert is $1 - 1/n_{K/\mathbb{Q}}$.

(19) Define

$$A(u) = \sum_{n \le e^u} (1 + \mu(n)) = [e^u] + M(e^u),$$

and note that $A(u)$ is nonnegative and nondecreasing. We calculate the Laplace transform

$$\begin{aligned} L(s) &= \int_0^\infty A(u)e^{-su}\,du = \int_0^\infty \left(\sum_{n \le e^u} (1 + \mu(n)) \right) e^{-su}\,du \\ &= \sum_{n=1}^\infty (1 + \mu(n)) \int_{\log(n)}^\infty e^{-su}\,du = \sum_{n=1}^\infty (1 + \mu(n)) \frac{n^{-s}}{s} \\ &= \frac{\zeta(s)}{s} + \frac{1}{s\zeta(s)} = \frac{1}{s-1} + h(s), \end{aligned}$$

where $h(s)$ is holomorphic on a domain containing $\sigma \ge 1$, since $\zeta(s)$ has no zeros in this closed half plane. Thus

$$\sum_{n \le e^u} (1 + \mu(n))e^{-u} = ([e^u] + M(e^u))e^{-u} \to 1$$

as $u \to +\infty$. Then $M(e^u)e^{-u} \to 0$, and we are finished.

(20) We have

$$h(d) = \frac{w|d_K|^{1/2}}{2\pi} L(1, \chi_d),$$

where the discriminant d_K equals d or $4d$, the number w of units equals 2 unless $d = -1$ or $d = -3$, and the character $\chi_d(m) = (d_K|m)$ has modulus at most $4|d|$. Now

$$L(1, \chi_d) = \lim_{x \to +\infty} \sum_{n \le x} \frac{\chi_d(n)}{n} = \lim_{x \to +\infty} \left(\frac{X_d(x)}{x} + \int_1^x X_d(u)u^{-2} \right) du$$

with

$$X_d(x) = \sum_{n \le x} \chi_d(n),$$

by partial summation. Then

$$L(1, \chi_d) = \int_1^\infty X_d(u)u^{-2}\,du,$$

since $X_d(x)$ is bounded. Now

$$\begin{aligned}|L(1,\chi_d)| &\le \int_1^b |X_d(u)|u^{-2}\,du + \int_b^\infty |X_d(u)|u^{-2}\,du \\ &\le \int_1^b uu^{-2}\,du + \int_b^\infty \sqrt{8/3}\sqrt{4|d|}\log(4|d|)u^{-2}\,du\end{aligned}$$

by the Pólya-Vinogradov inequality. We should choose b to make the right-hand side small. The choice $b = \sqrt{4|d|}$ is reasonable, though not quite optimal. We obtain

$$|L(1,\chi_d)| \le \frac{1}{2}\log(4|d|) + \sqrt{8/3}\log(4|d|) \le \pi\log(4|d|),$$

and so the desired inequality follows.

(21) By the theorem of von Koch, the error term is $O(\log^2(x))$ on the Riemann Hypothesis. No improvement on this is known. So we should take T large enough in terms of x that $R(x,T) = O(x^{1/2}\log^2(x))$, and no larger. From the last of the three terms in our bound for $R(x,T)$ we see that $T = x^{1/2}$ works for that term, and it clearly works for the first term too. While the middle term is no larger than $\log(x)$. So we choose $T = x^{1/2}$.

Bibliography

[Abe26] N. H. Abel, *Untersuchungen über die Reihe...*, J. Reine Angew. Math. **1** (1826), 311–339.

[Apo76] Tom M. Apostol, *Introduction to Analytic Number Theory*, Undergraduate Texts in Mathematics, Springer, Berlin, 1976.

[Art23] E. Artin, *Über eine neue Art von L-Reihen*, Abh. Math. Sem. Univ. Hamburg **3** (1923), 89–108.

[Art30] ———, *Zur Theorie der L-Reihen mit allgemeimnen Gruppencharakteren*, Abh. Math. Sem. Univ. Hamburg **8** (1930), 292–306.

[Bac94] Paul Bachmann, *Die Analytische Zahlentheorie*, Bibliotheca Mathematica Teubneriana, vol. 16, B. G. Teubner, Leipzig, 1894.

[Bac29] R. J. Backlund, *Über die Differenzen zwischen den Zahlen, die zu den ersten n Primzahlen teilerfremd sind*, Ann. Acad. Sci. Fenn. **32** (1929), no. 2, 1–9.

[Bar81] Klaus Barner, *On A. Weil's explicit formula*, J. Reine Angew. Math. **323** (1981), 139–152.

[Bau03a] M. Bauer, *Über einen Satz von Kronecker*, Arch. der Math. u. Phys. (3) **6** (1903), 218–219.

[Bau03b] ———, *Über zusammengesetzte Kőrper*, Arch. der Math. u. Phys. (3) **6** (1903), 221–222.

[BD66] E. Bombieri and H. Davenport, *Small differences between prime numbers*, Proc. Roy. Soc. Ser. A **293** (1966), 1–18.

[BDD86a] Ramachandra Balasubramanian, Jean-Marc Deshouillers, and François Dress, *Problème de Waring pour les bicarrés. I. Schéma de la solution*, C. R. Acad. Sci. Paris **303** (1986), no. 4, 85–88.

[BDD86b] ———, *Problème de Waring pour les bicarrés. II. Résultats pour le théorème asymptotique*, C. R. Acad. Sci. Paris **303** (1986), no. 5, 161–163.

[Bel15] E. T. Bell, *An Arithmetical Theory of Certain Numerical Functions*, Publications in Mathematical and Physical Sciences, no. 1, The University of Washington, Seattle, 1915.

[Ber00] Michael C. Berg, *The Fourier-analytic Proof of Quadratic Reciprocity*, Pure and Applied Mathematics, Wiley-Interscience, New York, NY, 2000.

[BEW98] Bruce C. Berndt, Ronald J. Evans, and Kenneth S. Williams, *Gauss and Jacobi Sums*, CMS series of monographs and advanced texts, vol. 21, Wiley-Interscience, New York, 1998.

[BH62] P. T. Bateman and R. A. Horn, *A heuristic asymptotic formula concerning the distribution of prime numbers*, Math. Comp. **16** (1962), 363–367.

[BH96] R. C. Baker and G. Harman, *The difference between consecutive primes*, Proc. London Math. Soc. **72** (1996), no. 2, 261–280.

[BHP01] R. C. Baker, G. Harman, and J. Pintz, *The difference between consecutive primes. II*, Proc. London Math. Soc. **83** (2001), no. 3, 532–562.

[Bir57] B. Birch, *Homogenous forms of odd degree in a large number of variables*, Mathematika **4** (1957), 102–105.

[Boc59] Salomon Bochner, *Lectures on Fourier Integrals*, Annals of Mathematics Studies, no. 42, Princeton University Press, Princeton, NJ, 1959, translated from the original by Morris Tenenbaum and Harry Pollard.

[Bor99] E. Borel, *Sur les zéros des fonctions entièrs*, Acta Math. **20** (1899), 357–396.

[Bra45] R. D. Brauer, *A note on systems of homogenous algebraic equations*, Bull. Amer. Math. Soc. **51** (1945), 749–755.

[Bru15] V. Brun, *Űber das Goldbachsche Gesetz und die Anzahl der Primzahlpaare*, Archiv for Math. og Natur. **34** (1915), no. 8, 1–19.

[Bug73] N. W. Bugaev, *Theory of number-theoretical derivatives*, Mathematical collection, Moscow Math. Soc., Moscow, 1870 and 1872-1873 (Russian).

[Bur00] William Burnside, *On group-characteristics*, Proc. London Math. Soc. **33** (1900), 146–162.

[Bur03] ———, *On the representation of a group of finite order as an irreducible group of linear substitutions and the direct establishment of the relations between the group characteristics*, Proc. London Math. Soc. (2) **1** (1903), 117–123.

[Bur04] ———, *On the reduction of a group of homogenous linear substitutions of finite order*, Acta Math. **28** (1904), 369–387.

[Bur57] D. A. Burgess, *The distribution of quadratic residues and non-residues*, Mathematika **4** (1957), 106–112.

[Bur62] ———, *On character sums and primitive roots*, Proc. London Math. Soc. **12** (1962), no. 3, 179–192.

[BZ30] A. Brauer and H. Zeitz, *Űber eine zahlentheoretische Behauptung von Legendre*, Sitz. Berliner Math. Ges. **29** (1930), 116–125.

[Cah94] E. Cahen, *Sur la fonction $\zeta(s)$ de Riemann et sur des fonctions analogues*, Ann. Sci. Ec. Norm. Sup. (3) **11** (1894), 75–164.

[Cau13] A. L. Cauchy, *Recherches sur les nombres*, J. Ecole Polytech. **9** (1813), 99–116.

[CE59] E. D. Cashwell and C. J. Everett, *The ring of number-theoretic functions*, Pacific J. Math. **9** (1959), 975–985.

[Cés88] E. Césaro, *Sur une fonction arithmétique*, C. R. Acad. Sci. Paris **106** (1888), 1340–1343.

[Che48] P. L. Chebyshev, *Sur la Fonction qui Détermine la Totalité des Nombres Premiers Inférieurs à une Limité Donnée*, Mémoires des savants étrangers de l'Acad. Sci. St. Pétersbourg **6** (1848), 1–19.

[Che50] ———, *Mémoire sur nombres premiers*, Mémoires des savants étrangers de l'Acad. Sci. St. Pétersbourg **7** (1850), 17–33.

[Che73] J. R. Chen, *On the representation of a larger even integer as the sum of a prime and the product of at most two primes*, Sci. Sinica **16** (1973), 157–176.

[Chi50] T. T. Chih, *The Dirichlet's divisor problem*, Sci. Rep. Nat. Tsing Hua Univ. Ser. A **5** (1950), 402–427.

[Cho35] Inder Chowla, *A theorem on the addition of residue classes: Application to the number $\Gamma(k)$ in Waring's problem*, Proc. Ind. Acad. Sci. **2** (1935), 242–243.

[Cho37] ———, *A theorem on the addition of residue classes: Application to the number $\Gamma(k)$ in Waring's problem*, Quarterly J. Math. **8** (1937), 99–102.

[Chu36] N. G. Chudakov, *On zeros of Dirichlet's L-functions*, Mat. Sbornik **43** (1936), 591–602.

[Cip15] M. Cipolla, *Sui principi del calculo arithmetico-integrale*, Atti Acc. Gioenia Catania **8** (1915), no. 5.

[CM06] A. C. Cojocaru and M. Ram Murty, *An introduction to Sieve Methods and their Applications*, London Mathematical Society Student Texts, no. 66, Cambridge University Press, Cambridge, 2006.

[Cor22] J. G. van der Corput, *Verschärfung der Abschätzung beim Teilerproblem*, Math. Ann. **87** (1922), 39–65, Berichtigungen: **89** (1923), 160.

[Cra36] H. Cramér, *On the order of magnitude of the difference between consecutive prime numbers*, Acta Arith. **2** (1936), 23–46.

[Cur99] Charles W. Curtis, *Pioneers of Representation Theory: Frobenius, Burnside, Schur, and Brauer*, History of Mathematics, vol. 15, American Mathematical Society and London Mathematical Society, Providence, RI, 1999.

[Dab84] H. Daboussi, *Sur le théorème des nombres premiers*, C. R. Acad. Sci. Paris Série I **298** (1984), no. 8, 161–164.

[Dav35] H. Davenport, *On the addition of residue classes*, J. London Math. Soc. **10** (1935), 30–32.

[Dav39] Harold Davenport, *On Waring's problem for fourth powers*, Ann. of Math. **40** (1939), 731–747.

[Dav63] H. Davenport, *Cubic forms in sixteen variables*, Proc. Royal Soc. Ser. A **272** (1963), 285–303.

[Dav00] Harold Davenport, *Multiplicative Number Theory*, third ed., Springer, New York, 2000, revised by Hugh L. Montgomery.

[Dav05] ———, *Analytic Methods for Diophantine Equations and Diophantine Inequalities*, second ed., Cambridge Mathematical Library, Cambridge University Press, Cambridge, 2005.

[DE80] Harold G. Diamond and Paul Erdős, *On sharp elementary prime number estimates*, Enseign. Math. (2) **26** (1980), 313–321.

[Ded57] R. Dedekind, *Abriss einer Theorie der hőhern Congruenzen in Bezug auf einen reellen Primzahl-Modulus*, J. Reine Angew. Math. **54** (1857), 1–26.

[Ded79] ———, *Űber die Theorie der ganzen algebraischen Zahlen*, Vorlesungen űber Zahlentheorie, Friedrich Vieweg und Sohn, Braunschweig, 3 ed., 1879, XI Supplement to Dirichlet's lectures on number theory.

[Des79] R. Descartes, *De la façon de trouver les nombres de parties aliquotes in ratione data*, Bull. Bibl. Storia Sc. Mat. e Fis. **12** (1879), 713–715, published by C. Henry using a mss. in the Bibliothèque Nationale, Paris.

[Des98] René Descartes, *Oeuvres et lettres*, vol. 2, L. Cerf, Paris, 1898, Charles Adam et Paul Tannery eds., 149.

[Dia82] Harold G. Diamond, *Elementary methods in the study of the distribution of prime numbers*, Bull. Amer. Math. Soc. (N. S.) **7** (1982), 553–589.

[Dic34] Leonard Eugene Dickson, *History of the Theory of Numbers*, Publications of the Carnegie Institution of Washington, vol. 256, G. E. Stechert, New York, NY, 1934, reprint of the first ed.

[Dir37a] G. Lejeune Dirichlet, *Beweis des Satzes, daß jede unbegrenzte arithmetische Progression, deren ersten Glied und Differenz ganze Zahlen ohne gemeinschaftlichen Faktor sind, unendlich vielen Primzahlen enthält*, Abhandl. Kgl. Preuß. Akad. Wiss. (1837), 45–81.

[Dir37b] ———, *Beweis eines Satzes über die arithmetischen Progression*, Ber. Verhandl. Kgl. Preuß. Akad. Wiss. (1837), 108–110.

[Dir38a] ———, *Sur l'usage des séries infinies dans la théorie des nombres*, J. Reine Angew. Math. **18** (1838), 259–274.

[Dir38b] ———, *Über die Bestimmung asymptotischer Gesetze in der Zahlentheorie*, Ber. Verhandl. Kgl. Preuß. Akad. Wiss. (1838), 13–15.

[Dir39] ———, *Recherches sur les diverses applications de l'analyse infinitésimale a la théorie des nombres*, J. Reine Angew. Math. **19** (1839), 324–369.

[Dir40] ———, *Über eine Eigenschaft der quadratischen formen*, Ber. Verhandl. Kgl. Preuß. Akad. Wiss. (1840), 49–52.

[Dir41a] ———, *Untersuchungen über die Theorie der complexen Zahlen*, Sitz. Kgl. Preuß. Akad. Wiss. Berlin (1841), 190–194.

[Dir41b] ———, *Untersuchungen über die Theorie der complexen Zahlen*, Abh. Kgl. Preuß. Akad. Wiss. Berlin (1841), 141–161.

[Dir42a] ———, *Recherches sur les formes quadratiques à coefficients et à indéterminés complexes*, J. Reine Angew. Math. **24** (1842), 291–371.

[Dir42b] ———, *Verallgemeinerung eines Satzes aus der Lehre von den Kettenbrüchen nebst einigen Anwendungen auf die Theorie der Zahlen*, Ber. Verhandl. Kgl. Preuß. Akad. Wiss. (1842), 93–95.

[Dir42c] P. G. Lejeune Dirichlet, *Vorlesungen über Zahlentheorie*, Vieweg, Braunschweig, 1842.

[Dir46] G. Lejeune Dirichlet, *Zur Theorie der complexen Einheiten*, Ber. Verhandl. Kgl. Preuß. Akad. Wiss. (1846), 103–107.

[Dir49] ———, *Über die Bestimmung der mittleren Werte in der Zahlentheorie*, Abhandl. Kgl. Preuß. Akad. Wiss. (1849), 69–83.

[DL63] H. Davenport and D. J. Lewis, *Homogenous additive equations*, Proc. Royal Soc. Ser. A **274** (1963), 443–460.

[dlVP96] C. J. de la Vallé Poussin, *Recherches analytiques sur la théorie des nombres premiers, I-III*, Ann. Soc. Sci. Bruxelles **20** (1896), 183–256, 281–362, 363–397.

[dlVP99] ———, *Sur la fonction $\zeta(s)$ de Riemann et le nombre des nombres premiers inférieurs à une limite donnée*, Mem. couronnés de l'Acad. Sci. Bruxelles **59** (1899), 183–256, 281–362, 363–397.

[dP51] A. de Polignac, *Recherches nouvelles sur les nombres premiers*, Bachelier, Paris, 1851.

[Edw01] H. M. Edwards, *Riemann's Zeta Function*, Dover Publishing, Mineola, NY, 2001.

[Erd35] P. Erdős, *On the difference of consecutive primes*, Quarterly J. Math. Oxford Ser. **6** (1935), 124–128.

[Erd40] ______, *The difference of consecutive primes*, Duke Math. J. **6** (1940), 438–441.

[Erd49] Paul Erdős, *On a new method in elementary number theory which leads to an elementary proof of the prime number theorem*, Proc. Nat. Acad. Sci. USA **35** (1949), 374–384.

[Eul37] L. Euler, *Variae observationes circa series infinitas*, Comm. Acad. Sci. Petropol. **9** (1737), 160–188.

[Eul38] ______, *Methodus generalis summandi progressiones*, Comm. Acad. Sci. Petropol. 1732/1733 **6** (1738), 68–97.

[FG86] É. Fouvry and F. Grupp, *On the switching principle in sieve theory*, J. Reine Angew. Math. **370** (1986), 101–126.

[FI98] J. Friedlander and H. Iwaniec, *The polynomial $X^2 + Y^4$ captures its primes*, Annals of Math. **148** (1998), no. 3, 945–1040.

[Fil90] M. Filaseta, *Short interval results for squarefree numbers*, J. Number Theory **35** (1990), no. 2, 128–149.

[Fog41] E. Fogels, *On the average values of arithmetic functions*, Proc. Cambridge Phil. Soc. **37** (1941), 358–372.

[Fra99] J. Franel, *Sur une formule utile dans la détermination de certaines valeurs asymptotiques*, Math. Ann. **51** (1899), no. 3, 369–387.

[Fro96] F. G. Frobenius, *Űber Gruppencharactere*, Sitz. Kgl. Preuß. Akad. Wiss. Berlin (1896), 985–1021.

[Fro97] ______, *Űber die Darstellung der endlichen Gruppen durch lineare Substitutionen*, Sitz. Kgl. Preuß. Akad. Wiss. Berlin (1897), 944–1015.

[Fro98] ______, *Űber Relationen zwischen den Charakteren einer Gruppe und denen ihrer Untergruppen*, Sitz. Kgl. Preuß. Akad. Wiss. Berlin (1898), 501–515.

[FT90] M. Filaseta and O. Trifonov, *On gaps between squarefree numbers*, Analytic number theory (Allerton Park, IL, 1989) (Boston, MA) (B. C. Berndt, H. G. Diamond, H. Halberstam, and A. J. Hildebrand, eds.), Birkhäuser Boston, 1990, pp. 235–253.

[FT92] ______, *On gaps between squarefree numbers. II*, J. London Math. Soc. (2) **45** (1992), no. 2, 215–221.

[Gau11] C. F. Gauss, *Summatio quarumdam serierum singularium*, Comm. Soc. Reg. Sci. Gottingensis **1** (1811).

[Gau33] ______, *Werke*, Dieterichschen Universitätsdruckerei, Gőttingen, 1863-1933.

[Geg85] L. Gegenbauer, *Einige asymptotische Gesetze der Zahlentheorie*, Sitz. Kais. Akad. Wiss. (Wien) IIa **92** (1885), 1290–1306.

[Gál61] I. S. Gál, *Lectures on Algebraic and Analytic Number Theory*, Jones Letter Service, Minneapolis, 1961.

[GK88] S. W. Graham and G. Kolesnik, *On the difference between consecutive squarefree integers*, Acta Arith. **49** (1988), 435–447.

[Gol74] D. M. Goldfeld, *A Simple Proof of Siegel's Theorem*, Proc. Natl. Acad. Sci. USA **71** (1974), no. 4, 1055.

[Gol76] D. Goldfeld, *The class number of quadratic fields and the conjectures of Birch and Swinnerton-Dyer*, Ann. Scuola Norm. Sup. Pisa Cl. Sci. **3** (1976), no. 4, 624–663.

[Gon93] S. M. Gonek, *An explicit formula of Landau and its applications to the theory of the zeta-function*, A Tribute to Emil Grosswald: Number Theory and Related Analysis (Providence, RI) (M. Knopp and M. Sheingorn, eds.), Contemp. Math., vol. 143, Amer. Math. Soc., 1993, pp. 395–413.

[GPY09] Daniel A. Goldston, János Pintz, and Cem Y. Yıldırım, *Primes in tuples I*, Annals Math. **170** (2009), no. 2, 819–862.

[Gra86] J. P. Gram, *Undersøgelser angaaende Mængden af Primtal under en given Grænse*, Det Kongelige Danske Videnskabernes Selskabs Skrifter. Naturvitenskabelig og mathematisk Afdeling **6** (1881-1886), no. 2, 183–308 (Danish).

[Gra95] A. Granville, *Harald Cramér and the distribution of prime numbers*, Scand. Actuar. J. (1995), no. 1, 12–28.

[Gra98] ———, *ABC allows us to count squarefrees*, Internat. Math. Res. Notices **19** (1998), 991–1009.

[Gro13] T. H. Gronwall, *Sur les séries de Dirichlet correspondant á des caractéres complexes*, Rend. Circ. Mat. di Palermo **35** (1913), 145–159.

[Gui42] A. P. Guinand, *Summation formulae and self-reciprocal functions. III.*, Quart. J. Math., Oxford Ser. **13** (1942), 30–39.

[GZ83] B. Gross and D. Zagier, *Points de Heegner et dérivées de fonctions L*, C. R. Acad. Sci. Paris Sér. I Math. **297** (1983), 85–87.

[Had93] J. Hadamard, *Étude sur les proprietés des fonctions entières et en particulier d'une fonction considérée par Riemann*, J. Math. Pures Appl. **9** (1893), 171–215.

[Had96] ———, *Sur la distribution des zéros de la fonction $\zeta(s)$ et ses conséquences arithmétiques*, Bull. Soc. Math. France **24** (1896), 199–220.

[Had08] ———, *Sur les séries de Dirichlet*, Rend. Circ. Mat. di Palermo **25** (1908), 326–330, 395–396.

[Har15a] G. H. Hardy, *On Dirichlet's divisor problem*, Proc. London Math. Soc. (2) **15** (1915), 1–25.

[Har15b] ———, *Sur le problème des diviseurs de Dirichlet*, C. R. Acad. Sci. Paris **160** (1915), 617–619.

[Har66] ———, *Collected Papers*, Oxford University Press, Oxford, 1966.

[HB83] D. R. Heath-Brown, *Cubic forms in ten variables*, Proc. London Math. Soc. (3) **47** (1983), no. 2, 225–257.

[HB92] ———, *Zero-free regions for Dirichlet l-functions, and the least prime in an arithmetic progression*, Proc. London Math. Soc. (3) **32** (1992), 265–338.

[HB01] ———, *Primes represented by $x^3 + 2y^3$*, Acta Math. **186** (2001), no. 1, 1–84.

[HB07] ———, *Cubic forms in* 14 *variables*, Invent. Math. **170** (2007), no. 1, 199–230.

[HBI79] D. R. Heath-Brown and H. Iwaniec, *On the difference between consecutive primes*, Invent. Math. **55** (1979), no. 1, 49–69.

[HBM02] D. R. Heath-Brown and B. Z. Moroz, *Primes represented by binary cubic forms*, Proc. London Math. Soc. (3) **84** (2002), no. 2, 257–288.

[HBM04] ———, *On the representation of primes by cubic polynomials in two variables*, Proc. London Math. Soc. (3) **88** (2004), no. 2, 289–312.

[HBP79] D. R. Heath-Brown and S. J. Patterson, *The distribution of Kummer sums at prime arguments*, J. Reine Angew. Math. **310** (1979), 111–130.

[Hec17] E. Hecke, *Über die Zetafunktion beliebiger algebraischer Zahlkörper*, Göttinger Nachrichten (1917), 77–89.

[Hec81] Erich Hecke, *Lectures on the Theory of Algebraic Numbers*, Graduate Texts in Mathematics, Springer-Verlag, New York, NY, 1981.

[Hei33] H. Heilbronn, *Über den Primzahlsatz von Herrn Hoheisel*, Math. Z. **36** (1933), no. 1, 394–423.

[Hei67] H. A. Heilbronn, *Zeta-Functions and L-Functions*, Algebraic Number Theory (London) (J. W. S. Cassels and A. Fröhlich, eds.), Academic Press, 1967, Prepared by D. A. Burgess and H. Halberstam, pp. 204–230.

[Hel05] Henry Helson, *Dirichlet Series*, Berkeley, 2005.

[Hil09] D. Hilbert, *Beweis für die Darstellbarkeit der ganzen Zahlen durch eine fester Anzahl nter Potenzen (Waringsche Problem)*, Göttinger Nach. (1909), 17–36.

[Hil86] A. Hildebrand, *The prime number theorem via the large sieve*, Mathematika **33** (1986), 23–30.

[Hil88] Adolf Hildebrand, *On the constant in the Pólya-Vinogradov inequality*, Canad. Math. Bull. **31** (1988), no. 3, 347–352.

[HL20a] G. H. Hardy and J. E. Littlewood, *A new solution of Waring's Problem*, Quarterly J. Math. **48** (1920), 272–293.

[HL20b] ———, *Some problems of 'Partitio Numerorum': I. A new solution of Waring's Problem*, Nachr. Akad. Wiss. Göttingen. Math.-Phys. Kl. (1920), 33–54.

[HL21] ———, *Some problems of 'Partitio Numerorum': II. proof that every large number is the sum of at most 21 biquadrates*, Math. Z. **9** (1921), 14–27.

[HL22a] ———, *Some problems of 'Partitio Numerorum': III. On the expression of a number as a sum of primes*, Acta Math. (1922), 1–70.

[HL22b] ———, *Some problems of 'Partitio Numerorum': IV. The singular series in Waring's Problem and the value of the number* $G(k)$, Math. Z. **12** (1922), 161–188.

[Hoh30] G. Hoheisel, *Primzahlprobleme in der Analysis*, S.-B. Preuß. Akad. Wiss. Phys.-Math. Kl. **33** (1930), 580–588.

[Hoo76] C. Hooley, *Applications of sieve methods to the theory of numbers*, Cambridge Tracts in Mathematics, vol. 70, Cambridge University Press, Cambridge, 1976.

[Hoo88] ———, *On nonary cubic forms*, J. Reine Angew. Math. **386** (1988), 32–98.

[Hoo91] ———, *On nonary cubic forms. II*, J. Reine Angew. Math. **415** (1991), 95–165.

[HR17a] G. H. Hardy and S. Ramanujan, *The normal number of prime factors of a number n*, Quart. J. Math. **48** (1917), 76–92.

[HR17b] ———, *Une formule asymptotique pour le nombre des partitions de n*, C. R. Acad. Sci. Paris **164** (1917), 35–38.

[HR18] ———, *Asymptotic formulae in combinatory analysis*, Proc. London Math. Soc (2) **18** (1918), 75–115.

[Hua38] L. K. Hua, *On Waring's problem*, Quarterly J. Math. **9** (1938), 199–202.

[Hur82] A. Hurwitz, *Einige Eigenschaften der Dirichlet'schen Functionen die bei der Bestimmung der Classenanzahlen binärer quadratischer Formen auftreten*, Z. Math. Phys. **27** (1882), 86–101.

[Hux72] M. N. Huxley, *On the difference between consecutive primes*, Invent. Math. **15** (1972), 164–170.

[Hux73] ———, *Small differences between consecutive primes*, Mathematika **20** (1973), 229–232.

[Hux77] ———, *Small differences between consecutive primes. II*, Mathematika **24** (1977), no. 2, 142–152.

[Hux84] ———, *An application of the Fouvry-Iwaniec theorem*, Acta Arith. **43** (1984), no. 4, 441–443.

[Hux93] ———, *Exponential sums and lattice points. II*, Proc. London Math. Soc. (3) **66** (1993), no. 2, 279–301, corrigenda: Proc. London Math. Soc. (3) **68** (1994), no. 2, 264.

[Hux02] ———, *Integer points, exponential sums and the Riemann zeta function*, Number theory for the millennium, II (Urbana, IL 2000) (Natick, MA) (M. A. Bennett, B. C. Berndt, N. Boston, H. G. Diamond, A. J. Hildebrand, and W. Philipp, eds.), A K Peters, 2002, pp. 275–290.

[Hux03] ———, *Exponential sums and lattice points. III*, Proc. London Math. Soc. (3) **87** (2003), no. 3, 591–609.

[HW08] G. H. Hardy and E. M. Wright, *An Introduction to the Theory of Numbers*, sixth ed., Oxford University Press, Oxford, 2008, revised by D. R. Heath-Brown and J. H. Silverman.

[IJ79] H. Iwaniec and M. Jutila, *Primes in short intervals*, Ark. Mat. **17** (1979), no. 1, 167–176.

[IK04] Henryk Iwaniec and Emmanuel Kowalski, *Analytic number theory*, Colloqium Publications, vol. 53, American Mathematical Society, Providence, RI, 2004.

[Ike31] S. Ikehara, *An extension of Landau's Theorem in the Analytic Theory of Numbers*, J. Math. Phys. M.I.T. **10** (1931), 1–12.

[IM88] H. Iwaniec and C. J. Mozzochi, *On the divisor and circle problems*, J. Number Theory **29** (1988), no. 1, 60–93.

[Ing37] A. E. Ingham, *On the difference between consecutive primes*, Quart. J. Math. **8** (1937), 255–266.

[Ing90] ———, *The Distribution of Prime Numbers*, Cambridge Mathematical Library, Cambridge University Press, Cambridge, 1990, reissue of the 1932 ed. with a Foreword by R. C. Vaughan.

[IP84] H. Iwaniec and J. Pintz, *Primes in short intervals*, Monatsh. Math. **98** (1984), no. 2, 115–143.

[Iwa78] H. Iwaniec, *Almost-primes represented by Quadratic Polynomials*, Invent. Math. **47** (1978), no. 2, 171–188.

[Jak29] C. G. J. Jakobi, *Fundamenta Nova Theoriae Functionum Ellipticarum*, Bornträger, Königsberg, 1829.

[Jak32] ———, *Observatio arithmetica de numero classium divisorum quadraticorum formae aa + azz, designante a numerum primum formae 4n + 3*, J. Reine Angew. Math. **9** (1832), 189–192.

[Jen84] J. L. W. V. Jensen, *Om rækkers konvergens*, Tidsskrift for Mathematik (Kbh.) (5) **2** (1884), 63–72 (Danish).

[Jen88] ———, *Sur une généralisation d'un théorème de Cauchy*, C. R. Acad. Sci. Paris **106** (1888), 833–836.

[Jen99] J. L. V. W. Jensen, *Sur un nouvel et important théorème de la théorie des fonctions*, Acta Math. **22** (1899), 359–364.

[Kem12] Aubrey Kempner, *Bemerkungen zum Waringschen Problem*, Math. Ann. **72** (1912), 387–399.

[Ker73] John Kersey, *The elements of that mathematical art, commonly called algebra*, vol. 1, Thomas Passinger and Benjamin Hurlock, London, 1673, page 199.

[Kin62] H. Kinkelin, *Allgemeine Theorie der harmonischen Reihen, mit Anwendungen auf die Zahlentheorie*, Programm der Gewerbeschule Basel 1861/62 (Basel), Schweighauserische Buchdruckerei, 1862, pp. 1–32.

[Klo26] H. D. Kloosterman, *On the representation of numbers in the form $ax^2 + by^2 + cz^2 + dt^2$*, Acta Math. **49** (1926), 407–464.

[Kno41] K. Knopp, *Funktionentheorie ii*, 5 ed., Sammlung Gőschen, no. 703, de Gruyter, Berlin, 1941.

[Kol26] A. N. Kolmogorov, *Une série de Fourier-Lebesgue divergente partout*, C. R. **183** (1926), 1327–1328.

[Kol69] G. A. Kolesnik, *An improvement of the remainder term in the divisor problem*, Mat. Zametki **6** (1969), 545–554 (Russian), English translation: Math. Notes **6** (1969), 784-791.

[Kol74] ______, *An estimate for certain trigonometric sums*, Acta Arith. **25** (1973/1974), 7–30 (Russian), errata insert.

[Kol82] ______, *On the order of $\zeta(\frac{1}{2} + it)$ and $\Delta(R)$*, Pacific J. Math. **98** (1982), 107–122.

[Kol85] ______, *On the method of exponent pairs*, Acta Arith. **45** (1985), no. 2, 115–143.

[Kor58] Nikolai M. Korobov, *Estimates of trigonometric sums and applications*, Uspekhi Mat. Nauk **13** (1958), no. 4, 185–192 (Russian).

[Kor06] Jaap Korevaar, *The Wiener-Ikehara theorem by complex analysis*, Proc. Amer. Math. Soc. **134** (2006), no. 4, 1107–1116.

[Kro80] Leopold Kronecker, *Űber die Irreduktibilität von Gleichungen*, Mon. Ber. Kgl. Preuss. Akad. Wiss. Berlin (1880), 155–163.

[Kro01] ______, *Vorlesungen űber Zahlentheorie*, B. G. Teubner, Leipzig, 1901, bearbeitet und ausgegeben von Dr. Kurt Hensel.

[Lag70] J. L. Lagrange, *Démonstration d'un théorème d'arithmétique*, Nouv. Mém. Acad. Roy. Sc. de Berlin (1770), 123–133.

[Lan93] G. Landsberg, *Zur Theorie der Gauss'schen Summen und der linearen Transformation der Thetafunktionen*, J. Reine Angew. Math. **111** (1893), 234–253.

[Lan00] Edmund Landau, *Sur quelques problémes rélatifs à la distribution des nombres premiers*, Bull. Soc. Math. France **28** (1900), 25–38.

[Lan01] ______, *Ueber die asymptotische Werthe einiger Zahlentheoretischer Functionen*, Math. Ann. **54** (1901), 570–591.

[Lan03a] ______, *Neuer Beweis des Primzahlsatzes und Beweis des Primidealsatzes*, Math. Ann. **56** (1903), 645–670.

[Lan03b] ______, *Űber die Primzahlen einer arithmetischen Progression*, Sitz. Kais. Akad. Wiss. Wien IIa **112** (1903), 493–535.

[Lan05a] ______, *Sur quelques inégalités dans la théorie de la fonction $\zeta(s)$ de Riemann*, Bull. Soc. Math. France **33** (1905), 229–241.

[Lan05b] ______, *Űber einen Satz von Tschebyschef*, Math. Ann. **61** (1905), 527–550.

[Lan06] ———, *Über den Zusammenhang einiger neuerer Sätze der analytischen Zahlentheorie*, Sitz. Kais. Akad. Wiss. (Wien) IIa **115** (1906), 589–632.

[Lan08] ———, *Nouvelle démonstration pour la formule de Riemann sur le nombre des nombres premiers inférieurs à une limite donné, et démonstration d'une formule plus générale pour le cas des nombres premiers d'une progression arithmétique*, Ann. Sci. Ec. Norm. Sup. (3) **25** (1908), 399–442.

[Lan09] ———, *Über eine Anwendung der Primzahlen auf das Waringsche Problem in der elementaren Zahlentheorie*, Math. Ann. **66** (1909), 102–105.

[Lan11] ———, *Über die Äquivalenz zweier Häuptsätze der analytischer Zahlentheorie*, Sitz. Kais. Akad. Wiss. Wien IIa **120** (1911), 973–988.

[Lan12] ———, *Über einige Summen, die von den Nullstellen der Riemannschen Zetafunktion abhängen*, Acta Math. **35** (1912), 271–294.

[Lan18a] ———, *Abschätzungen von Charaktersummen, Einheiten und Klassenzahlen*, Göttinger Nachrichten (1918), 79–97.

[Lan18b] ———, *Über die Klassenzahl imaginär quadratischer Zahlkörper*, Göttinger Nachrichten (1918), 285–295.

[Lan24a] ———, *Über die Wurzeln der Zetafunktion*, Math. Z. **20** (1924), 98–104.

[Lan24b] ———, *Über die ζ-funktion und die L-funktionen*, Math. Z. **20** (1924), 105–125.

[Lan27] ———, *Über die Zetafunktion und die Hadamardsche Theorie der ganzen Funktionen*, Math. Z. **26** (1927), 170–175.

[Lan30] ———, *Über die neue Winogradoffsche Behandlung des Waringschen Problem*, Math. Z. **31** (1930), 319–338.

[Lan35] E. Landau, *Bemerkungen zum Heilbronnschen Satz*, Acta. Arith. **1** (1935), 1–18.

[Lan70] Serge Lang, *Algebraic Number Theory*, Addison-Wesley Series in Mathematics, Addison Wesley, Reading, MA, 1970.

[Lan74] Edmund Landau, *Handbuch der Lehre von der Verteilung der Primzahlen*, third ed., Chelsea, New York, 1974, with an Appendix by Dr. Paul T. Bateman.

[Leg85] A.-M. Legendre, *Recherches d'analyse indéterminée*, Histoire de l'Académie Royale des Sciences (Paris) (1785), 465–559.

[Leg08] ———, *Essai sur la Théorie des Nombres*, second ed., Courcier, Paris, 1808.

[Leg11] A. M. Legendre, *Exercises de calcul intégral sur divers ordres de transcendentes et sur les quadratures*, Courcier, Paris, 1811.

[Lew52] D. J. Lewis, *Cubic homogenous polynomials over p-adic number fields*, Ann. of Math. (2) **56** (1952), 473–478.

[Lew57] ———, *Cubic forms over algebraic number fields*, Mathematika **4** (1957), 97–101.

[Lib32] G. Libri, *Mémoire sur la théorie des nombres*, J. Reine Angew. Math. **9** (1832), no. 1, 54–80, part one of the memoir.

[Lin08] E. Lindelőf, *Quelques remarques sur la croissance de la fonction $\zeta(s)$*, Bull. Sci. Math. (2) **32** (1908), 341–356.

[Lin42] Yuri V. Linnik, *On the representation of large numbers as sums of seven cubes*, Dokl. Akad. Nauk SSSR **35** (1942), 162 (Russian).

[Lin43] ———, *On the representation of large numbers as sums of seven cubes*, Mat. Sbornik **12** (1943), 218–224 (Russian).

[Lio57] J. Liouville, *Sur l'expression $\varphi(n)$, qui marque combien la suite $1, 2, 3, \ldots, n$ contient de nombres premiers à n*, J. Math. Pures Appl. (2) **2** (1857), 110–112.

[Lip89] Rudolf Lipschitz, *Untersuchungen der Eigenschaften einer Gattung von unendlichen Reihen*, J. Reine Angew. Math. **105** (1889), 127–156.

[Lit22] J. E. Littlewood, *Researches in the theory of the Riemann ζ-function*, Proc. London Math. Soc. (2) **20** (1922), Records xxii–xxvii.

[Lit86] ______, *Littlewood's Miscellany*, Cambridge University Press, Cambridge, 1986, Bela Bollobás ed.

[LMS10] Florian Luca, Anirban Mukhopaday, and Kotyada Srinivas, *Some results on Oppenheim's "factorisatio numerorum" function*, Acta Arith. **142** (2010), no. 1, 41–50.

[LY92] S. T. Lou and Q. Yao, *A Chebyshev's type of prime number theorem in a short interval. II*, Hardy-Ramanujan J. **15** (1992), 1–33.

[LY93] ______, *The number of primes in a short interval*, Hardy-Ramanujan J. **16** (1993), 21–43.

[Mac42] C. Maclaurin, *Treatise of Fluxions*, T. W. and T. Ruddimans, Edinburgh, 1742.

[Mai85] H. Maier, *Primes in short intervals*, Michigan Math. J. **32** (1985), no. 2, 221–225.

[Mai88] ______, *Small differences between prime numbers*, Michigan Math. J. **35** (1988), no. 3, 323–344.

[Man97] H. von Mangoldt, *Beweis der Gleichung...*, Sitz. Kgl. Preuss. Akad. Wiss. Berlin (1897), 835–852.

[Man72] Szolem Mandelbrojt, *Dirichlet Series: Principles and Methods*, Reidel, Dordrecht, 1972.

[Mas99] H. Maschke, *Beweis des Satzes, dass dejenigen endlichen linearen Substitutionsgruppen, in welchen einige durchgehends verschwindende Coefficienten auftreten, intransitiv sind*, Math. Ann. **52** (1899), 363–368.

[Mei54] E. Meissel, *Observationes quaedam in theoriae numerorum*, J. Reine Angew. Math. **48** (1854), 301–316, also A. G. Haynii, Berlin 1850.

[Mel00] Hjalmar Mellin, *Eine Formel fűr den Logarithmus transcendenter Funktionen von endlicher Geschlect*, Acta Soc. Sci. Fenn. **29** (1900), no. 4, 1–50.

[Mer74a] F. Mertens, *Ein Beitrag zur analytischen Zahlentheorie*, J. Reine Angew. Math. **78** (1874), 46–62.

[Mer74b] ______, *Ueber einige asymptotische Gesetze der Zahlentheorie*, J. Reine Angew. Math. **77** (1874), 289–338.

[Mer95a] ______, *Űber das Nichtverschwinden Dirichletschen Reihen mit reellen Gliedern*, Sitz. Kais. Akad. Wiss. (Wien) IIa **104** (1895), 1158–1166.

[Mer95b] ______, *Űber Dirichletschen Reihen*, Sitz. Kais. Akad. Wiss. (Wien) IIa **104** (1895), 1093–1153.

[Mer98] ______, *Űber eine Eigenschaft der Riemannscher ζ-Funktion*, Sitz. Kais. Akad. Wiss. (Wien) IIa **107** (1898), 1429–1434.

[Mey88] A. Meyer, *Űber einen Satz von Dirichlet*, J. Reine Angew. Math. **103** (1888), 98–117.

[Min91a] H. Minkowski, *Théorèmes arithmétiques*, C. R. Acad. Sci. Paris **112** (1891), 261–263.

[Min91b] ———, *Über die positiven quadratischen Formen und über kettenbruchähnliche Algorithmen*, J. reine Angew. Math. **107** (1891), 278–297.

[MO07] M. Ram Murty and Jeanine Van Order, *Counting integral ideals in a number field*, Expo. Math. **10** (2007), 53–66.

[Möb31] A. F. Möbius, *Über eine besondere Art von Umkehrung der Reihen*, J. Reine Angew. Math **10** (1831), 105–123.

[Mon71] Hugh L. Montgomery, *Topics in Multiplicative Number Theory*, Lecture Notes in Math., no. 227, Springer-Verlag, Berlin-New York, 1971.

[Moo98] E. H. Moore, *An universal invariant for finite groups of linear transformations: with applications in the theory of the canonical form of a linear substitution of finite period*, Math. Ann. **50** (1898), 213–219.

[Mor29] L. J. Mordell, *Poisson's summation formula and the Riemann zeta-function*, J. London Math. Soc. **4** (1929), 285–291.

[Mor37] ———, *A remark on indeterminate equations in several variables*, J. London Math. Soc. **12** (1937), 127–129.

[Moz86] C. J. Mozzochi, *On the difference between consecutive primes*, J. Number Theory **24** (1986), no. 2, 181–187.

[MP90] H. Maier and C. Pomerance, *Unusually large gaps between consecutive primes*, Trans. Amer. Math. Soc. **322** (1990), no. 1, 201–237.

[MS87] M. Ram Murty and N. Saradha, *On the Sieve of Eratosthenes*, Canad. J. Math. **39** (1987), no. 5, 1107–1122.

[MV77] H. Montgomery and R. C. Vaughan, *Exponential sums with multiplicative coefficients*, Invent. Math. **43** (1977), 69–82.

[MV07] Hugh L. Montgomery and Robert C. Vaughan, *Multiplicative Number Theory I. Classical Theory*, Cambridge Studies in Advanced Mathematics, no. 97, Cambridge University Press, Cambridge, 2007.

[Nar00a] W. Narkiewicz, *The Development of Prime Number Theory*, Springer, Berlin, 2000.

[Nar00b] ———, *Elementary and Analytic Theory of Algebraic Numbers*, third ed., Springer Monographs in Mathematics, Springer, Berlin, 2000.

[New48] F. W. Newman, *On $\Gamma(a)$ especially when a is negative*, Cambridge and Dublin Math. J. **3** (1848), 57–60.

[Noe29] E. Noether, *Hypercomplexe Grössen und Darstellungstheorie*, Math. Z. **30** (1929), 641–692.

[Oes85] J. Oesterlé, *Le probléme de Gauss sur le nombre de classes*, Enseign. Math. **34** (1985), 43–67.

[Pag35] A. Page, *On the number of primes in an arithmetic progression*, Proc. London Math. Soc. (2) **39** (1935), 116–141.

[Pal32] R. E. A. C. Paley, *A theorem on characters*, J. London Math. Soc. **7** (1932), 28–32.

[Per08] O. Perron, *Zur Theorie der Dirichletschen Reihen*, J. reine Angew. Math. **134** (1908), 95–143.

[Phr04] E. Phragmén, *Sur une extension d'un théorème de la théorie des fonctions*, Acta Math. **28** (1904), 351–368.

[Pil72] G. Z. Pil'tjaĭ, *The magnitude of the difference between consecutive primes*, Studies in number theory (D. N. Lenskoĭ, ed.), vol. 4, Saratov Univ., Saratov, 1972, pp. 73–79 (Russian).

[Pin84] J. Pintz, *On primes in short intervals. II*, Studia Sci. Math. Hungar. **19** (1984), no. 1, 89–96.

[Pin97] ______, *Very large gaps between consecutive primes*, J. Number Theory **63** (1997), no. 2, 286–301.

[PL08] E. Phragmén and E. Lindelőf, *Sur une extension d'un principe classique d'analyse et sur quelques propriétés du fonctions monogènes dans le voisinage d'un point singulier*, Acta Math. **31** (1908), 381–406.

[Pól18] G. Pólya, *Űber die Verteilung der der quadratische Reste und Nichtreste*, Gőttinger Nachrichten (1918), 21–29.

[PS53] I. I. Piatetski-Shapiro, *On the distribution of prime numbers in sequences of the form* $[f(n)]$, Mat. Sbornik N. S. **33 (75)** (1953), 559–566 (Russian).

[PW27] F. Peter and H. Weyl, *Die vollstǎndigkeit der primitiven Darstellungen einer geschlossenen kontinuerlichen Gruppe*, Math. Ann. **97** (1927), 737–755.

[Rad38] Hans Rademacher, *On the partition function* $p(n)$, Proc. London Math. Soc. (2) **43** (1938), 241–254.

[Rad73] ______, *Topics in Analytic Number Theory*, Die Grundlehren der mathematischen Wissenschaften, vol. 169, Springer, Berlin, 1973.

[Ram15] S. Ramanujan, *Highly composite numbers*, Proc. London Math. Soc. **14** (1915), no. 2, 347–409.

[Ram18] ______, *On certain trigonometrical sums and their applications in the theory of numbers*, Trans. Camb. Phil. Soc. **22** (1918), 259–76.

[Ram19] ______, *A proof of Bertrand's postulate*, J. Indian Math. Soc. **11** (1919), 181–182.

[Ran38] R. A. Rankin, *The difference between consecutive prime numbers*, J. London Math. Soc. **13** (1938), 242–247.

[Ran50] ______, *The difference between consecutive prime numbers. IV*, Proc. Amer. Math. Soc. **1** (1950), 143–150.

[Ran55] ______, *Van der Corput's method and the theory of exponent pairs*, Quart. J. Math. Oxford Ser. 2 **6** (1955), 147–153.

[Ran63] ______, *The difference between consecutive prime numbers, V*, Proc. Edinburgh Math. Soc. (2) **13** (1962/1963), 331–332.

[Ric34] G. Ricci, *Ricerche arithmetiche sui polinomi II (Intorno a una proposizione non vera di Legendre)*, Rend. Circ. Mat. di Palermo **58** (1934), 190–208.

[Ric54a] ______, *Sull'andamento della differenza di numeri primi consecutivi*, Riv. Mat. Univ. Parma **5** (1954), 3–54.

[Ric54b] H.-E. Richert, *On the difference between consecutive squarefree numbers*, J. London Math. Soc. (2) **29** (1954), 16–20.

[Rie59] B. Riemann, *Ueber die Anzahl der Primzahlen unter einer gegebenen Grősse*, Monatsber. Kgl. Preuß. Akad. Wiss. (Berlin) (1859), 671–680.

[RM95] P. Alessandra Maccioni Ruju and Marco Mostert, *The Life and Times of Guglielmo Libri*, Verloren Publishers, Hilversum, 1995.

[Rot51] K. F. Roth, *On the gaps between squarefree numbers*, J. London Math. Soc. (2) **26** (1951), 263–268.

[RR09] Morten S. Risager and Zeev Rudnick, *On the statistics of the minimal solution of a linear diophantine equation and uniform distribution of the real parts of orbits in hyperbolic spaces*, Spectral analysis in geometry and number theory

(Providence, RI) (M. Kotani, H. Naito, and T. Tate, eds.), Contemp. Math., vol. 484, American Mathematical Society, 2009, pp. 187–194.

[RW01] J. Rivat and J. Wu, *Prime numbers of the form* $[n^c]$, Glasg. Math. J. **43** (2001), no. 2, 237–254.

[Sch43] O. Schlőmilch, *Einiges űber die Eulerischen Iintegrale der zweiten Art*, Arch. Math. Phys. **4** (1843), 167–174.

[Sch49] M. Schaar, *Mémoire sur la théorie des résidus quadratiques*, Mémoires de l'Académie Royale des Sciences, des lettres et des Beaux-arts de Belgique **24** (1849), 1–14.

[Sch50] ———, *Recherches sur la théorie des résidus quadratiques*, Mémoires de l'Académie Royale des Sciences, des lettres et des Beaux-arts de Belgique **25** (1850), 1–20.

[Sch60] W. Scheibner, *Űber unendliche Reihen und deren Convergenz*, Hirzel, Leipzig, 1860.

[Sch05] Issai Schur, *Neue Begrűndung der Theorie der Gruppencharaktere*, Sitz. Kgl. Preuss. Akad. Wiss. Berlin (1905), 406–432.

[Sch18] I. Schur, *Einige Bemerkungen zu der Vorstehenden Arbeit des Herrn G. Pólya: Űber die Verteilung der der quadratische Reste und Nichtreste*, Gőttinger Nachrichten (1918), 30–36.

[Sch63] A. Schőnhage, *Eine Bemerkung zur Konstruktion grosser Primzahllűcken*, Arch. Math. (Basel) **14** (1963), 29–30.

[Sch64] P. G. Schmidt, *Abschắtzungen bei unsymmetrischen Gitterpunktproblemen*, doctoral dissertation, Georg-August-Universität zu Gőttingen, 1964.

[Sel49] A. Selberg, *An elementary proof of the prime-number theorem*, Ann. of Math. (2) **50** (1949), 305–313.

[Sha50] H. N. Shapiro, *On primes in arithmetic progressions II*, Ann. of Math. (2) **52** (1950), 231–243.

[Sie35] C. L. Siegel, *Űber die Classenzahl quadratischer Zahlkőrpern*, Acta Arith. **1** (1935), 83–86.

[Sie61] Carl Ludwig Siegel, *Lectures on Advanced Analytic Number Theory*, Lecture Notes, Tata Institute of Fundamental Research, Bombay, 1961.

[SS58] A. Schinzel and W. Sierpiński, *Sur certaines hypothèses concernant les nombres premiers*, Acta Arith. **4** (1958), 185–208, Erratum: ibid. **5**, 259.

[Sta74] H. M. Stark, *Some effective cases of the Brauer-Siegel theorem*, Inventiones Math. **23** (1974), 135–152.

[Sta75] ———, *The analytic theory of algebraic numbers*, Bull. Amer. Math. Soc. **81** (1975), no. 6, 961–972.

[Sta95] ———, *Galois Theory, Algebraic Number Theory, and Zeta Functions*, From Number Theory to Physics (Berlin) (J.-M. Luck M. Waldschmidt, P. Moussa and C. Itzykson, eds.), Springer, 1995, pp. 313–393.

[Sti85] T. J. Stieltjes, *Sur une loi asymptotique dans la théorie des nombres*, C. R. Acad. Sci. Paris **101** (1885), 368–370.

[Sti99] ———, *Sur le développement de* $\log(\Gamma(a))$, J. Math. Pur. Appl. (4) **5** (1899), 425–444.

[Syl81] J. J. Sylvester, *On Tchebycheff's theory of the totality of the prime numbers comprised within given limits*, Amer. J. Math. **4** (1881), 230–247.

[Ter85] Audrey Terras, *Harmonic Analysis on Symmetric Spaces and Applications*, Graduate Texts in Mathematics, Springer-Verlag, New York, NY, 1985, in two volumes.

[TF00] Gérald Tenenbaum and Michel Mendès France, *The Prime Numbers and Their Distribution*, Student Mathematical Library, vol. 6, American Mathematical Society, Providence, RI, 2000.

[Tit30] E. C. Titchmarsh, *A divisor problem*, Rend. Circ. Mat. di Palermo **54** (1930), 414–429.

[Tit39] ______, *The Theory of Functions*, Oxford University Press, Oxford, 1939.

[Tit86] ______, *The Theory of the Riemann Zeta-function*, second ed., Oxford University Press, Oxford, 1986.

[Tri88] O. Trifonov, *On the squarefree problem*, C. R. Acad. Bulgare Sci. **41** (1988), no. 12, 37–40.

[Tri89] ______, *On the squarefree problem. II*, Math. Balkanica (N.S.) **3** (1989), no. 3-4, 284–295.

[Tur34] P. Turán, *On a theorem of Hardy and Ramanujan*, J. London Math. Soc. **9** (1934), 274–276.

[Vau85] Robert C. Vaughan, *Sur le problème de Waring des cubes*, C. R. Acad. Sci. Paris **301** (1985), no. 6, 253–255.

[Vau86] R. C. Vaughan, *On Waring's problem for cubes*, J. Reine Angew. Math. **365** (1986), 122–170.

[Vau89] ______, *A new iterative method in Waring's problem*, Acta Math. **162** (1989), 1–71.

[Vau97] ______, *The Hardy-Littlewood Method*, second ed., Cambridge Tracts in Mathematics, vol. 125, Cambridge University Press, Cambridge, 1997.

[Vin18a] I. M. Vinogradov, *Sur la valeur moyenne du nombre des classes des formes proprement primitives du déterminant négatif*, Comm. Soc. Math. Kharkov (2) **16** (1918), 10–38 (Russian).

[Vin18b] Ivan M. Vinogradov, *On the distribution of power residues and non-residues*, J. Soc. Phys. Math. Univ. Permi **1** (1918), 94–98 (Russian).

[Vin19] ______, *On the distribution of quadratic residues and non-residues*, J. Soc. Phys. Math. Univ. Permi **2** (1919), 1–16 (Russian).

[Vin28] I. M. Vinogradov, *On Waring's theorem*, Izvestia Akad. Nauk SSSR, Otd. Fiz.-Mat. Nauk (1928), no. 4, 393–400 (Russian).

[Vin34] Ivan M. Vinogradov, *A new estimate for $G(n)$ in Waring's problem*, Doklady Akad. Nauk SSSR **5** (1934), no. 5-6, 249–253.

[Vin37] I. M. Vinogradov, *Representation of an odd number as a sum of three primes*, Doklady Akad. Nauk SSSR **15** (1937), no. 6-7, 291–294 (Russian).

[Vin58] Ivan M. Vinogradov, *A new estimate of the function $\zeta(1+it)$*, Iz. Akad. Nauk. SSSR Ser. Mat. **22** (1958), 161–164 (Russian).

[Vin80] I. M. Vinogradov, *The Method of Trigonometric Sums in Number Theory*, second ed., Nauka, Moscow, 1980.

[vK01] H. von Koch, *Sur la distribution des nombres premiers*, Acta Math. **24** (1901), 159–182.

[vK10] ______, *Contributions à la théorie des nombres premiers*, Acta Math. **33** (1910), 293–320.

[vM95] H. von Mangoldt, *Zu Riemann's Abhandung "Ueber die Anzahl der Primzahlen unter einer gegebenen Grősse"*, J. Reine Angew. Math. **114** (1895), 255–305.

[vM05] ———, *Zur Verteilung der Nullstellen der Riemannscher Funktion* $\xi(t)$, Math. Ann. **60** (1905), 1–19.

[Vor03] G. Voronoi, *Sur un problème du calcul des fonctions asymptotiques*, J. Reine Angew. Math. **126** (1903), no. 4, 241–282.

[Vor04] ———, *Sur un fonction transcendante et ses applications a la sommation de quelques séries*, Ann. Sci. Ecole Norm. Sup. **21** (1904), 203–267,459–533.

[VW02] R. C. Vaughan and T. D. Wooley, *Integer points, exponential sums and the Riemann zeta function*, Number theory for the millennium, III (Urbana, IL 2000) (Natick, MA) (M. A. Bennett, B. C. Berndt, N. Boston, H. G. Diamond, A. J. Hildebrand, and W. Philipp, eds.), A K Peters, 2002, pp. 301–340.

[Wal36] A. Walfisz, *Zur additiven Zahlentheorie. II*, Math. Z. **40** (1936), 592–607.

[Wal63] A. Z. Walfisz, *Weylschen Exponentialsummen in der neueren Zahlentheorie*, Mathematische Forschungsberichte, vol. 15, VEB Deutscher Verlag der Wissenschaften, Berlin, 1963.

[War70] Edward Waring, *Meditationes Algebraicae*, J. Woodyer, Cambridge, 1770.

[Web82] H. Weber, *Beweis des Satzes, dass jede eigentlich primitive quadratische Form unendlich viele Primzahlen darzustellen fāhig ist*, Math. Ann. **20** (1882), 301–329.

[Web08] ———, *Lehrbuch der Algebra*, Friedrich Vieweg und Sohn, Braunschweig, 1908, Zweite Auflage.

[Wei56] K. Weierstrass, *Űber die Theorie der analytischen Facultāten*, J. Reine Angew. Math. **51** (1856), 1–60.

[Wei49] André Weil, *Numbers of solutions of equations in finite fields*, Bull. Amer. Math. Soc. **55** (1949), 497–508.

[Wei52] ———, *Sur les "formules explicites" de la théorie des nombres premiers*, Comm. Sém. Math. Univ. Lund (1952), 252–265, Tome Supplémentaire.

[Wes31] E. Westzynthius, *Űber die Verteilung der Zahlen, die zu den n ersten Primzahlen teilerfremd sind*, Comm. Phys.-Math. Helsingfors **25** (1931), no. 5, 1–37.

[Wey16] H. Weyl, *Űber die Gleichverteilung von Zahlen mod. eins*, Math. Ann. **77** (1916), 313–352.

[Wie09] Arthur Wieferich, *Beweis des Satzes, dass sich eine jede ganze Zahl als Summe von hőchstens neun positiven Kuben darstellen lāsst*, Math. Ann. **66** (1909), 95–101.

[Wig07] S. Wigert, *Sur l'ordre de grandeur du nombre des diviseurs d'un entier*, Arkiv f. Mat. Astr. Fys. **3** (1906/1907), no. 18, 1–9.

[Woo92] T. D. Wooley, *Large improvements in Waring's problem*, Ann. Math. **135** (1992), 131–164.

[Zha14] Y. Zhang, *Bounded gaps between primes*, Ann. of Math. (2) **179** (2014), no. 3, 1121–1174.

List of Notations

$\mathbf{A}$	limit superior of $\psi(x)/x$, page 4
$\mathbf{a}$	limit inferior of $\psi(x)/x$, page 4
$\mathfrak{a}$	nonzero ideal, page 255
$A(x)$	counting function $A(x) = \sum_{A \ni n \leq x} 1$ of a sequence A, page 43
$a_K(m)$	the number of ideals in $\mathcal{O}_K$ whose norm equals m, page 255
$A_n(q)$	an exponential sum in the Circle Method, page 120
$\mathfrak{b}'$	the Dedekind complement of $\mathfrak{b}$, page 260
$\mathbb{C}$	the complex plane, page 3
Cl_K	the class group of $\mathcal{O}_K$, page 257
$C_{k,s}$	a constant in the Circle Method, page 121
$c_q(n)$	Ramanujan sum, page 89
$\deg(\rho)$	degree of a group representation ρ, page 71
$\mathbb{D}$	unit disk, page 206
$\mathrm{d}A$	asymptotic density of a sequence A, page 43
D	discriminant $D = b^2 - 4ac$ of a binary quadratic form, page 100
$d(n)$	divisor function, page 16
$D(x)$	summatory function $D(x) = \sum_{n \leq x} d(n)$, page 23
d_K	the discriminant of a number field K, page 257
$d_k(n)$	number of ways of writing n as a product of k factors, page 103
$\mathrm{e}(n)$	neutral element under Dirichlet convolution, page 17
$\mathrm{e}_k(n)$	indicator function of $\{k\}$, page 20
$e(x)$	exponential function $e(x) = \exp(2\pi i x)$ of period 1, page 68

$E_J(z)$	Weierstrass primary factor, page 235
$\hat{f}$	Fourier transform of f, page 213
$\hat{f}_n$	n-th Fourier coefficient of f, page 210
$\mathrm{Frob}_{\mathfrak{P}}$	Frobenius element, page 296
$\mathfrak{f}_\chi$	conductor of the Dirichlet character χ, page 91
$f(\alpha)$	an exponential sum in the Circle Method, page 117
$f_j(\alpha)$	an exponential sum in the Circle Method, page 117
F_n	Fermat number, page 108
$\hat{G}$	dual of a finite group G, page 79
$\mathrm{GL}(V)$	general linear group, page 71
$G(k)$	least s in Waring's problem for k-th powers for large n, page 112
$g(k)$	least s in Waring's problem for k-th powers for all n, page 112
$\mathbb{H}$	upper half plane of points τ with $\mathrm{Im}(\tau) > 0$, page 216
$\mathrm{Ind}_H^G(\rho)$	the representation induced on G by ρ on a subgroup H, page 293
$\mathrm{I}_A(x)$	indicator function of a sequence A, page 43
$\mathrm{I}_{q,a}(n)$	indicator function of the arithmetic progression $a + q\mathbb{Z}$, page 84
$\mathrm{id}(n)$	identity function $\mathrm{id}(n) = n$, page 16
J_K	the group of nonzero fractional ideals of $\mathcal{O}_K$, page 257
$J_\rho(n)$	an integral in the Circle Method, page 120
K	an algebraic number field, page 255
$\liminf$	limit inferior, page 4
$\limsup$	limit superior, page 4
$\mathrm{Li}(x)$	integral logarithm $\mathrm{Li}(x) = \mathrm{PV} \int_{[0,x]} du/\log(u)$, page 246
$L(s,\chi)$	Dirichlet L-function $L(s,\chi) = \sum_{n\in\mathbb{N}} \chi(n)n^{-s}$, page 85
$\mathrm{li}(x)$	integral logarithm $\mathrm{li}(x) = \int_{[2,x]} du/\log(u)$, page 1
$(\mathcal{M}f)(s)$	Mellin transform of f, page 227
$\mathfrak{M}$	the major arcs intervals in the Circle Method, page 117
$\mathfrak{m}$	the minor arcs intervals in the Circle Method, page 117
$\mathfrak{M}(q,a)$	a single major arcs interval in the Circle Method, page 117
$M(x)$	summatory function $M(x) = \sum_{n\le x} \mu(n)$, page 50
$M_n(q)$	number of solutions of a congruence for a diagonal form, page 113
$N(\alpha)$	norm of an algebraic number α, page 255
$N(\mathfrak{a})$	norm of an ideal $\mathfrak{a}$, page 255
$N(\sigma,T)$	number of zeros ρ of $\zeta(s)$ with $\beta > \sigma$ and $0 < \gamma \le T$, page 323

$n(p)$	least positive quadratic nonresidue modulo p, page 94
$N(T)$	number of zeros of $\zeta(s)$ with $0 \le \sigma \le 1$ and $0 < t \le T$, page 309
$N_0(T)$	number of zeros $s = 1/2 + it$ of $\zeta(s)$ with $0 < t \le T$, page 245
n_K	the degree of K over $\mathbb{Q}$, page 255
$\mathcal{O}_K$	ring of algebraic integers in a number field K, page 255
$O(g)$	big-O notation of Bachmann, page 7
$o(g)$	small-o notation of Landau, page 7
$\mathfrak{p}$	nonzero prime ideal, page 255
$p(n)$	partition function, page 104
$P^+(n)$	the largest prime factor of n, page 37
$P^-(n)$	the smallest prime factor of n, page 37
P_K	the subgroup of J_K of principal fractional ideals, page 257
$\mathbb{Q}$	the field of rational numbers, page 149
$Q(x)$	counting function $Q(n) = \sum_{n \le x} \lvert\mu(n)\rvert$ of the squarefrees, page 42
$Q(x,y)$	binary quadratic form $Q(x,y) = ax^2 + bxy + cy^2$, page 100
$\mathbb{R}$	the field of real numbers, page 68
$\mathrm{Res}_H^G(\rho)$	the restriction of ρ from G to a subgroup H, page 295
r	$= r_1 + r_2 - 1$ is the rank of the group of units of $\mathcal{O}_K$, page 264
$R(n)$	number of solutions in the Circle Method, page 116
r_1	the number of real embeddings of a number field, page 257
r_2	half the number of complex embeddings, page 257
R_K	the regulator of $\mathcal{O}_K$, page 267
$\mathrm{rad}(n)$	radical $\mathrm{rad}(n) = \prod_{p \mid n} p$ of n, page 16
$\mathfrak{S}(n)$	the singular series in the Circle Method, page 123
S	Weyl sum, page 130
s	notation $s = \sigma + it$ for a complex variable, page 62
$S(q,a)$	an exponential sum in the Circle Method, page 114
$S(x)$	sawtooth function $S(x) = x - [x] - 1/2$, page 11
t	imaginary part of the complex variable $s = \sigma + it$, page 62
$T(x)$	summatory function $T(x) = \sum_{n \le x} \log(n)$, page 3
$T_{\mathfrak{P}}$	inertia group, page 296
$\mathfrak{U}$	an interval of integration in the Circle Method, page 117
U_K	the group of units of $\mathcal{O}_K$, page 264
$v(\beta)$	an integral in the Circle Method, page 117

$v_j(\beta)$	an integral in the Circle Method, page 120
V_ρ	representation space of a group representation ρ, page 71
$W(\chi)$	root number in the functional equation of $L(s,\chi)$, page 231
w_K	the number of units of finite order in $\mathcal{O}_K$, page 264
$Z(s)$	completed zeta function $Z(s) = \pi^{-s/2}\Gamma(s/2)\zeta(s)$, page 230
$Z_K(s)$	the completed Dedekind zeta function, page 257
$Z_K(s,C)$	the completed zeta function of an ideal class C, page 257
$Z_{\mathfrak{P}}$	decomposition group, page 296
β	real part of a zero ρ of $\zeta(s)$ in $\sigma > 0$, page 170
χ	Dirichlet character, page 84
χ_0	principal Dirichlet character, page 85
χ_0	trivial character, page 76
χ_ρ	character of a group representation ρ, page 76
δA	logarithmic density of a sequence A, page 44
$\Delta(x)$	error term in $D(x) = x\log(x) + (2\gamma - 1)x + \Delta(x)$, page 24
δ_{il}	Kronecker delta $\delta_{il} = 1$ if $i = l$, zero otherwise, page 75
γ	Euler-Mascheroni constant, page 12
γ	imaginary part of a zero ρ of $\zeta(s)$ in $\sigma > 0$, page 170
$\Gamma(s)$	gamma function, page 223
$\gamma_K(s)$	the gamma factor for the number field K, page 257
λ	rank of an entire function, page 238
$\Lambda(n)$	von Mangoldt function, page 2
$\lambda(n)$	Liouville function $\lambda(n) = \prod_{p^k \mid n}(-1)$, page 16
μ	exponent of convergence for an entire function, page 237
$\mu(n)$	Möbius function, page 18
$\Omega(n)$	total number $\Omega(n) = \sum_{p^k \mid n} 1$ of prime factors of n, page 16
$\omega(n)$	number of distinct prime factors $\omega(n) = \sum_{p \mid n} 1$ of n, page 16
$\Phi(\lambda,\alpha,s)$	Lerch zeta function, page 251
$\Phi(x)$	summatory function $\Phi(x) = \sum_{n \le x} \phi(n)$, page 21
$\phi_2(q)$	number of primitive characters modulo q, page 92
$\Pi(x)$	weighted counting function $\Pi(x) = \sum_{p^k \le x} k^{-1}$, page 246
$\pi(x)$	counting function $\pi(x) = \sum_{p \le x} 1$ of the primes, page 1
$\Psi(s;q,a)$	a Dirichlet series, page 183
$\psi(x)$	Chebyshev function $\psi(x) = \sum_{p^k \le x} \log(p)$, page 2

$\psi(x;q,a)$	$=\sum_{x\geq n\equiv a \bmod q}\Lambda(n)$, page 183
$\psi_1(x)$	smoothed Chebyshev function, page 173
ρ	group representation, page 71
ρ	zero of $\zeta(s)$ in $\sigma>0$, page 170
ρ_0	trivial group representation, page 71
σ	real part of the complex variable $s=\sigma+it$, page 62
$\sigma(n)$	sum-of-divisors function $\sigma(n)=\sum_{d\mid n}d$, page 17
σ_a	abscissa of absolute convergence of a Dirichlet series, page 66
$\sigma_a(A)$	abscissa of absolute convergence of the Dirichlet series of A, page 67
σ_c	abscissa of convergence of a Dirichlet series, page 65
$\sigma_c(A)$	abscissa of convergence of the Dirichlet series of A, page 67
$\tau(\chi,n)$	Gauss sum of a Dirichlet character χ, page 89
τ_p	classical Gauss sum, page 70
ϱ	order of an entire function, page 237
$\vartheta(x)$	weighted counting function $\vartheta(x)=\sum_{p\leq x}\log(p)$ of the primes, page 2
$\vartheta(z,\tau)$	Jacobi theta function, page 216
$\vartheta_{jk}(z,\tau)$	Jacobi theta function, page 249
$\widetilde{\rho}$	the inflation of ρ from G/H to G, page 296
$\xi(s)$	Riemann xi function $\xi(s)=s(s-1)\pi^{-s/2}\Gamma(s/2)\zeta(s)/2$, page 230
$\Xi(z)$	Riemann Xi function $\Xi(z)=\xi(1/2+iz)$, page 230
$\zeta(s)$	Riemann zeta function $\zeta(s)=\sum_{n\in\mathbb{N}}n^{-s}$, page 63
$\zeta_K(s)$	Dedekind zeta function of a number field K, page 256
$\zeta_K(s,C)$	the zeta function of an ideal class C, page 257
$\delta_{A,B}$	Dirichlet density of A in B, page 88
$1(n)$	constant arithmetic function $1(n)\equiv 1$, page 17
$[x]$	integer part function $[x]=\max\{n\in\mathbb{Z}\mid n\leq x\}$, page 2
$*$	Dirichlet convolution $(f*g)(n)=\sum_{d\mid n}f(d)g(n/d)$, page 17
$\langle\alpha\vert\beta\rangle_{L^2(\hat{G})}$	Hermitian inner product of functions on $\hat{G}$, page 79
$\langle\alpha\vert\beta\rangle_{L^2(G)}$	Hermitian inner product of functions on G, page 77
$\ll$	Vinogradov notation, page 7
$\overline{n}$	multiplicative inverse modulo q, page 70
$\sim$	asymptotic equality $f(x)\sim g(x)$, page 1

$\varphi \mapsto \check{\varphi}$	inverse Fourier transform operator, page 79
$f \mapsto \hat{f}$	Fourier transform operator, page 79
$f \mapsto \hat{f}$	finite Fourier transform operator on $\mathbb{Z}/q\mathbb{Z}$, page 68

Index

Selected Published Titles in This Series

160 **Marius Overholt,** A Course in Analytic Number Theory, 2014
159 **John R. Faulkner,** The Role of Nonassociative Algebra in Projective Geometry, 2014
158 **Fritz Colonius and Wolfgang Kliemann,** Dynamical Systems and Linear Algebra, 2014
157 **Gerald Teschl,** Mathematical Methods in Quantum Mechanics, 2014
156 **Markus Haase,** Functional Analysis, 2014
155 **Emmanuel Kowalski,** An Introduction to the Representation Theory of Groups, 2014
154 **Wilhelm Schlag,** A Course in Complex Analysis and Riemann Surfaces, 2014
153 **Terence Tao,** Hilbert's Fifth Problem and Related Topics, 2014
152 **Gábor Székelyhidi,** An Introduction to Extremal Kähler Metrics, 2014
151 **Jennifer Schultens,** Introduction to 3-Manifolds, 2014
150 **Joe Diestel and Angela Spalsbury,** The Joys of Haar Measure, 2013
149 **Daniel W. Stroock,** Mathematics of Probability, 2013
148 **Luis Barreira and Yakov Pesin,** Introduction to Smooth Ergodic Theory, 2013
147 **Xingzhi Zhan,** Matrix Theory, 2013
146 **Aaron N. Siegel,** Combinatorial Game Theory, 2013
145 **Charles A. Weibel,** The K-book, 2013
144 **Shun-Jen Cheng and Weiqiang Wang,** Dualities and Representations of Lie Superalgebras, 2012
143 **Alberto Bressan,** Lecture Notes on Functional Analysis, 2013
142 **Terence Tao,** Higher Order Fourier Analysis, 2012
141 **John B. Conway,** A Course in Abstract Analysis, 2012
140 **Gerald Teschl,** Ordinary Differential Equations and Dynamical Systems, 2012
139 **John B. Walsh,** Knowing the Odds, 2012
138 **Maciej Zworski,** Semiclassical Analysis, 2012
137 **Luis Barreira and Claudia Valls,** Ordinary Differential Equations, 2012
136 **Arshak Petrosyan, Henrik Shahgholian, and Nina Uraltseva,** Regularity of Free Boundaries in Obstacle-Type Problems, 2012
135 **Pascal Cherrier and Albert Milani,** Linear and Quasi-linear Evolution Equations in Hilbert Spaces, 2012
134 **Jean-Marie De Koninck and Florian Luca,** Analytic Number Theory, 2012
133 **Jeffrey Rauch,** Hyperbolic Partial Differential Equations and Geometric Optics, 2012
132 **Terence Tao,** Topics in Random Matrix Theory, 2012
131 **Ian M. Musson,** Lie Superalgebras and Enveloping Algebras, 2012
130 **Viviana Ene and Jürgen Herzog,** Gröbner Bases in Commutative Algebra, 2011
129 **Stuart P. Hastings and J. Bryce McLeod,** Classical Methods in Ordinary Differential Equations, 2012
128 **J. M. Landsberg,** Tensors: Geometry and Applications, 2012
127 **Jeffrey Strom,** Modern Classical Homotopy Theory, 2011
126 **Terence Tao,** An Introduction to Measure Theory, 2011
125 **Dror Varolin,** Riemann Surfaces by Way of Complex Analytic Geometry, 2011
124 **David A. Cox, John B. Little, and Henry K. Schenck,** Toric Varieties, 2011
123 **Gregory Eskin,** Lectures on Linear Partial Differential Equations, 2011
122 **Teresa Crespo and Zbigniew Hajto,** Algebraic Groups and Differential Galois Theory, 2011

For a complete list of titles in this series, visit the AMS Bookstore at **www.ams.org/bookstore/gsmseries/**.